MEDIUM AND HIGH TEMPERATURE SOLAR PROCESSES

ENERGY SCIENCE AND ENGINEERING:
RESOURCES, TECHNOLOGY, MANAGEMENT
An International Series
EDITOR
JESSE DENTON
Belton, Texas

Medium and High Temperature Solar Processes

Jan F. Kreider
Jan F. Kreider and Associates
Boulder, Colorado

ACADEMIC PRESS   *New York   San Francisco   London   1979*
A Subsidiary of Harcourt Brace Jovanovich, Publishers

ACADEMIC PRESS, INC.
111 Fifth Avenue, New York, New York 10003

United Kingdom Edition published by
ACADEMIC PRESS, INC. (LONDON) LTD.
24/28 Oval Road, London NW1   7DX

Library of Congress Cataloging in Publication Data

Kreider, Jan F   Date
    Medium and high temperature solar processes.

    (Energy science and engineering)
    Bibliography: p.
    Includes index.
    1. Solar energy. 2. Solar heating. 3. Solar
radiation. I. Title. II. Series.
TJ810.K72   621.47   79–51694
ISBN 0–12–425980–4

PRINTED IN THE UNITED STATES OF AMERICA

79 80 81 82   9 8 7 6 5 4 3 2 1

Who will prevent the clouds in their heavenly movement
Who will stop the winds and the swaying willows
Who can threaten the sunrise and
The light falling under our feet.

Ivan Rabuzin

CONTENTS

3 SELECTED TOPICS IN THERMODYNAMICS
AND HEAT TRANSFER FOR ELEVATED-
TEMPERATURE SOLAR PROCESSES

4 MEDIUM-TEMPERATURE SOLAR
COLLECTORS AND ANCILLARY
COMPONENTS

5 MEDIUM-TEMPERATURE SOLAR
PROCESSES

6 HIGH-TEMPERATURE SOLAR PROCESSES

7 *ECONOMIC ANALYSIS OF SOLAR-THERMAL SYSTEMS*

PREFACE

Many processes in the industrial and commercial sectors of developed economies require energy at elevated temperature with high thermodynamic availability. The sun is one source of such energy, and it is the purpose of this book to describe in an engineering context many medium- and high-temperature solar processes that are or will become technically viable during the remainder of this century. In addition, methods of assessing the economic viability of elevated temperature solar processes are developed in detail, and expected environmental impacts are summarized.

The use of solar heat for many processes is attractive since solar systems (1) are safe and benign environmentally, (2) are relatively simple to design and fabricate, (3) conserve fossil fuels for other essential applications such as pharmaceuticals, liquid fuels, and chemical feedstocks, and (4) have no significant effect on the global heat balance. Of course, there are impediments to the near-term use of solar-thermal processes for high temperatures, including (1) relatively high cost, (2) large materials requirements for metals, insulation, and glass, (3) lack of industrial experience with solar processes, and (4) lack of familiarity with the principles of solar system design. One of the purposes of this book is to eliminate this final barrier.

The subject matter of this book is concerned with those solar-thermal processes operating above 100°C. A broad division between medium- and high-temperature processes is made at 300–400°C, corresponding to practical operating temperatures achieved by single- and compound-curvature solar concentrators.

Further, only concentrating collectors are efficacious for temperatures above 100°C; therefore, flat plate collectors are not considered in this book.

xi

Elevated-temperature solar processes is a relatively new field of solar-thermal applications, and field experience is only beginning to become available. It is, therefore, premature to present detailed design guidelines and costs for some of the applications described in this book. Consequently, the principal or first-order design parameters of solar systems which use basic mechanical and chemical engineering concepts applied to solar heat are described; but details required for final system design will be left to future editions. A unique feature of this book is the emphasis on the second law of thermodynamics and its use in matching solar collection methods to thermal processes. We also provide more details for small- and intermediate-scale applications than for large-scale ones since large-scale plants such as central power facilities require very detailed design studies to account for features unique to each project, whereas smaller-scale applications will be much greater in number and general design guidelines, for them, will be more useful.

It is assumed that the user of this book has a background in engineering heat transfer, thermodynamics, fluid mechanics, and mathematics through the calculus. It is not necessary for the reader to have any prior knowledge of solar radiation, of collectors and conversion concepts, or of engineering economics. However, references to other works are made for the reader who wishes to reexamine some of the basic concepts of solar engineering. Chapter 1 is an overview of energy use patterns in the U.S. and of the various solar-thermal processes considered in detail later in the book. The concepts of economics of solar systems and possible environmental impacts are also summarized. Chapter 2 deals with solar radiation—its quantity, geographic availability, and quality. Particular emphasis is placed on beam or direct radiation since it has the highest thermodynamic availability and is used by most elevated-temperature collectors. The trigonometry of various solar tracking modes and optical properties of materials are also described. Chapter 3 considers selected topics in thermodynamics and heat transfer, including various heat engine designs and their first and second law efficiencies, radiation heat transfer, and the properties of selective surfaces usable at high temperature.

Chapter 4 considers components and systems for medium temperature processes. Many concentrating collectors are described along with types of thermal storage, heat exchangers, and energy transport systems. Chapter 5 treats systems for power production, shaft power, industrial process heat, and total energy (shaft power plus thermal power). Chapter 6 includes engineering design data for high-temperature collectors and their use in solar furnaces, central solar power plants, distributed power plants, and solar thermionics. Finally, Chapter 7 considers the economics of the foregoing systems with emphasis on methods and principles of analysis. Costs of specific systems are not the major focus since they are subject to uncertainty resulting from inflation and lack of mass production experience. Throughout, the emphasis of the book is on engineering design and all methods and tools currently available are described by example or analytical equations and charts. Since the field is in a rapid state of development, however, the reader can remain completely current only by study of the technical literature.

The author acknowledges the essential assistance provided by his typist Janet Hagood in assembling the manuscript. James Snow prepared the original graphics. Discussions with Dr. Fred Hoffman on the topic of solar thermionics; with Drs. Manuel Collares-Pereira, Ari Rabl, and James Leonard on concentrator design, insolation modeling, and performance prediction; with Bim Gupta on the performance of high-temperature selective surfaces; with Dr. L. R. Bush on solar total energy systems; and with Dr. Malcolm Fraser on process heat systems were very valuable in assuring that the most current information on these topics was included. The support of Dottie Lang through the course of the book's preparation was greatly appreciated. The anonymous reviewers of the manuscript also added to the quality of the book by their views and detailed comments from perspectives different from those of an author.

MEDIUM AND HIGH TEMPERATURE SOLAR PROCESSES

1 | PROLOGUE

Eyes, though not ours, shall see
Sky high a signal flame,
The sun returned to power above
A world, but not the same.

C. D. Lewis

I. INTRODUCTION

The subject of this book—medium- and high-temperature solar processes—is the synthesis of established mechanical and chemical engineering design principles with the developing technology of solar-thermal conversion above 100°C. The high thermodynamic quality of sunlight is comparable to the high quality (low entropy) of fossil fuels, so it is expected that the substitution of the former for the latter can be achieved successfully from a technical point of view. The question of economic viability is less clear and is governed by exogenous variables including competing energy prices, solar flux levels, and consistency and specific system output requirements. Both technical and economic topics are covered in detail in the succeeding chapters of the book.

Solar energy has certain positive and negative attributes

when applied to thermal processes (K2). A few of the major areas are reviewed herein. The major technical obstacle for solar-thermal applications is the intermittent nature of the resource both on daily and shorter time scales. The intermittency leads to a storage requirement not present in nonsolar systems. The extra cost and complexity of storage (solar energy cannot be stored in its primary form) is a negative attribute of solar systems. Solar systems also require a good solar "climate" for efficient operation. This implies a regional or geographic effect which may be at odds with the geographical distribution of energy demands and the land which can be devoted to solar collectors to meet those demands. Finally, solar system must meet the criterion of economic competitiveness in order to be widely adopted.

Solar energy has several unique features which place it in an advantageous position vis-à-vis progressively more scarce fossil fuels. Most of the thermal end use demands of the U.S. are achievable by relatively simple and fairly unsophisticated solar systems. The majority of the components of solar systems treated in this book are off-the-shelf, commercial items. Collectors and controls are the only special, new components and neither is particularly complex. In addition, by the proper application of solar technology, an excellent thermodynamic match between the solar energy resource and many end uses can be achieved. On the contrary, the match between fossil or nuclear fuels and most industrial end uses is notoriously bad with second law efficiencies of 10% or less.

This book attempts to present an analytical framework for comparing the many features of thermal, solar, and nonsolar systems in an unbiased manner so that technically and economically optimal systems for performing a given task may be designed.

A. Solar Systems to Meet U.S. Energy Needs

The use of heat energy in the U.S. has been estimated by various investigators. The principal uses expected in the commercial and industrial sectors are shown in Fig. 1.1 for 1985. It is seen that the largest uses lie in space heating (can be done effectively below 100°C), direct and steam process heating, and various electrical and shaft drive end uses.

The topics to be covered in this book were selected to treat the major end uses of heat above 100°C in the commercial and industrial sectors. Specifically, the solar conversion systems discussed are

industrial process heating;
shaft work production;

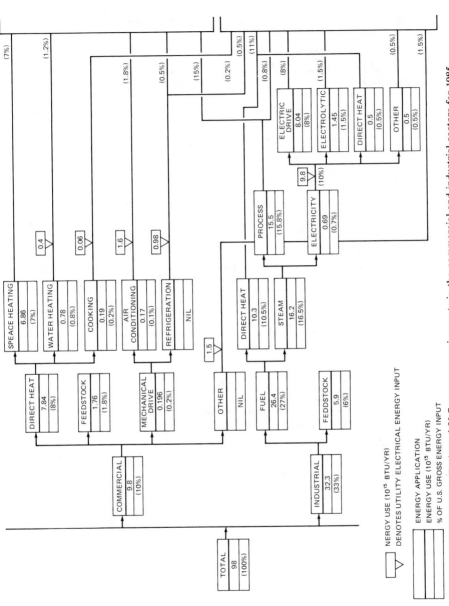

FIGURE 1.1 Projected U.S. energy requirements in the commercial and industrial sectors for 1985.

total energy systems—meeting electrical and thermal loads;
electric power production—
    distributed thermal systems,
    central receiver thermal systems,
    distributed thermionic systems.

In addition, solar furnaces are discussed. Each of these principal end uses is briefly described below.

    Industrial process heat is used to process and prepare finished goods from raw materials. Process heat can be in the form of direct heat, hot liquids, hot gases, or steam depending on the process temperature and heat rate specifications. Solar systems are readily adaptable to these several process heat types except for direct heat, which is frequently at very high temperature, for example, in a blast furnace. Industrial processes frequently operate for 24 h per day, indicating the requirement for solar heat storage if this energy source is to provide more than about one-third of a continuous demand. Early solar process heat systems have not used storage, however, to minimize capital cost and system complexity. This sun-follower mode can be economical but annual solar load fractions are relatively small for continuous processes. Detailed descriptions of process heat energy demands and solar process heat systems are given in Chapter 5.

    Shaft work or shaft drive energy refers to the output of a heat engine which is used in many applications in industry. Shaft drive is conventionally produced by turbines or electric motors. Since terrestrial solar flux is of high thermodynamic quality, work production can be achieved efficiently using heat engines of various types including Rankine, Stirling, or Brayton engines. Since Rankine cycles are the most common heat engines in industry, they are treated most completely of the several engine types in this book.

    Solar total energy systems (STES's) represent an integrated series of energy conversion subsystems which are powered in turn by progressively lower grade heat energy. For example, a solar-powered turbine may be operated from a high-temperature collector field. The exhaust of the turbine could then be used for an intermediate temperature process load followed by a low-temperature feedstream preheat application. Properly designed, an STES can provide an excellent second law match between energy inputs and process requirements. Therefore, the use of the thermodynamic availability of sunlight is maximized.

    Electric power production can be accomplished by two generic types of systems. The first, called the distributed concept, uses large fields of solar-thermal collectors producing either heat or power at many local points in the field. This heat or power is transmitted to a central plant which in turn is connected to a grid. The second type of power plant produces heat and power at only one large, central plant, the solar flux being reflected by large fields of mirrors to a central receiver. This concept has the advantage that large networks of pipes for heat transport or cables for power transport are not required. However, a large central tower with a solar flux collector at its top is used instead. The decision as to which power plant concept is most appropriate has been shown to be a matter of size. For plant sizes below 10 MW_e, the distributed collector concept seems most economic, whereas for larger plants the central receiver is more economic using current technology. Several thermodynamic and thermionic solar power production schemes are described in Chapters 5 and 6. Solar furnaces used to produce very high temperatures for research are also described in Chapter 6.

B. Common Features of All Solar-Thermal Systems

All solar-thermal systems used in a continuous mode (batch process solar furnace not included) consist of five principal components:

solar collector;
storage—thermal, chemical, or mechanical;
energy transport; .
end use converter;
controls.

Solar collectors convert radiant solar flux to heat at a temperature compatible with the system's end use. In this book only focusing or concentrating devices are considered since processes above 100°C are emphasized. About a dozen solar collector types are considered in analytical detail in Chapters 4 and 6. First-order optical and thermal characteristics are described for most practical concentrators which have been built of both the single- and compound-curvature types. Of course, adequate solar flux is required to energize a solar system and Chapter 2 contains detailed analyses of solar flux on monthly, daily, and hourly time scales. The use of historical data is emphasized and methods of converting historical, horizontal flux data to other time scales and collector surfaces of arbitrary orientation are given.

Energy storage is usually present in solar systems to buffer periods of solar outage. Storage serves as a thermal capacitance to provide a relatively more uniform output from a solar system subject to a relatively nonuniform solar flux input owing to decimal geometric or shorter term weather effects. The precise size of storage is an economic and reliability question and varies drastically by application. In addition, the temperature of storage as well as the size is a principal design variable. The higher the required storage temperature the smaller the selection of economic and reliable materials which are available. The thermodynamic quality or entropy level of storage must be compatible with other components of the solar system. Sensible heat, latent heat, and chemical storage are described in Chapters 4 and 5 with additional guidelines provided for system application type.

Energy transport occurs by means of heat or chemical energy in fluids, or by electric power, in most solar-thermal systems. The design of thermal transport systems is well established and only important results are summarized in this book in Chapter 3. Since thermal transport in solar and nonsolar systems is not substantially different, no new results are needed for solar systems. Transport components include pumps, pipes and their insulation, heat exchangers, valves, and other small components.

Solar heat end uses determine the type of end use converter. For example, if a hot process gas is needed, a solar-heated liquid-to-gas heat exchanger is used; but if a pumped fluid is to be the output, the end use converter would be a heat engine–gearbox–pump assembly. The end use converter is clearly specific to each process and is described in Chapters 5 and 6 by process type.

Solar system controls are critical to the efficient performance of the system. They perform many tasks including making decisions on collector turn-on or turn-off, collector tracking, mode selection, backup system operation, fluid stream temperature control and modulation, fluid makeup, and fail-safe system protection. Although most of these tasks other than collector tracking are familiar in the process and utility industries, solar systems require careful design since their operation is subject not only to process variables, which are the most common control parameters for nonsolar systems, but also to climatic and temporal variables.

C. The Solar Design Process

The structure of the solar system design process also contains certain fundamental steps which are the same for all solar processes. Three main steps can be identified:

develop a system design,
estimate system cost,
estimate system performance.

If the performance per unit cost is better than for competing systems performing the same task, the solar system is viable and will probably be selected over its competitors. Several iterations of the three-step process are required. As system design is refined, cost and performance estimates improve. It is the purpose of this book to provide enough information for the designer to properly carry out the three-step design process at the design development level. Final design can also be carried out for the most part except for the final step of estimating performance in detail, the tools for which do not currently exist.

Throughout the book the importance of the second law of thermodynamics to solar design is emphasized. Although final system selection is based on economic criteria, the second law provides valuable insights regarding which components cause first order performance penalties or limits and which components are less important. These conclusions frequently have economic consequences.

II. SOLAR SYSTEM ECONOMICS

Solar energy in its natural state is without cost and without value, as is the case for every energy source. It is the cost of equipment and labor for collection, conversion, and distribution which causes a solar cost to exist. The final chapter of this book analyzes solar economic methodology in detail in a general form applicable to any conversion system. Discounted cash flow principles are used to assess the comparative economics of solar and nonsolar energy forms to perform a given task.

Solar costs include many cost components which occur at the start of or during the expected life of a project: capital, land, taxes, debt interest, operating costs, insurance, maintenance, and repairs. These and other costs are compared to the costs of a competing process or to the value of fuel displaced by solar heat. If the benefits outweigh the solar costs, solar is the system of choice.

Solar system selection has an added feature not present in fuel-operated systems since most solar-thermal systems are subject to the law of diminishing returns. It is usually found that one specific solar system comprised of a unique set of physical components exhibits the smallest cost–benefit ratio of all possible configurations. As a result it is not sufficient to compare only one solar system to the nonsolar alternative, but the "best" or most economic solar system must be used in the comparison. The identification of the optimal solar system requires the use of optimization methodology as described in Chapter 7.

III. ENVIRONMENTAL IMPACTS OF SOLAR SYSTEMS

Solar systems are usually considered to be rather benign environmentally. One of the reasons for the use of solar energy is to reduce environmental pollution and costs for its control. However, there are several negative impacts which have been identified. These are described briefly below.

A. Air Pollution

The replacement of fossil fuel combustion with solar heat will reduce the emission of pollutants such as CO, NO_x, SO_x, particulates, and unburned hydrocarbons. However, certain solar systems using chemical reactants for storage or organic fluids for heat transport pose the threat of release of CO, SO_2, SO_3, hydrocarbon vapors, and other noxious gases. Their release, of course, is not routine and not expected to occur often. A fire hazard may exist if overheated, organic working fluids are present near an ignition source. If an on-site, fossil-fueled backup is used, there will be emissions of pollutants but at a reduced level and depending upon the types of fuels used.

Mining of raw materials such as bauxite, copper ore, and iron ore for solar collectors will be increased. Most of the air pollutants from mining operations are airborne particulates and exhaust gases from machinery. The majority of materials used in solar systems can be recycled so the mining emissions can be a one-time event. Surface mines are subject to strict regulations on the amount of dust which can be released.

A form of air "pollution" exists near the focus of large solar reflectors or refractors. Human tissue would be destroyed upon only a short contact. At the focus of very large, central receiver power plants very high flux densities occur over appreciable areas (tens of meters). These could blind aircraft pilots or damage the aircraft itself. Planes can be constrained from flying through focal zones and solar tracking devices can be programmed in such a way that a focus cannot exist above the design focal zone.

B. Water Pollution

Water pollution can result from release or spillage of collector or process working fluids. Organic working fluids and eutectic heat transfer salts are specified for use in many elevated temperature processes. If these materials were released, for whatever reason, they could find their way into the hydrosystem by direct spillage into a waterway or by surface runoff. Several fluids are quite toxic and require special spillage containment such as dikes or ponds. In addition, some fluids requiring periodic replacement will require careful control during this operation.

Nitrates and nitrites present in eutectic heat transfer salts have well-established physiological impacts including cyanosis, carcinogenosis, mental impairment, etc. Nitrosamines formed by the reaction of nitrites with amino acids in the body are potent carcinogens. Certain glycol working fluids are toxic and heat transfer oils severely impact the potability of water and aquatic fauna. Trace elements for corrosion control in aqueous fluids include chromates, phosphates, and sulfates. All of these compounds have specific maximum acceptable concentrations in raw water to be used for drinking. Water quality is also deteriorated by runoff from mines for solar system raw materials as described above.

C. Land Use

One of the largest impacts of solar collection is the requirement of relatively large amounts of land. A typical 100-Mw$_e$ power plant will require about one square mile, for example. This represents a severe problem in industrialized areas such as the Northeast of the U.S. The land intensiveness of solar plants is offset to some extent because they do not generate solid or liquid wastes requiring land for disposal.

Collector fields produce shading not normally present over large areas. Impacts on the local ecosystem include an increase of

shade-seeking flora which may impact collector function. In desert areas
flora may cease to exist in shaded zones and some form of active dust control may be needed (oiling, paving, etc).

The amount of land required for appreciable solar energy
use is small when viewed from the macroscale, however. Weingart (W9)
has shown that less than 1% (0.5×10^6 km^2) of U.S. land area could provide all U.S. energy needs. This amounts to 225 m^2 per person. By comparison, croplands and pastures consume 41% of all U.S. land (W9).

D. Thermal Pollution

Solar systems eliminate the local thermal pollution produced by fossil fuel combustion. Solar flux, formerly reflected or absorbed by the local environment, is partially collected and transported to
a nearby solar-thermal conversion facility. Hence, the thermal effects of a
solar plant are minor. If electric power produced by a solar plant is used
hundreds of miles away in a city, some reduction in the local environmental heat budget would occur. If, for example, the power plant has an
overall efficiency of 20%, about one-tenth of the solar heat incident on the
power plant site is exported, assuming a ground cover ratio of 50%.

E. Noise Pollution

Solar systems are not expected to add any new noise impacts beyond those normally present at industrial or utility sites. Collector trackers, pumps, and other components are not major noise pro-
ducers.

F. Summary

Most of the environmental impacts of solar-thermal systems
arise during the mining and construction phases. Pollutants arising during
routine operation are relatively minor and can be controlled by methods
routinely in use in industry. The overall level of environmental impacts
for most solar systems seems to be significantly lower than for comparable nonsolar systems.

2 | PRINCIPLES
OF SOLAR RADIATION
AND OPTICS

Above all Brother Sun
Who Brings us the day
And Sends us his light.

St. Francis of Assisi

I. INTRODUCTION

Calculation of the performance and economics of medium-
and high-temperature solar systems requires an accurate analysis of both
the temporal and geographical variation of solar intensity at the earth's
surface. The quantitative description of the sun's virtual motion as seen
by earth is, therefore, the first subject of this chapter. Second, methods of
calculation of radiation incident upon surfaces of arbitrary orientation and
tracking mode are given. The final topic covered in this chapter is the optical analysis of reflecting and transparent media which are always used in
solar thermal systems.

The majority of high-temperature solar systems rely on the
existence of beam radiation which may be concentrated to achieve high
efficiency and high operating temperature. Therefore, the emphasis in this
chapter is on direct or beam radiation. Diffuse radiation is considered

11

briefly since it can be partially captured by some concentrators. The following definitions and symbols are used for beam, diffuse, and total solar radiation:

A. Beam Radiation

Solar flux (energy rate per unit area) incident on a surface without significant direction change or scattering in the atmosphere. Beam radiation is also called *direct radiation*. $\bar{B}$ denotes monthly averaged, daily beam radiation (kW h/m² day or kJ/m² day); $\bar{I}_b$ the daily (kW h/m² day or kJ/m² day) and I_b either the instantaneous or hourly beam radiation (W/m² or kJ/m² h).

B. Diffuse Radiation

Solar flux dispersed by various scattering mechanisms into the sky dome without a specific incidence angle on a terrestrial surface. $\bar{D}$ denotes monthly averaged, daily diffuse radiation, $\bar{I}_d$ the daily, and I_d the instantaneous of hourly diffuse radiation with units being the same as for beam radiation.

C. Total Radiation

The total of beam and diffuse radiation, sometimes referred to as *global radiation*. On a horizontal surface $\bar{H}_h$ denotes the monthly averaged, daily total radiation, $\bar{I}_h$ the daily total, and I_h the instantaneous or hourly total. For collector surfaces other than horizontal the symbols $\bar{I}_c$, $\bar{I}_c$, and I_c are used, respectively, the subscript c denoting *collector*. The distinction between monthly averaged, daily, and total daily quantities is obvious by the context.

II. TERRESTRIAL AND EXTRATERRESTRIAL BEAM RADIATION

All solar radiation beyond the atmosphere is beam radiation. Its magnitude and spectral distribution are well known from an extensive series of measurements made from satellites and from solar observatories on earth. This section describes extraterrestrial radiation and its attenuation in the atmosphere.

A.   The Solar Constant

The annual mean solar flux at the "edge" of the atmosphere is called the *solar constant* I_{sc}. Its value is 1353 W/m², 429 Btu/h ft² or 1.94 cal/cm² min ($\pm 1.6\%$) (T7). The solar constant represents the maximum solar energy available for terrestrial use.

Since the earth's orbit about the sun is slightly elliptical (but with small eccentricity), extraterrestrial radiation shows a $\pm 3.4\%$ variation through a year. Table 2.1 shows the yearly extraterrestrial intensity

TABLE 2.1

Monthly Variation of Extraterrestrial Solar Radiation Owing to Orbital Eccentricity[a]

Date		Sun–earth distance in astronomical units	Irradiance due to the sun (W m⁻²)
Jan	1	0.983 34	1399
	4	0.983 32	1399
Feb	1	0.985 31	1393
	12	0.987 03	1389
Mar	1	0.990 97	1378
	20	0.995 94	1364
Apr	1	0.999 41	1355
	4	0.999 98	1353
May	1	1.007 70	1332
	14	1.010 72	1324
Jun	1	1.014 13	1316
	21	1.016 30	1310
Jul	1	1.016 72	1309
	3	1.016 73	1309
	26	1.015 66	1312
Aug	1	1.014 95	1313
Sep	1	1.009 14	1329
	23	1.003 37	1344
Oct	1	1.001 07	1350
	5	0.999 89	1353
Nov	1	0.992 39	1347
	3	0.991 87	1375
Dec	1	0.985 98	1392
	22	0.983 68	1398

[a] From (T1).

variation. An adequate analytical representation of this effect is

$$I_0(N)/I_{sc} = 1 + 0.034 \cos(2\pi N/365), \qquad (2.1)$$

where $I_0(N)$ is the extraterrestrial intensity and N the day number counted from January 1.

B.   The Extraterrestrial Solar Spectrum

The solar constant described above is the integral over all wavelengths of the extraterrestrial solar *irradiance* expressed as W/m² (μm), for example. However, the careful evaluation of many optical properties—index of refraction, extinction coefficient, reflectance— requires information on the spectral nature of sunlight. The extraterrestrial solar spectrum is shown in Fig. 2.1 as the "air mass zero" line (the air mass concept will be described shortly). It is seen that the peak intensity is in the visible wavelength range at about 0.45 μm.

Also shown in Fig. 2.1 is a blackbody spectral curve corresponding to a source at 5762 K located at the mean sun–earth distance. The wavelength integral of this curve is the same as the solar constant. Therefore, the effective temperature of the sun can be taken as 5762 K for engineering purposes. Of course, the temperature of the sun ranges from 10^7 K to 10^3 K from the center to the outer layer so the effective value is only an artifice useful for engineering analysis.

Table 2.2 contains a numerical tabulation of the standard solar spectrum published by NASA in 1971 (T1). This table is used for careful engineering calculation of optical properties of materials used in the solar environment.

C.   Terrestrial Beam Radiation Estimation

Air Mass   A convenient method of estimating the attenuation of beam radiation due to absorption or scattering is by means of a Beer's law or Bouger's law exponential decay model such as

$$I_{b,\lambda} = I_{0,\lambda} e^{-c_\lambda m}, \qquad (2.2)$$

where $I_{b,\lambda}$ is the beam irradiance on a terrestrial surface at wavelength λ, c_λ is the attenuation coefficient, and m is a dimensionless atmospheric path length called the air mass.

The air mass ratio is defined in Fig. 2.2 and is seen to be the ratio of optical path length BP to the nominal atmospheric layer thickness AP. For small air mass values the approximation $m = \sec \alpha$ (where α is

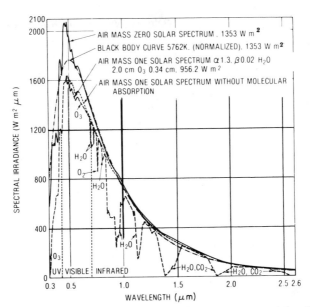

FIGURE 2.1 The extraterrestrial solar spectrum and the effective spectrum at 5762 K. Also shown are two standard terrestrial spectra described later. [From (T1).]

TABLE 2.2

The Standard Solar Irradiance Spectrum[a]

λ (μm)	$E_\lambda{}^c$ (W/m²·μm)	$E_\lambda{}^c$ (Btu/hr·ft²·μm)	$D_\lambda{}^d$ (%)	λ (μm)	E_λ (W/m²·μm)	E_λ (Btu/hr·ft²·μm)	D_λ (%)	λ (μm)	E_λ (W/m²·μm)	E_λ (Btu/hr·ft²·μm)	D_λ (%)
0.115	0.007	0.002	1×10^{-4}	0.43	1639	520	12.47	0.90	891	283	63.37
0.14	0.03	0.010	5×10^{-4}	0.44	1810	574	13.73	1.00	748	237	69.49
0.16	0.23	0.073	6×10^{-4}	0.45	2006	636	15.14	1.2	485	154	78.40
0.18	1.25	0.397	1.6×10^{-3}	0.46	2066	655	16.65	1.4	337	107	84.33
0.20	10.7	3.39	8.1×10^{-3}	0.47	2033	645	18.17	1.6	245	77.7	88.61
0.22	57.5	18.2	0.05	0.48	2074	658	19.68	1.8	159	50.4	91.59
0.23	66.7	21.2	0.10	0.49	1950	619	21.15	2.0	103	32.7	93.49
0.24	63.0	20.0	0.14	0.50	1942	616	22.60	2.2	79	25.1	94.83
0.25	70.9	22.5	0.19	0.51	1882	597	24.01	2.4	62	19.7	95.86
0.26	130	41.2	0.27	0.52	1833	581	25.38	2.6	48	15.2	96.67
0.27	232	73.6	0.41	0.53	1842	584	26.74	2.8	39	12.4	97.31
0.28	222	70.4	0.56	0.54	1783	566	28.08	3.0	31	9.83	97.83
0.29	482	153	0.81	0.55	1725	547	29.38	3.2	22.6	7.17	98.22
0.30	514	163	1.21	0.56	1695	538	30.65	3.4	16.6	5.27	98.50
0.31	689	219	1.66	0.57	1712	543	31.91	3.6	13.5	4.28	98.72
0.32	830	263	2.22	0.58	1715	544	33.18	3.8	11.1	3.52	98.91
0.33	1059	336	2.93	0.59	1700	539	34.44	4.0	9.5	3.01	99.06
0.34	1074	341	3.72	0.60	1666	528	35.68	4.5	5.9	1.87	99.34
0.35	1093	347	4.52	0.62	1602	508	38.10	5.0	3.8	1.21	99.51
0.36	1068	339	5.32	0.64	1544	490	40.42	6.0	1.8	0.57	99.72
0.37	1181	375	6.15	0.66	1486	471	42.66	7.0	1.0	0.32	99.82
0.38	1120	355	7.00	0.68	1427	453	44.81	8.0	0.59	0.19	99.88
0.39	1098	348	7.82	0.70	1369	434	46.88	10.0	0.24	0.076	99.94
0.40	1429	453	8.73	0.72	1314	417	48.86	15.0	0.0048	0.015	99.98
0.41	1751	555	9.92	0.75	1235	392	51.69	20.0	0.0015	0.005	99.99
0.42	1747	554	11.22	0.80	1109	352	56.02	50.0	0.0004	0.0001	100.00

[a] Adapted from Thekaekara (T7).

[b] Solar constant = 429 Btu/hr · ft² = 1353 W/m².

[c] E_λ is the solar spectral irradiance averaged over a small bandwidth centered at λ.

[d] D_λ is the percentage of the solar constant associated with wavelengths shorter than λ.

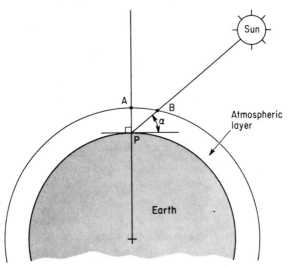

FIGURE 2.2   At sea level air mass ratio is defined as $m = BP/AP$ and the solar altitude angle is α. The atmosphere is modeled as a uniform thickness layer. For other altitudes above sea level see Eq. (2.4).

the solar altitude; see Eq. (2.8) and associated discussion) can be used. A more accurate equation for m is (W6)

$$m(0,\alpha) = [1229 + (614 \sin \alpha)^2]^{1/2} - 614 \sin \alpha, \qquad (2.3)$$

where $m(0,\alpha)$ is the air mass at sea level ($z = 0$). At altitudes z above sea level the air mass is

$$m(z,\alpha) = m(0,\alpha)[p(z)/p(0)], \qquad (2.4)$$

where $p(z)$ is the atmospheric pressure at altitude z.

Attenuation Coefficients   The attenuation coefficient c_λ in Eq. (2.2) includes three effects—Rayleigh scattering, c_1; ozone absorption, c_2; aerosol and turbidity, c_3. There are other effects present as well (W6) but the three above are the first order effects. Table 2.3 tabulates the c_i versus wavelength. Values of c_1 and c_2 are calculated for the U.S. standard atmosphere and c_3 represents typical urban–industrial atmospheric turbidity.

It should be noted that additional absorption phenomena exist in the infrared ($\lambda > 0.72$ μm) which are not included in the c_1, c_2, and c_3 coefficients (T1). Therefore the calculated value of the wavelength integral of Eq. (2.2) may be larger than the measured terrestrial beam radiation. Additional attenuation coefficients c_i may be added to include the infrared (ir) effects.

TABLE 2.3

Atmospheric Attenuation Coefficients for Rayleigh Scattering (c_1),
Ozone Absorption (c_2), and Turbidity (c_3)[a]

Wavelength, λ (μm)	c_1	c_2	c_3 ($\mu = 0.66; \beta = 0.17$)
0.270	1.928	70.956	0.4034
0.280	1.645	35.816	0.3938
0.300	1.222	3.413	0.3763
0.320	0.927	0.303	0.3606
0.340	0.717	0.022	0.3465
0.360	0.564	0.001	0.3336
0.380	0.450	0.000	0.3220
0.400	0.364	0.000	0.3112
0.450	0.223	0.001	0.2880
0.500	0.145	0.012	0.2686
0.550	0.098	0.031	0.2522
0.600	0.069	0.045	0.2382
0.650	0.050	0.021	0.2259
0.700	0.037	0.008	0.2151
0.800	0.021	0.003	0.1970
0.900	0.013	0.000	0.1822
0.026	0.007	0.000	0.1671
1.060	0.003	0.000	0.1636
1.670	0.001	0.000	0.1212
2.170	0.000	0.000	0.1020
3.500	0.000	0.000	0.0744
4.000	0.000	0.000	0.0681

[a] From (Tl).

Standard Irradiance Curves for Various Air Mass Ratios A number of spectral irradiance curves may be calculated by assuming values of atmospheric ozone and water content and turbidity levels. The integrals of these curves are the ground-level beam radiation. Table 2.4 summarizes such calculations for four levels of turbidity. The turbidity attenuation coefficient c_3 used in Table 2.4 has been represented as

$$c_3 = \beta/\lambda^\mu. \qquad (2.5)$$

Therefore large β and small μ values correspond to a polluted atmosphere. Clear atmosphere flux is represented in Table 2.4 by $\mu = 1.3$ and $\beta = 0.02$. A polluted industrial environment is represented by $\mu = 0.66$ and $\beta = 0.17$. The full spectral irradiance table for this industrial atmosphere is contained in Appendix Table A.1.

TABLE 2.4

Beam Radiation I_b as a Function of Air Mass for Various Turbidity Levels[a]

Air mass	Solar zenith angle (deg)	Turbidity factors μ	Turbidity factors β	Total flux (W m⁻²)	Ratio of total flux to solar constant (%)	Fraction of the total energy in the uv, λ < 0.4 μm (%)	Fraction of the total energy in the Visible, 0.4 μm < λ < 0.72 μm (%)	Fraction of the total energy in the Infrared, λ > 0.72 μm (%)
0	0			1353.0	100.0	8.7	40.1	51.1
1	0	1.30	0.02	956.2	70.7	4.8	46.9	48.3
4	75.5	1.30	0.02	595.2	44.0	1.23	44.2	54.5
7	81.8	1.30	0.02	413.6	30.6	0.35	39.4	60.3
10	84.3	1.30	0.02	302.5	22.4	0.102	34.7	65.2
1	0	1.30	0.04	924.9	68.4	4.6	46.4	49.0
4	75.5	1.30	0.04	528.9	39.1	1.04	42.1	56.9
7	81.8	1.30	0.04	324.0	25.3	0.26	35.9	63.8
10	84.3	1.30	0.04	234.5	17.3	0.065	30.3	69.6
1	0	0.66	0.085	889.2	65.7	4.7	46.4	48.9
4	75.5	0.66	0.085	448.7	33.2	1.14	42.4	56.5
7	81.8	0.66	0.085	255.2	18.9	0.30	36.3	63.4
10	84.3	0.66	0.085	153.8	11.4	0.08	30.7	69.2
1	0	0.66	0.17	800.2	59.1	4.5	45.4	50.1
4	75.5	0.66	0.17	303.1	22.4	0.88	38.3	60.8
7	81.8	0.66	0.17	133.3	9.85	0.14	30.0	69.8
10	84.3	0.66	0.17	63.4	4.69	0.039	22.9	77.1

[a] U.S. standard atmosphere: H_2O. 20 mm; O_3. 3.4 mm; the zenith angle is the complement of the solar altitude angle α.

[From T1).]

TABLE 2.5

Beam Radiation I_b for Various Levels of Atmospheric Water Vapor and Ozone[a]

Atmospheric parameters				Air mass				
				1	1.5	2	3	5
				For solar zenith angles				
H_2O (cm)	O_3 (cm)	β	μ	0	48.2	60	70.5	78.5
2.0	0.20	0.02	1.3	961	876	805	693	534
2.0	0.20	0.17	0.66	804	672	568	414	232
2.0	0.34	0.02	1.3	956	870	798	684	523
2.0	0.34	0.17	0.66	800	668	563	409	228
2.0	0.55	0.02	1.3	949	861	788	672	508
2.0	0.55	0.17	0.66	795	661	556	402	222
0.5	0.34	0.02	1.3	1024	943	875	767	612
0.5	0.34	0.17	0.66	859	727	622	464	273
1.0	0.34	0.02	1.3	992	909	840	729	571
1.0	0.34	0.17	0.66	832	700	595	439	252
5.0	0.34	0.02	1.3	898	807	732	615	454
5.0	0.34	0.17	0.66	750	616	513	363	193

[a] From (T1). Units are in W m^{-2}. Solar zenith angle is the complement of the solar altitude angle.

Table 2.5 shows the effects of atmospheric water vapor and ozone upon the beam radiation level I_b. The amount of water vapor is given in units of the depth (cm) of a liquid layer containing all the water vapor, as condensate, in a column above a surface. The data show that increased water vapor and ozone levels cause reduced beam radiation levels. It is seen that from 25 to 40% of the extraterrestrial radiation at noon is absorbed by the typical atmospheres considered. This absorption is an effect apart from attenuation and blockage of sunlight by clouds. A detailed description of atmospheric physics and the attenuation of sunlight is beyond the purview of this book. The reader may refer to (W6) for an introduction.

D. Solar Disk Intensity Distribution

The beam radiation levels I_b discussed above are assumed to be from a point-source sun. However, the sun is not a point source but rather subtends a 32' (9.3 mrad) angle. For careful analysis of the distribution of the sun's radiation on the absorbers of solar collectors, the intensity distribution of solar flux must be known over the sun's disk. One non-uniform solar disk model is represented by (J1)

$$I_b(r)/I_b(0) = (1 + 1.5641\sqrt{1 - (r/r_0)^2})/2.5641, \qquad (2.6)$$

where $I_b(r)$ is the intensity at radius r and r_0 is the radius of the sun (or its image on an absorber surface).

III. SOLAR RADIATION GEOMETRY

The preceding section has described a method for calculating the beam component of radiation. However, many solar collectors do not face the sun directly at all times and hence cannot capture the full I_b radiation amount since they suffer the incidence angle *cosine loss*. That is, the effective portion of beam radiation which can be intercepted is the direct-normal radiation I_b times the cosine of the incidence angle (i.e., the scalar product of the solar flux vector and the surface-normal unit vector). This section analyzes solar motion and its effect upon collector flux interception and incidence angle.

A. The Sun's Location

It is convenient to adopt the convention that the sun moves on an imaginary spherical surface called the celestial sphere. The celestial sphere is sufficiently large so that the sun can be considered as a point. The celestial sphere also is assumed to be centered at the site of analysis on the earth's surface. Since an object moving on a sphere has two degrees of freedom, only two angles are needed to specify the instantaneous position of the sun. The solar altitude and azimuth angles are used for this purpose. (*All angles and arguments of trigonometric functions in this book are expressed in degrees unless otherwise stated.*)

The *solar altitude* angle α is measured from the local horizontal plane upward to the center of the sun. It is zero at sunrise and sunset and is at its maximum at local solar noon. Figure 2.3 shows the altitude angle.

The *solar azimuth* angle is measured in the horizontal plane between the north–south line and the projection of the sun's rays onto the

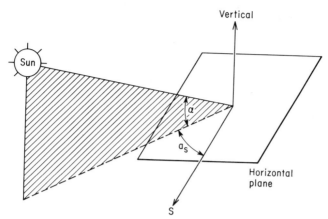

FIGURE 2.3   Solar altitude α and azimuth a_s angles with reference to the local horizontal plane for the Northern Hemisphere.

horizontal plane. The azimuth angle is zero at local solar noon, positive before noon, and negative after noon in accordance with the right-hand rule convention.

The local north–south direction can be determined by sighting Polaris in the Northern Hemisphere or by use of a magnetic compass with proper account taken of local magnetic declination. See (B1) for a shadow technique for finding the local meridian.

The solar altitude and azimuth angles are related to the solar hour angle h_s, the latitude L, and the solar declination δ_s. The *hour angle* is given by 15° times the number of hours from local solar noon. It is zero at noon, positive before noon, and negative after noon. The signs of azimuth and hour angles are based on the usual positive angular displacement rule in righthand coordinate systems.

The *latitude* is the angle between the earth's equatorial plane and a line from the center of the earth to the site of study. It is positive north of the equator and negative south. The value of the latitude can be read from an atlas.

The *solar declination* is the angle at noon on the equator between the overhead (zenith) direction and a line collinear with the sun's rays. Alternatively, it has the same value as the latitude at which the sun is overhead at local solar noon. Declination values are positive when the sun is north of the equator and negative when south. Table 2.6 contains a tabulation of the declination for 5-day periods for a full year. For engineering purposes the declination is constant over a day and is given by (P7)

TABLE 2.6

The Average Solar Ephemeris[a]

Date	Declination Deg	Declination Min	Equation of time Min	Equation of time Sec	Date	Declination Deg	Declination Min	Equation of time Min	Equation of time Sec
Jan. 1	−23	4	− 3	14	Feb. 1	−17	19	−13	34
5	22	42	5	6	5	16	10	14	2
9	22	13	6	50	9	14	55	14	17
13	21	37	8	27	13	13	37	14	20
17	20	54	9	54	17	12	15	14	10
21	20	5	11	10	21	10	50	13	50
25	19	9	12	14	25	9	23	13	19
29	18	8	13	5					
Mar. 1	− 7	53	−12	38	Apr. 1	+ 4	14	− 4	12
5	6	21	11	48	5	5	46	3	1
9	4	48	10	51	9	7	17	1	52
13	3	14	9	49	13	8	46	− 0	47
17	1	39	8	42	17	10	12	+ 0	13
21	− 0	5	7	32	21	11	35	1	6
25	+ 1	30	6	20	25	12	56	1	53
29	3	4	5	7	29	14	13	2	33
May 1	+14	50	+ 2	50	June 1	+21	57	+ 2	27
5	16	2	3	17	5	22	28	1	49
9	17	9	3	35	9	22	52	1	6
13	18	11	3	44	13	23	10	+ 0	18
17	19	9	3	44	17	23	22	− 0	33
21	20	2	3	34	21	23	27	1	25
25	20	49	3	16	25	23	25	2	17
29	21	30	2	51	29	23	17	3	7
July 1	+23	10	− 3	31	Aug. 1	+18	14	− 6	17
5	22	52	4	16	5	17	12	5	59
9	22	28	4	56	9	16	6	5	33
13	21	57	5	30	13	14	55	4	57
17	21	21	5	57	17	13	41	4	12
21	20	38	6	15	21	12	23	3	19
25	19	50	6	24	25	11	2	2	18
29	18	57	6	23	29	9	39	1	10
Sep. 1	+ 8	35	− 0	15	Oct. 1	− 2	53	+10	1
5	7	7	+ 1	2	5	4	26	11	17
9	5	37	2	22	9	5	58	12	27
13	4	6	3	45	13	7	29	13	30
17	2	34	5	10	17	8	58	14	25
21	+ 1	1	6	35	21	10	25	15	10
25	− 0	32	8	0	25	11	50	15	46
29	2	6	9	22	29	13	12	16	10
Nov. 1	−14	11	+16	21	Dec. 1	−21	41	+11	16
5	15	27	16	23	5	22	16	9	43
9	16	38	16	12	9	22	45	8	1
13	17	45	15	47	13	23	6	6	12
17	18	48	15	10	17	23	20	4	17
21	19	45	14	18	21	23	26	2	19
25	20	36	13	15	25	23	25	+ 0	20
29	21	21	11	59	29	23	17	− 1	39

[a]Since each year is 365.25 days long, the precise value of declination varies from year to year. *The American Ephemeris and Nautical Almanac* published each year by the U.S. Government Printing Office contains precise values for each day of each year.

$$\delta_s = \sin \mu \cos[2\pi(10.5 + N)/365.25], \qquad (2.7)$$

where N is the day number from January 1 and $\mu = -23° \ 27'$, the angle between the earth's axis and the ecliptic plane.

The solar altitude angle α (Fig. 2.3) is given by

$$\sin \alpha = \sin L \sin \delta_s + \cos L \cos \delta_s \cos h_s \qquad (\sin \alpha \geq 0) \qquad (2.8)$$

and the azimuth angle a_s is

$$a_s = \sin^{-1}(\cos \delta_s \sin h_s/\cos \alpha). \qquad (2.9)$$

B. Sunrise and Sunset Time

At sunrise and sunset the solar altitude angle is zero; therefore Eq. (2.8) can be used to find the sunrise hour angle $h_{sr}(\alpha = 0)$ and sunset hour angle $h_{ss}(\alpha = 0)$ by equating $\sin \alpha$ to zero:

$$\sin \alpha_{sr} = 0 = \sin L \sin \delta_s + \cos L \cos \delta_s \cos h_{sr}(\alpha = 0); \quad (2.10)$$

then

$$h_{sr}(\alpha = 0) = \cos^{-1}(-\tan L \tan \delta_s) \quad (2.11)$$

and

$$h_{ss}(\alpha = 0) = -h_{sr}(\alpha = 0). \quad (2.12)$$

The sunrise and sunset hour angles are expressed in radians or degrees depending upon the context. If $|\tan L \tan \delta_s| > 1.0$, the location is beyond the Arctic or Antarctic circles and

$$h_{ss} = h_{sr} = 180° \quad (2.13)$$

That is, the sun does not set in summer. In winter, the sun does not rise if the above inequality applies. Table 2.7 shows the average length of a day at various latitudes for each month.

The preceding sunset and sunrise times are measured relative to the intersection of the center of the sun and the horizon plane. However, true sunrise occurs when the upper edge or limb of the sun appears to cross the horizon. This extra period of daylight consists of 16′ of

TABLE 2.7

Average Length of a Day for Various Latitudes in the Northern Hemisphere

North latitude (deg)	Months of year											
	Jan	Feb	Mar	Apr	May	Jun	Jul	Aug	Sep	Oct	Nov	Dec
20	10.9	11.3	11.9	12.5	12.9	13.2	13.1	12.7	12.1	11.5	11.0	10.8
25	10.6	11.2	11.8	12.6	13.2	13.5	13.4	12.9	12.1	11.4	10.8	10.5
30	10.3	10.9	11.8	12.7	13.5	13.9	13.8	13.1	12.2	11.3	10.5	10.1
35	9.9	10.7	11.7	12.9	13.8	14.3	14.1	13.3	12.2	11.1	10.1	9.7
40	9.4	10.5	11.7	13.1	14.2	14.8	14.6	13.6	12.3	10.9	9.7	9.2
45	8.9	10.2	11.6	13.3	14.6	15.4	15.1	13.9	12.3	10.7	9.3	8.6
50	8.3	9.8	11.6	13.5	15.2	16.1	15.7	14.3	12.4	10.5	8.7	7.9
55	7.5	9.4	11.5	13.8	15.9	17.1	16.6	14.7	12.4	10.1	8.0	6.9
60	6.3	8.8	11.4	14.2	16.8	18.4	17.7	15.4	12.5	9.7	7.1	5.6

solar arc for the sun to move its center one-half solar diameter below the horizon plus an atmospheric refraction effect amounting to another 36' of arc (W6). Of course, when the sun is near the horizon, the flux level is below the threshhold of practical collectibility. The additional 52' of arc noted here is only of interest in the design of solar collector tracking devices.

C. Solar Time, the Equation of Time

A *solar day* is the length of time between two successive crossings of the local meridian plane, formed by extension of the local longitude great circle to the celestial sphere, by the sun. Since the earth moves forward in its orbit during a solar day, the time required for a 360° rotation of the earth is about 3.95' less than a solar day.

The length of a solar day varies because of the tilt of the earth's axis, the nonspherical shape of the earth, and the orbital eccentricity. Therefore, standard clock time and solar time differ. The difference is called the equation of time (ET). Solar time and clock time in hours are related by

$$\text{solar time (in hours)} \equiv h_s/15$$
$$= 12.0 - \text{corrected standard time} - \text{ET}/60, \quad (2.14)$$

where ET is expressed in minutes. Corrected standard time is standard clock time (not daylight savings time) corrected for differences between the standard meridian for a time zone and the local site meridian—

$$\text{corrected standard time} = \text{standard clock time (in the 24-h system)}$$
$$+ \text{(time zone meridian}$$
$$- \text{local site longitude)}/15. \quad (2.15)$$

The longitude values are measured west of Greenwich and the units of time are hours. The equation of time is tabulated in Table 2.6 and shown graphically by the analemma in Fig. 2.4.

Equations (2.8) and (2.9) include all first-order effects necessary for determining the position of the sun within one degree. However, for very precise tracking of the sun (without feedback, i.e., with an "open-loop" tracker) for a long period, more than ten second-order effects must be included. These are described in detail by Pitman and Vant-Hull (P5), who also mention several common sources of error in locating the sun exactly. If a feedback sun tracker is used, these second-order effects are not of practical interest.

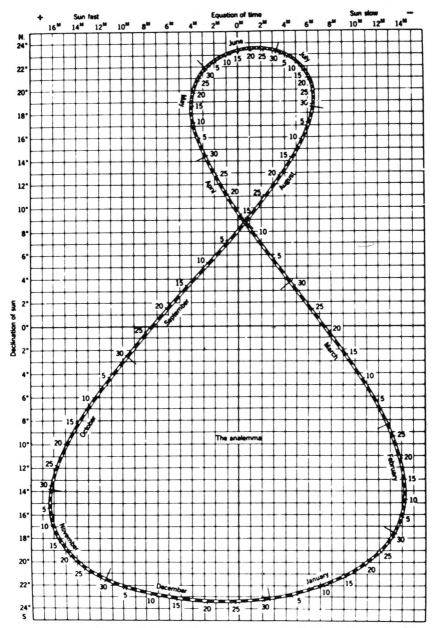

FIGURE 2.4   The analemma showing the declination and equation of time for each day of a year. Equation of time values to the right of the centerline are negative, those to the left, positive.

IV. BEAM RADIATION INTERCEPTED BY SURFACES

Many solar collectors are subject to the incidence angle cosine optical effect. This section summarizes incidence angles for many types of fixed and tracking solar collectors.

A. Incidence Angle Definition

The solar *incidence angle* is the angle between the normal to an irradiated surface and a line collinear with the sun's rays. Figure 2.5 shows the incidence angle i for an arbitrary solar collector aperture C. The beam radiation $I_{b,c}$ intercepted by collector C is then given by

$$I_{b,c} = I_b \cos i. \tag{2.16}$$

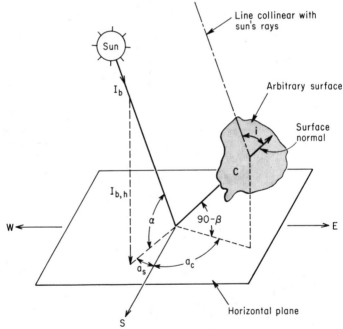

FIGURE 2.5 Solar incidence angle i of a collector surface C. Also shown are the complement of the surface tilt angel β, azimuth angle a_c, and horizontal component of beam radiation $I_{b,h}$. The tilt factor relating horizontal to tilted surface radiation $I_{b,c}/I_{b,h}$ is seen to be $\cos i / \sin \alpha$.

The calculation of cos i is required for all possible geometries to be used in medium- and high-temperature solar processes.

B. *Incidence Angles for Fixed Surfaces*

The solar incidence angle for any fixed surface is given by (B6)

$$\cos i = \sin \delta_s (\sin L \cos \beta - \cos L \sin \beta \cos a_c)$$
$$+ \cos \delta_s \cos h_s (\cos L \cos \beta + \sin L \sin \beta \cos a_c)$$
$$+ \cos \delta_s \sin \beta \sin a_c \sin h_s, \quad (2.17)$$

where β is the *surface tilt* angle from the horizontal and a_c is the *surface azimuth* angle defined analogously to the solar azimuth angle a_s. The surface tilt angle β is positive for south-facing surfaces and negative for north-facing ones. A negative value of cos i indicates that $i > 90°$ and the sun's rays do not strike the surface. An alternative form of Eq. (2.17) is given by Kondratyev and Fedorova (K13).

For special cases, equations simpler than Eq. (2.17) can be used. For a horizontal surface

$$\cos i = \sin L \sin \delta_s + \cos L \cos \delta_s \cos h_s. \quad (2.18)$$

For a south-facing surface

$$\cos i = \sin(L - \beta) \sin \delta_s + \cos(L - \beta) \cos \delta_s \cos h_s. \quad (2.19)$$

For a cylindrical tubular surface the incidence angle is defined as the angle between the sun's rays and a plane perpendicular to the long axis of the tube. For a tilted tubular collector with a north–south centerline

$$\cos i_t = \{1 - [\sin(\beta - L) \cos \delta_s \cos h_s$$
$$+ \cos(\beta - L) \sin \delta_s]^2\}^{1/2}. \quad (2.20)$$

Sunrise and sunset on a fixed surface occur when $i = 90°$. It is possible for collector surface sunset to occur before the sun moves below the horizon; therefore collector sunset dictates solar system operational hours under this condition. If $i = 90°$ is substituted into Eq. (2.19), the hour angle for *collector sunrise* $h_{sr}(i = 90)$ and *sunset* can be calculated:

$$\cos 90° = 0 = \sin(L - \beta) \sin \delta_s$$
$$+ \cos(L - \beta) \cos \delta_s \cos h_{sr}(i = 90). \quad (2.21)$$

Solving for the sunrise hour angle $h_{sr}(i = 90)$,

$$h_{sr}(i = 90) = \cos^{-1}[-\tan(L - \beta) \tan \delta_s] \qquad (2.22)$$

and the sunset hour angle is

$$h_{ss}(i = 90) = -h_{sr}(i = 90). \qquad (2.23)$$

The effective time of sunrise h_{sr} for a solar system is the smaller of the two hour angles $h_{sr}(\alpha = 0)$ from Eq. (2.11) and $h_{sr}(i = 90)$ from Eq. (2.22). That is,

$$h_{sr} = \min[h_{sr}(\alpha = 0), h_{sr}(i = 90)] \qquad (2.24a)$$

and

$$h_{ss} = -h_{sr}. \qquad (2.24b)$$

If a fixed collector does not face due south, a more complex equation than Eq. (2.22) must be used to find h_{sr} and h_{ss} for $i = 90°$ (K1).

C. Incidence Angles for Tracking Collectors

Most of the collectors considered in this book are concentrating devices which require some form of sun tracking to maintain the sun within their field of view and to minimize the incidence angle. Either one or two-degree-of-freedom modes are used. Two-degree-of-freedom tracking maintains the collector aperture normal to the sun's rays so that the minimum incidence angle is zero. Therefore, for perfect tracking

$$\cos i = 1.0. \qquad (2.25)$$

Single-axis tracking can involve either a horizontal, east–west axis or a horizontal, north–south axis of rotation. Also, a polar mount can be used which employs a north–south axis tilted up at an angle equal to the local latitude. For these three single-axis tracking modes, the incidence angles are

(1) E–W horizontal:

$$\cos i = (1 - \cos^2 \alpha \sin^2 a_s)^{1/2}; \qquad (2.26)$$

(2) N–S horizontal:

$$\cos i = (1 - \cos^2 \alpha \cos^2 a_s)^{1/2}; \qquad (2.27)$$

(3) Polar:

$$\cos i = \cos \delta_s. \qquad (2.28)$$

A less common tracking mode uses a collector rotating about a vertical axis. If the collector is tilted at an angle β from the horizontal, the incidence angle is

$$\cos i = \sin(\alpha + \beta). \tag{2.29}$$

V. THE SOLAR ENERGY RESOURCE

The first section of this chapter describes a method of predicting instantaneous clear sky radiation by means of a Bouger's law absorption model. However, the method requires the estimation of several empirical constants and cannot account for effects of clouds and other microclimatic phenomena. Therefore, it is often preferable to use historical radiation data when available. This section describes the historical solar radiation data base and its geographic distribution.

A. Data Measurement

Beam radiation is measured by a normal incidence *pyrheliometer* of the type shown in Fig. 2.6. The basic components include a radiation sensor mounted at the end of a tube and a tracking mechanism. The collimating tube is sufficiently long to block most of the diffuse, sky radiation from the sensor. One common pyrheliometer has a field of view of 5.7°. Since the sun subtends a 32′ angle, some circumsolar diffuse radiation is recorded as beam radiation by this device, however. The circumsolar portion of the measured radiation can vary from 5 to 15% of the total depending on atmospheric conditions.

The tracking device is usually a polar-mount, clock drive which requires occasional adjustment for solar declination. Radiation sensors can be either thermopiles or photocells. A thermopile has a flat wavelength response and is preferred for thermal applications.

Diffuse radiation data are measured in several ways the most common of which uses a horizontal *pyranometer*. Pyranometers of the type shown in Fig. 2.7 measure total or global radiation I_h on the horizontal plane. The horizontal diffuse component $I_{d,h}$ can be calculated by the difference

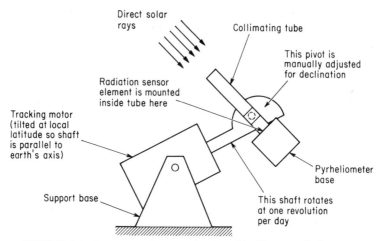

FIGURE 2.6   Schematic diagram of a normal incidence pyrheliometer.

$$I_{d,h} = I_h - I_b \sin \alpha \qquad (2.30)$$

if I_b is known from pyrheliometer data. Alternatively, the horizontal diffuse component can be measured directly by using an occulting disk or ring to block beam radiation from striking the pyranometer. A pyranometer can be tilted to measure global flux on a tilted surface. However, the free convection heat transfer regime within the transparent cover varies with tilt angle and the manufacturer must provide calibration data or correction factors for the tilt angles to be used.

B. Solar Radiation Records

Solar radiation data for the U.S. are measured by the National Weather Service (NWS) and are archived by the National Climatic Center (NCC) in Asheville, North Carolina. The records include data measured in many locations for widely varying periods of time. Many of these data are of questionable accuracy, however, since pyranometer calibrations were not made frequently and the sensor response showed a loss of accuracy with time.

Recently the NCC has rehabilited some hourly and daily time-scale data for 55 sites for the period 1950–1975. These rehabilitated data and not the original data should be used for solar design although problems remain because of the relatively simple method used for filling

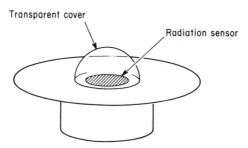

FIGURE 2.7   Schematic diagram of a pyranometer used to measure global radiation.

gaps and accounting for pyranometer loss of calibration. Hourly data recorded since 1977 are available under the standard SOLMET format and daily data under the SOLDAY format. Both formats also include standard surface weather data such as temperature, wind speed, and direction and cloud cover.

In addition to historical solar data, synthetic data have been generated for approximately 220 sites using meteorological regression models developed by NOAA. Hourly solar data were predicted using widely available sunshine hour and cloud cover data. These generated data are also available in the SOLMET format. All SOLMET tapes also include predicted direct-normal values. The NWS publishes solar and weather data in a monthly summary entitled "Monthly Summary—Solar Radiation Data." Historical monthly solar data are listed in Appendix Table A.2.

The present solar data measurement network of the NWS consists of 38 stations shown in Fig. 2.8. The network was established in 1977 and each station records both direct-normal (I_b) and total horizontal (I_h) data on a 1-min time scale. These data are then totaled for hourly amounts. Standard surface observations of wind, temperature, etc., are also recorded. Several research solar data stations are also planned. Additional data collected will include spectral irradiance, circumsolar radiation, sky radiation distribution, atmospheric turbidity, and humidity.

Solar data records elsewhere in the world have been summarized, as of 1966, in (L1). Nearly 1000 stations are listed which measure either solar flux or percent of possible sunshine. Monthly solar data for all sites are listed in (L1) along with a description of the algorithm used to compute average solar radiation from percent of possible sunshine measurements. In the years since 1966, many stations have been added so that the total is now well over 1000. The locations of the 1966 sites are shown in Fig. 2.9.

FIGURE 2.8   The U.S. NWS solar radiation data network. Not shown are Fairbanks, Alaska, Puerto Rico, Hawaii, and Guam.

C.   Solar Flux Maps

A convenient means of graphically displaying average solar flux data is by plotting isopleths of solar flux on a map. Many kinds of data may be plotted, e.g. direct normal, total horizontal, diffuse. Maps are constructed using standard computer contour routines with solar data and site locations as inputs. An example map showing average horizontal flux for the world for September is shown in Fig. 2.10.

Beam radiation levels are of the most interest in this book since concentrating collectors will be used for most applications. Figure 2.11 shows direct-normal flux maps for the U.S. for four seasons of the year. These maps were prepared by Boes (B1) using rehabilitated hourly data for 26 SOLMET stations for the period 1958–1962. The data on these maps show the availability of radiation to two-degree-of-freedon tracking collectors. The data from which Fig. 2.11 was prepared are contained in Appendix Table A.3.

Solar isoflux maps should not be used for careful system design since the resolution of the maps is usually poor and the data base used for contouring is sparse. Maps are useful for regional-level preliminary analysis but not for detailed engineering and economic analysis for a specific project.

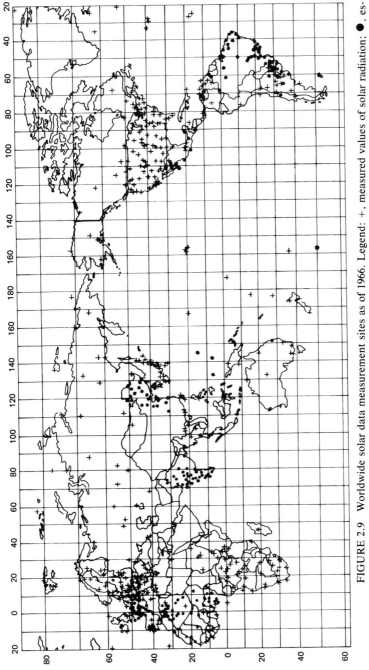

FIGURE 2.9 Worldwide solar data measurement sites as of 1966. Legend: +, measured values of solar radiation; ●, estimated values of solar radiation. [From (T2).]

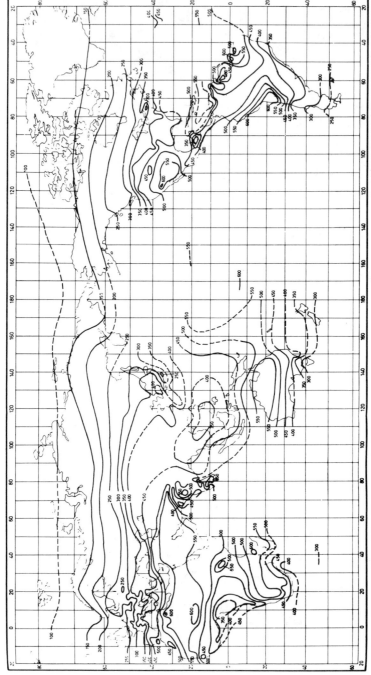

FIGURE 2.10 Monthly averaged, daily horizontal solar flux for September for the world in cal/cm². [From (D1).]

(a)

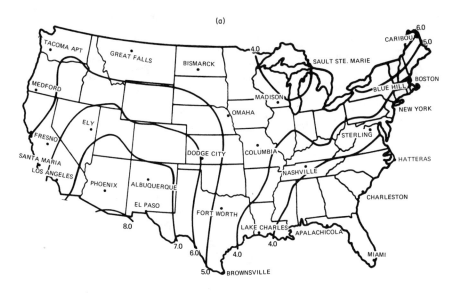

(b)

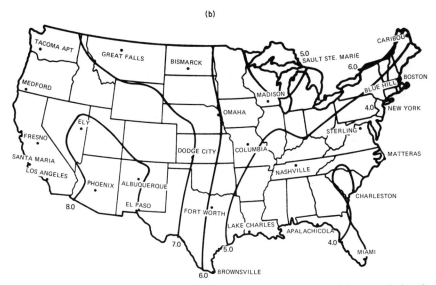

FIGURE 2.11 (a) Seasonally averaged direct-normal beam radiation (in kW h/m²) for March–May. [From (B1).] (b) Seasonally weighed beam radiation (in kW h/m²) for June–August. [From (B1).] (c) Seasonally averaged beam radiation (in kW h/m²) for September–November. [From (B1).] (d) Seasonally averaged beam radiation (in kW h/m²) for December–February. [From (B1).]

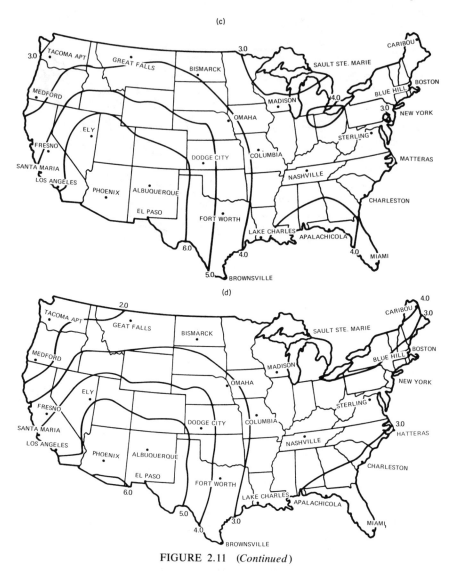

FIGURE 2.11 (*Continued*)

D. Circumsolar Radiation

          Since pyrheliometers in common use record circumsolar
radiation is a 5–6° cone, it is important to know what portion of these
records consist of flux from only within the solar disk. High concentration

solar collectors are designed to collect sunlight from only the solar disk, not the 5–6° pyrheliometer cone; hence, the assumption that pyrheliometric records represent only the solar disk flux will cause errors in the estimation of performance of such collectors, notably of the point focus type.

A substantial literature on solar aureole flux levels does not presently exist. It has been recently recognized that aureole data are required and the U.S. Lawrence Berkeley Laboratory has deployed four aureole telescopes to measure the aureole flux (G5). These data have not yet been processed into a uniform format for easy design use.

The Swedish Astrophysical Observatory (R14) made aureole measurements over several years and found that the ratio of aureole brightness i_{sky} to that of the solar disk i_{sun} near noon could be well represented for clear days by equations of the form

$$\log_{10} i_{sky}/i_{sun} = -(kR + l) \qquad (2.31)$$

where k and l are constants and R is the distance from the sun *center* measured in units of solar radii; the range of validity of Eq. (2.31) is $R = [1.25, 5.0]$, i.e., $R \approx [4', 66']$ of arc from the edge of the sun. The constants calculated from data in (R14) are

Climate type	k	l
Maritime polar	0.256	4.13
Continental polar	0.193	3.97
Maritime tropic	0.176	3.88
Continental tropic	0.133	3.46

These data show that aureole intensities are from 10^{-3} to 10^{-5} of the solar disk intensities on clear days. At times away from noon the relative aureole intensity increases but detailed data on this phenomenon are not yet available.

VI. INSTANTANEOUS AND HOURLY SOLAR RADIATION CALCULATIONS

Three time scales naturally suggest themselves in the thermal analysis and performance prediction of solar-thermal systems—hourly, daily, and monthly. Monthly time scales, as will be seen in subsequent chapters, are useful in rapidly estimating solar system performance. Daily and hourly time scales are useful if more accurate calculations are required by computer. This section describes instantaneous and hourly calculation. Subsequent sections consider daily and monthly time scales.

A. Hourly Extraterrestrial Radiation

Equations (2.1) and (2.18) can be combined to give the instantaneous, extraterrestrial *horizontal* solar flux $I_{0,h}$ at time t (or hour angle h_s) measured from local solar noon:

$$I_{0,h} = I_{sc}[1.0 + 0.034 \cos(2\pi N/365)]$$
$$\times (\sin L \sin \delta_s + \cos L \cos \delta_s \cos h_s). \qquad (2.32)$$

Equation (2.32) can be integrated over an hour centered at hour angle h_s to give the hourly horizontal extraterrestrial radiation $I_{0,h}(h_s)$:

$$I_{0,h}(h_s) = \{I_{sc}[1.0 + 0.034 \cos(2\pi N/365)]\}$$
$$\times (0.9972 \cos L \cos \delta_s \cos h_s + \sin L \sin \delta_s). \qquad (2.33)$$

For engineering purposes the numerical coefficient 0.9972 is usually replaced by 1.0. Therefore, for engineering purposes, the total hourly extraterrestrial radiation is the same as the instantaneous radiation evaluated at the hour's midpoint.

B. Hourly Terrestrial Radiation

Hourly beam radiation useful for solar thermal processes operating at elevated temperatures is not generally available over a long, historical period. Until such data become available, a predictive method relating hourly total radiation to beam radiation must be used. Boes (B1) has developed such a predictive scheme.

It has been found that the hourly *clearness index* k_T or atmospheric transmittance can be used to correlate beam and total radiation. The hourly clearness index is defined as the ratio of total terrestrial, horizontal radiation I_h to extraterrestrial, horizontal radiation $I_{0,h}(h_s)$; that is,

$$k_T = I_h/I_{0,h}(h_s) \qquad (2.34)$$

where h_s is the hour angle at the midpoint of the hour of interest.

The expression relating terrestrial beam radiation I_b to extraterrestrial radiation is (B1)

TABLE 2.8

*Coefficients for Hourly Bean
Radiation Model*[a]

Interval for k_T	$a(k_T)$	$b(k_T)$
0.00, 0.05	0.04	0.00
0.05, 0.15	0.01	0.002
0.15, 0.25	0.06	−0.006
0.25, 0.35	0.32	−0.071
0.35, 0.45	0.82	−0.246
0.45, 0.55	1.56	−0.579
0.55, 0.65	1.69	−0.651
0.65, 0.75	1.49	−0.521
0.75, 0.85	0.27	0.395

[a] From (B1).

$$I_b = [a(k_T) \times k_T + b(k_T)] \times I_0 \tag{2.35}$$

where I_0 is given by Eq. (2.1).

The piecewise linear model uses two coefficients $a(k_T)$ and $b(k_T)$ which depend upon the clearness index, as shown in Table 2.8. Equation (2.35) is plotted in Fig. 2.12.

Hourly, horizontal diffuse radiation $I_{d,h}$ can be calculated by difference from

$$I_{d,h} = I_h - I_b \sin \alpha. \tag{2.36}$$

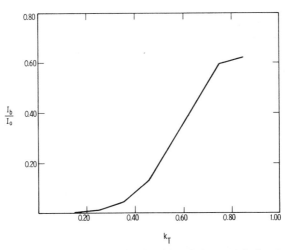

FIGURE 2.12 Graph of ratio of hourly beam radiation I_b to extraterrestrial, direct-normal radiation I_0 vs. clearness index k_T. [From (B1).]

Hourly beam radiation intercepted by a collector surface $I_{b,c}$ is given by

$$I_{b,c} = I_b \cos i \tag{2.37}$$

where the incidence angle i can be calculated from Eqs. (2.17) to (2.29) depending upon the tracking mode.

Hourly diffuse radiation on a collector surface $I_{d,c}$ is calculated from

$$I_{d,c} = I_{d,h} \cos^2(\beta/2) \tag{2.38}$$

where β is the tilt angle. Equation (2.38) assumes an isotropic, diffuse radiation distribution in the sky.

Hourly radiation $I_{r,c}$ diffusely reflected from a uniform foreground of reflectance ρ onto a collector aperture is given by

$$I_{r,c} = \rho I_h \sin^2(\beta/2) \tag{2.39}$$

If hourly horizontal data are not available for use in Eq. (2.36), I_h can be estimated from (R1)

$$I_h = I_{0,h}(h_s)[83.02 - 3.847m - 4.407(\text{CC}) + 1.1013(\text{CC})^2 - 0.1109(\text{CC})^3]/100 \tag{2.40}$$

where m is the air mass and CC is the opaque cloud cover recorded at many sites by the NWS (CC = 1.0 for full overcast and CC = 0.0 for a cloudless sky). Although this correlation was based on data for only three cities, the predictions were within 3.2% of measured solar data for an-

TABLE 2.9

Percent of Daily Beam Radiation Occurring during Each Hour of Clear Days[a]

Length of day (h)				am			Time
	5–6	6–7	7–8	8–9	9–10	10–11	11–12
8	—	—	—	3.2	10.2	16.4	19.9
9	—	—	—	5.2	10.8	15.5	18.1
10	—	—	2.0	6.2	11.0	14.1	16.6
11	—	0.4	3.1	7.2	10.5	13.4	15.3
12	—	1.1	4.0	7.7	10.4	12.6	14.1
13	0.4	1.7	4.8	7.9	10.0	11.8	13.3
14	0.7	2.5	5.4	6.9	9.8	11.3	12.3
15	1.3	3.2	5.6	7.8	9.6	10.7	11.5
16	1.8	3.9	5.9	7.5	9.2	10.3	11.0

[a] Hourly percentages are symmetric about noon.

other site. A method of estimating hourly solar data if only daily totals are known is described in (L2). Table 2.9 shows percentages of daily radiation which occur during each hour of a clear day.

One method of estimating *average* hourly, beam solar flux if the commonly measured *daily* total (direct plus diffuse) is known, was developed by Collares-Pereira and Rabl (C9). If $\bar{r}_b(h_s)$ is the ratio of *hourly horizontal beam* radiation $I_{b,h}$ to the *daily total of beam and diffuse* flux $\bar{H}_H$ (see Section VIII), *on the average,* then

$$\bar{r}_b(h_s- = \frac{\pi}{24}\left[\frac{\cos h_s - \cos h_{sr}(\alpha = 0)}{\sin h_{sr}(\alpha = 0) - h_{sr}(\alpha = 0)\cos h_{sr}(\alpha = 0)}\right]$$
$$\times [a + b\cos h_s] - (I_{d,h}/\bar{H}_h), \tag{2.41}$$

where

$$a = 0.409 + 0.5016\sin(h_s - 60),$$
$$b = 0.6609 - 0.4767\sin(h_s - 60).$$

The ratio $I_{d,h}/\bar{H}_h$ is the ratio of daily horizontal diffuse to daily horizontal total radiation, on the average, described in the Section VIII.

VII.   DAILY SOLAR RADIATION CALCULATIONS

A.   Daily Extraterrestrial Radiation

Equation (2.32) can be integrated over a day to give daily total, extraterrestrial solar flux on a horizontal surface $\bar{I}_{0,h}$, that is,

$$\bar{I}_{0,h} = \int_{h_{sr}(\alpha=0)}^{h_{ss}(\alpha=0)} I_{0,h}(t)\, dt, \tag{2.42}$$

where sunrise and sunset times are given by Eq. (2.24) and $t = h_s/15$ in hours. Carrying out the integration gives

$$\bar{I}_{0,h} = (24/\pi)I_{sc}\{[1 + 0.034\cos(2\pi N/365)]\}$$
$$\times (\cos L\cos\delta_s\sin h_{sr} + h_{sr}\sin L\sin\delta_s). \tag{2.43}$$

The declination δ_s is evaluated at midday. h_{sr} *is in units of radians when not the argument of a trigonometric function in this and following equations.* Values of $\bar{I}_{0,h}$ are plotted in Fig. 2.13.

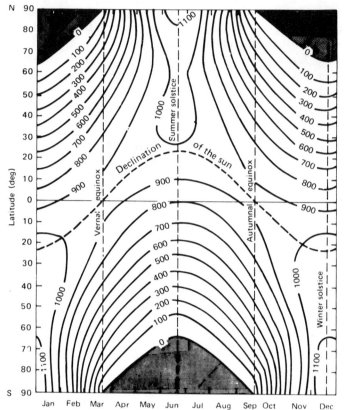

FIGURE 2.13 Daily total extraterrestrial horizontal radiation $\bar{I}_{0,h}$ for various latitudes and times of the year.

B. Daily Terrestrial Radiation

Daily beam radiation is not generally available from meteorological data services. As a result, a predictive method similar to that used for hourly radiation is required. A *daily clearness index* K_T is used as a measure of daily atmospheric transmittance. K_T is defined as

$$K_T = \bar{I}_h / \bar{I}_{0,h} \tag{2.44}$$

for each day, where $\bar{I}_h$ is the daily total horizontal radiation on earth. Terrestrial beam radiation *on a horizontal surface* $\bar{I}_{b,h}$ can be calculated using a correlation developed by Collares-Pereira and Rabl (C9) based on earlier work of Liu and Jordan (L3,L4,L5,L6).

$$\bar{I}_{b,h}/\bar{I}_h = -0.188 + 2.272K_T - 9.473K_T^2$$
$$+ 21.856K_T^3 - 14.648K_T^4 \quad (2.45)$$

for $K_T = [0.17, 0.80]$ and for $K_T < 0.17$, $\bar{I}_{b,h} \sim 0.0$.

Beam radiation on a south-facing, *fixed*, tilted surface $\bar{I}_{b,c}$ is given by

$$\bar{I}_{b,c} = \bar{I}_{b,h} \left[\frac{\cos(L - \beta) \cos \delta_s \sin h_{sr} + h_{sr} \sin(L - \beta) \sin \delta_s}{\cos L \cos \delta_s \sin h_{sr}(\alpha = 0) + h_{sr}(\alpha = 0) \sin L \sin \delta_s} \right]. \quad (2.46)$$

The factor in brackets is the integrated daily value of the tilt factor (see Fig. 2.5) relating instantaneous horizontal beam radiation $I_{b,h}$ to that on a tilted surface $I_{b,c}$, outside the atmosphere. It is assumed that the same tilt factor applies to terrestrial beam radiation.

Daily beam and total radiation on a *tracking* collector can be evaluated using the methods given in the next section. Daily diffuse radiation on a horizontal surface $\bar{I}_{d,h}$ can be evaluated from

$$\bar{I}_{d,h} = \bar{I}_h - \bar{I}_{b,h}. \quad (2.47)$$

VIII.   MONTHLY RADIATION CALCULATIONS

In subsequent chapters monthly time scale calculations will be used to estimate the performance of many solar-thermal systems. Monthly calculations have the attraction that only twelve calculations need be performed to evaluate average, annual performance. With the monthly methods used in Chapters 5 and 6, engineering accuracy can be achieved without resort to lengthy daily or hourly computations.

A.   *Monthly Extraterrestrial Radiation*

Instead of using monthly totals of solar flux, the usual practice is to divide monthly totals by the actual number of days in each month to calculate *monthly averaged, daily* amounts usually called "monthly" for short. Equation (2.48) can be used to calculate $\bar{H}_{0,h}$, the monthly horizontal extraterrestrial radiation from

$$\bar{H}_{0,h} = \frac{I_{sc}}{\Delta N} \int_{N_i}^{N_f} \left(1.0 + 0.034 \cos \frac{2\pi N}{365}\right)$$

$$\times \int_{h_{sr}(\alpha=0)}^{h_{ss}(\alpha=0)} (\sin L \sin \delta_s(N)$$

$$+ \cos L \cos \delta_s(N) \cos h_s) \, dt' \, dN, \qquad (2.48)$$

where N_i is the initial day number for a month, N_f is the final day number, $dt' = dh_s/15$, and the declination $\delta_s(N)$ is given by Eq. (2.7). Table 2.10 contains values of $\bar{H}_{0,h}$ calculated from Eq. (2.48) for various latitudes. A simpler method for calculating $\bar{H}_{0,h}$ is to evaluate the argument of Eq. (2.43) at the "design day number" N_{des} which gives a value of $\bar{I}_{0,h}$ which most closely agrees with value of $\bar{H}_{0,h}$ for the month. Klein (K3) recommends for this purpose the design day numbers shown in Table 2.11.

B. Monthly Terrestrial Radiation

Since long-term beam radiation data are not available, a predictive method using a monthly clearness index $\bar{K}_T$ is used. The monthly clearness index is

$$\bar{K}_T = \bar{H}_h/\bar{H}_{0,h}, \qquad (2.49)$$

where $\bar{H}_h$ is the monthly total solar radiation on a terrestrial horizontal surface. Using a correlation developed by Collares-Pereira and Rabl (C9) the diffuse radiation on horizontal surface $\bar{D}_h$ is given by (h_{sr}, radians)

$$\frac{\bar{D}_h}{\bar{H}_h} = 0.775 + 0.347[h_{sr}(\alpha = 0) - \pi/2]$$

$$- \{0.505 + 0.261[h_{sr}(\alpha = 0) - \pi/2]\} \cos \left[(\bar{K}_T - 0.9) \frac{360}{\pi}\right]. \quad (2.50)$$

Using Eq. (2.50), the total radiation $\bar{I}_c$ incident on surfaces of several tracking types can be calculated from

$$\bar{I}_c = [r_T - r_d(\bar{D}_h/\bar{H}_h)]\bar{H}_h. \qquad (2.51)$$

Equations for r_T and r_d are given in Table 2.12 for various tracking geometries. The collection start (or stop) time h_{coll} used in Table 2.12 represents the time at which the collector starts (or stops) operating measured from solar noon. The collection time is either determined by an optical cutoff or by the system controller which decides that collector operation is no longer worthwhile. Cutoff time is described more thoroughly in Chapter 5.

TABLE 2.10

Monthly Averaged, Extraterrestrial Solar Radiation on a Horizontal Surface $\bar{H}_{o,h}^{a,b}$

Latitude (deg)	Jan	Feb	Mar	Apr	May	Jun	Jul	Aug	Sep	Oct	Nov	Dec
20	7415	8397	9552	10422	10801	10868	10794	10499	9791	8686	7598	7076
25	6656	7769	9153	10312	10936	11119	10988	10484	9494	8129	6871	6284
30	5861	7087	8686	10127	11001	11303	11114	10395	9125	7513	6103	5463
35	5039	6359	8153	9869	10995	11422	11172	10233	8687	6845	5304	4621
40	4200	5591	7559	9540	10922	11478	11165	10002	8184	6129	4483	3771
45	3355	4791	6909	9145	10786	11477	11099	9705	7620	5373	3648	2925
50	2519	3967	6207	8686	10594	11430	10981	9347	6998	4583	2815	2100
55	1711	3132	5460	8171	10358	11352	10825	8935	6325	3770	1999	1320
60	963	2299	4673	7608	10097	11276	10657	8480	5605	2942	1227	623
65	334	1491	3855	7008	9852	11279	10531	8001	4846	2116	544	97

[a] In W h/m² day.

Latitude (deg)	Jan	Feb	Mar	Apr	May	Jun	Jul	Aug	Sep	Oct	Nov	Dec
20	2346	2656	3021	3297	3417	3438	3414	3321	3097	2748	2404	2238
25	2105	2458	2896	3262	3460	3517	3476	3316	3003	2571	2173	1988
30	1854	2242	2748	3204	3480	3576	3516	3288	2887	2377	1931	1728
35	1594	2012	2579	3122	3478	3613	3534	3237	2748	2165	1678	1462
40	1329	1769	2391	3018	3455	3631	3532	3164	2589	1939	1418	1193
45	1061	1515	2185	2893	3412	3631	3511	3070	2410	1700	1154	925
50	797	1255	1963	2748	3351	3616	3474	2957	2214	1450	890	664
55	541	991	1727	2585	3277	3591	3424	2826	2001	1192	632	417
60	305	727	1478	2407	3194	3567	3371	2683	1773	931	388	197
65	106	472	1219	2217	3116	3568	3331	2531	1533	670	172	31

[b] In Btu/day ft².

TABLE 2.11

Recommended Design Day Numbers N_{des}
for Monthly Calculations[a]

Month	Day number	Date
Jan	17	1/17
Feb	47	2/16
Mar	75	3/16
Apr	105	4/15
May	135	5/15
Jun	162	6/11
Jul	198	7/17
Aug	228	8/16.
Sep	258	9/15
Oct	288	10/15
Nov	318	11/14
Dec	344	12/16

[a] From (K3).

In Table 2.12 the value of solar collector concentration ratio CR enters since concentrating collectors accept $(CR)^{-1}$ of the diffuse radiation at their apertures.

To determine only the beam radiation on a tracking collector, set $(CR)^{-1} = 0.0$ in Table 2.12. The concentration ratio described in detail in Chapter 4 is the ratio of solar collector aperture area to collector receiver area. Boes (B1) recommends an alternative method of calculating $\bar{B}$, the monthly beam radiation:

$$\bar{B} = -3.17 + 1.31 \bar{H}_h \sin \bar{\alpha} \qquad (2.52)$$

in units of kw h/m² day. The average altitude angle $\bar{\alpha}$ is evaluated for the fifteenth day of the month at hour angle $h_s = 0.25 h_{sr}(\alpha = 0)$.

Note that Eq. (2.51) and expressions for r_T and r_d can used on a daily time scale to find $\bar{I}_c$ on a daily basis by using the sunset time on the collector and on a horizontal surface for a specific day.

If the value of $\bar{H}_h$ has not been measured for a specific site where needed, (K2) contains a method of estimating it from other climatic data.

C. Summary

Table 2.13 summarizes the equations developed in this section for finding (1) horizontal extraterrestrial, (2) horizontal terrestrial, (3)

TABLE 2.12

Parameters r_T and r_d Used to Calculate Monthly Solar Flux Incident on Various Collector Types[a]

Collector type	$r_T^{b,c,d}$	r_d^{e}
Fixed aperture concentrators which do not view the foreground	$[\cos(L - \beta)/(d \cos L)]\{[-ah_{coll} \cos h_{sr}(i = 90)]$ $+ [a - b \cos h_{sr}(i = 90)] \sin h_{coll}$ $+ (b/2)(\sin h_{coll} \cos h_{coll} + h_{coll})\}$	$(\sin h_{coll}/d)\{[\cos(L + \beta)/\cos L] - [1/(CR)]\}$ $+ (h_{coll}/d)\{[\cos h_{sr}(\alpha = 0)/(CR)]$ $- [\cos(L - \beta)/\cos L] \cos h_{sr}(i = 90)\}$
East–west axis tracking[f]	$\frac{1}{d} \int_0^{h_{coll}} \{[(a + b \cos x)/\cos L]$ $\times \sqrt{\cos^2 x + \tan^2 \delta_s}\}\, dx$	$\frac{1}{d} \int_0^{h_{coll}} \{(1/\cos L)\sqrt{\cos^2 x + \tan^2 \delta_s}$ $- [1/(CR)][\cos x - \cos h_{sr}(\alpha = 0)]\}\, dx$
Polar tracking	$(ah_{coll} + b \sin h_{coll})/(d \cos \delta_s \cos L)$	$(h_{coll}/d)\{(1/\cos L) + [\cos h_{sr}(\alpha = 0)/(CR)]$ $- \sin h_{coll}/[d(CR)]\}$
Two-axis tracking	$(ah_{coll} + b \sin h_{coll})/(d \cos \delta_s \cos L)$	$(h_{coll}/d)\{1/(\cos \delta_s \cos L) + [\cos h_{sr}(\alpha = 0)/(CR)]$ $- \sin h_{coll}/[d(CR)]\}$

[a] From (C1); the collection hour angle value h_{coll} not used as the argument of trigonometric functions is expressed in radians; note that the total interval $2h_{coll}$ is assumed to be centered about solar noon.

[b] $a = 0.409 + 0.5016 \sin[h_{sr}(\alpha = 0) - 60°]$.

[c] $b = 0.6609 - 0.4767 \sin[h_{sr}(\alpha = 0) - 60°]$.

[d] $d = \sin h_{sr}(\alpha = 0) - h_{sr}(\alpha = 0) \cos h_{sr}(\alpha = 0)$.

[e] CR is the collector concentration ratio; see Chapters 4 and 6.

[f] Use the identity $\cos \delta_s = \sin(90° - \delta_s)$ and multiply the integral by $\cos \delta_s$, a constant. For computer implementation a numerical method can be used. For hand calculations, use Weddle's rule or Cote's formula. Elliptic integral tables to evaluate terms of the form of $\int_0^h \sqrt{\cos^2 x + \tan^2 \delta_s}\, dx$ contained in r_T and r_d are given in the Appendix.

TABLE 2.13

Equations for Extraterrestrial and Terrestrial Beam and Terrestrial Total Radiation from Hourly, Daily, and Monthly Time Scales[a]

	Hourly	Daily	Monthly
Extraterr. horiz. flux	$I_{o,h}(h_s) = \{I_{sc}[1.0 + 0.034\cos(2\pi N/365)]\}$ $\times (0.9972\cos L\cos\delta_s\cos h_s + \sin L\sin\delta_s)$	$\bar{I}_{o,h} = (24I_{sc}/\pi)\{[1 + 0.034(2\pi N/365)]\}$ $\times (\cos L\cos\delta_s\sin h_{sr} + h_{sr}\sin L\sin\delta_s)$	$\bar{H}_{o,h} = (24I_{sc}/\pi)\{[1 + 0.034(2\pi N_{des}/365)]\}$ $\times [\cos L\cos\delta_s(N_{des})\sin h_{sr} + h_{sr}\sin L\sin\delta_s(N_{des})]$
Horiz. total rad.	$I_h = [83.02 - 3.847m - 4.07(CC) + 1.1013(CC)^2 - 0.1109(CC)^3][I_{o,h}(h_s)/100]$	$\bar{I}_h$ assumed available from meteorological services	$\bar{H}_h$ assumed available from meteorological services
Horiz. beam rad.	$I_{b,h} = \{\pi/24[\cos h_s - \cos h_{sr}(\alpha=0)/\sin h_{sr}(\alpha=0)$ $- h_{sr}(\alpha=0)\cos h_{sr}(\alpha=0)]$ $\times [a + b\cos h_s] - I_{d,h}/\bar{H}_h\} \times I_h$ if only daily data I_h are available: or $([a(k_T) \times k_T + b(k_T)]\sin\alpha) \times I_o$ if hourly solar data I_h are available	$\bar{I}_{b,h} = (-0.188 + 2.272K_T - 9.473K_T^2 + 21.856K_T^3 - 14.648K_T^4] \times \bar{I}_h$	$\bar{B}_h = \{0.225 - 0.347[h_{sr}(\alpha=0) - \pi/2] + (0.505 + 0.261[h_{sr}(\alpha=0) - \pi/2]) \times (\cos[(\bar{K}_T - 0.9)(360/\pi)]\} \times \bar{H}_h$
Total rad. on a surface	$I_c = [a(k_T) \times k_T + b(k_T)]$ $\times \{\cos i - [\sin\alpha\cos^2(\beta/2)/(CR)]\}$ $\times I_o + I_h[\cos^2(\beta/2)/(CR)]$	$\bar{I}_c = [r_T - r_d(\bar{I}_{d,h}/\bar{I}_h)]\bar{I}_h$ $(\bar{I}_{d,h} \equiv \bar{I}_h - \bar{I}_{b,h})$ (see Table 2.12)	$\bar{I}_c = [r_T - r_d(\bar{D}_h/\bar{H}_h)]\bar{H}_h$ $(\bar{D}_h \equiv \bar{H}_h - \bar{B}_h)$

[a] Abbreviations in table are, extraterr.: extraterrestrial, horiz.: horizontal, rad.: Radiation.

horizontal beam and (4) total radiation on surfaces. The equations are given for hourly, daily, and monthly time scales.

IX. SELECTED TOPICS IN OPTICS

In the context of this book optics is the descriptive method which quantitatively predicts the interaction of a surface and a beam of incident solar radiation. The previous sections of this chapter have described the calculation of beam radiation intercepted by a surface. This section analyzes the interaction of radiation and reflective and transparent elements of solar collectors. Radiation may either be reflected from, transmitted through, or absorbed by a substance as shown in Fig. 2.14. The first law of thermodynamics requires that

$$\tau + \alpha + \rho = 1. \tag{2.53}$$

That is, the three components must total the incident radiation amount. The transmittance τ, absorptance α, and reflectance ρ are, respectively, the ratios of radiation transmitted, absorbed, or reflected to the incident total.

In solar conversion systems three special optical phenomena are of interest:

transmission through a transparent material;

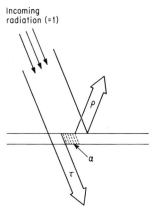

FIGURE 2.14   Schematic representation of transmittance τ, absorptance α, and reflectance ρ.

reflection from a mirrorlike surface;
absorption of radiation to produce heat.

These three cases apply to many solar collection devices. The first two
cases are described in turn in this section, the third in Chapter 3.

A. Transmission and Reflection in Nonopaque Media

Many high-temperature solar collectors use a transparent
envelope around the absorber to control heat losses from the solar-
absorbing surface. A material which is to serve this purpose well should
have three properties:

high transmittance of sunlight;
long-term durability under long exposure to sunlight and possible high
      temperature;
dimensional stability.

Other properties such as low-weight, low-cost, and low-infrared transmittance are also desirable.

Index of Refraction The transmission of light through a
smooth, homogeneous transparent medium with no internal scattering is
governed by principles laid down long ago by Fresnel, Snell, and Stokes
and outlined below. The optical behavior of a substance can be characterized by two wavelength-dependent physical properties—the index of refraction n and the extinction coefficient K. The index of refraction determines the amount of light reflected from a single interface; the extinction
coefficient, the amount absorbed in a substance in the course of a single
pass of radiation through the material.

Figure 2.15 defines the angles used in analyzing reflection
and transmission of light. The angle i is called the incidence angle as defined earlier. The angle at which a beam is reflected from a specular surface is equal to the incidence angle but opposite in sign. Angle θ_r is the
angle of refraction, which is defined as shown in the figure. The incidence
and refraction angles are related by Snell's law

$$\sin i/\sin \theta_r = n_i'/n_i' = n_r, \tag{2.54}$$

where n_i' and n_r' are the two refractive indices and n_r is the index ratio for
the two substances forming the interface. The refractive index n_r for
glass, the only material of interest for elevated temperature use, is
1.50–1.53 depending on the specific composition.

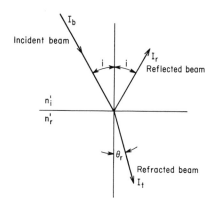

FIGURE 2.15   Incident, reflected, and refracted beams of light and incidence and refraction angles for a transparent medium.

The reflectance ρ from a surface of a transparent substance is related to the refractive index indirectly by the values of i and θ_r according to the Fresnel equations. The reflectance has two components corresponding to the two components of polarization resolved parallel and perpendicular to the plane of incidence (plane of Fig. 2.15). The perpendicular ($\perp$) and parallel ($\parallel$) components of reflectance, respectively, are given by the relations

$$\rho'_\perp = \sin^2(i - \theta_r)/\sin^2(i + \theta_r), \tag{2.55}$$

$$\rho'_\parallel = \tan^2(i - \theta_r)/\tan^2(i + \theta_r). \tag{2.56}$$

For normal incidence the two components are equal and

$$\rho'_\perp = \rho'_\parallel = (n_r - 1)^2/(n_r + 1)^2. \tag{2.57}$$

The reflectance of a glass–air interface may be reduced by a factor of four by an etching process. If glass is immersed in a silica-supersaturated, fluosilicic acid solution, the acid attacks the glass and leaves a porous silica surface layer. This layer has an index of refraction intermediate between glass and air. Consequently, reflectance losses are reduced significantly.

The efficacy of the process depends upon solution temperature and composition, immersion time, and surface pretreatment. Mar *et al*. (M1) have studied these effects in detail and have devised a repeatable process for producing glass with reflectance of 1% per interface (2% for a pane of glass). They also found that heat treatment at 100°C enhances the durability of the coating significantly. Figure 2.16 shows the spectral reflectance of a pane of glass before and after etching.

A number of empirical equations have been suggested to

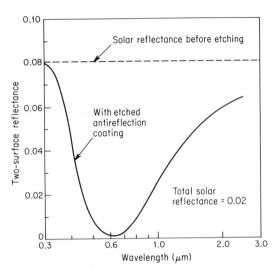

FIGURE 2.16   Reflection spectra for a sample of glass before and after etching. [From (M1).]

describe the wavelength variation of refractive index of glass. An early and simple form due to Cauchy is (H1)

$$n_r = A + B\lambda^{-2} + C\lambda^{-4}. \qquad (2.58)$$

For soda lime glass $A = 1.4$, $B \sim 10^{-14}$ m^2, $C \sim 10^{-28}$ m^4. This equation is satisfactory for the engineering analysis of glass in the solar environment.

Extinction Coefficient   For microscopically homogeneous, transparent media it has been verified experimentally that the light intensity decreases exponentially with layer thickness, as predicted by Beer's law. In equation form

$$I_\lambda(L) = I_\lambda(0)e^{-K_\lambda L}, \qquad (2.59)$$

where L is the layer thickness and K_λ is the extinction coefficient at wavelength λ. Then the transmittance is

$$\tau'_\lambda \equiv I_\lambda(L)/I_\lambda(0) = e^{-K_\lambda L}. \qquad (2.60)$$

The single-pass absorptance α'_λ of homogeneous substances is then

$$\alpha'_\lambda = 1 - e^{-K_\lambda L}. \qquad (2.61)$$

The optical path length L is the thickness of the material t divided by the cosine of the angle of refraction (see Fig. 2.15), i.e.,

$$L = t/\cos \theta_r. \qquad (2.62)$$

Over all wavelengths

$$\alpha' = 1 - e^{-KL} \tag{2.63}$$

where K is the wavelength-averaged value of extinction coefficient K_λ.

It is difficult to measure the extinction coefficient for materials used in solar energy since they are expressly selected to have very small values of K. In order to measure small values of K, a large material thickness L is required. However, thick sections made for measurement may have properties different from those for thin layers used in solar collectors. The value of K for clear glass with low iron content ($<0.01\%$ Fe_2O_3) is about 0.04 cm^{-1}.

Stokes' Equations The quantities ρ', τ', and α' apply to single surfaces, single passes and single reflections. In practice, a layer of glass or other transparent material will have multiple interreflections of radiation which must all be accounted for in calculating the total reflectance ρ, transmittance τ, and absorptance α as shown in Fig. 2.17.

The total wavelength-averaged reflectance ρ, the flux leaving the top surface divided by the incident flux, in Figure 2.17 can be calculated by summing the infinite series of components making up the total flux. For simplicity of notation $\tau' = e^{-KL}$ and ρ' is either $\rho'_\perp$ or $\rho'_\parallel$ based on the wavelength-averaged refractive index. The two series to be summed can be written from the figure; they are

$$\rho = \rho(\rho',\tau')$$
$$= \rho' + \rho'\tau'^2(1 - \rho')^2[1 + \rho'^2\tau'^2 + \rho'^4\tau'^4 + \cdots], \tag{2.64}$$

$$\tau = \tau(\rho',\tau') = (1 - \rho')^2\tau'[1 + \rho'^2\tau'^2 + \rho'^4\tau'^4 + \cdots]. \tag{2.65}$$

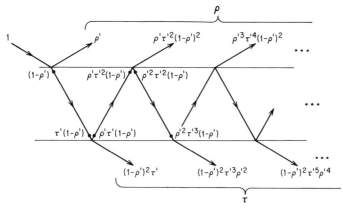

FIGURE 2.17 Ray trace diagram for a transparent medium showing the reflected and transmitted fractions accounting for multiple interreflections. Total fraction reflected and transmitted denoted by ρ and τ, respectively; $\tau' \equiv e^{-KL}$.

The terms in brackets are geometric series with first term unity and ratio $\rho'^2\tau'^2$. Using the formula for the sum of an infinite geometric series from elementary algebra we have

$$\rho(\rho',\tau') = \rho' \left[1 + \frac{\tau'^2(1 - \rho')^2}{1 - \rho'^2\tau'^2} \right], \tag{2.66}$$

$$\tau(\rho',\tau') = \tau' \left(\frac{(1 - \rho')^2}{1 - \rho'^2\tau'^2} \right). \tag{2.67}$$

The total absorptance is simply unity decreased by $(\rho + \tau)$ from Eq. (2.53) or

$$\alpha(\rho',\tau') = 1 - \rho(\rho',\tau') - \tau(\rho',\tau'). \tag{2.68}$$

Equations (2.64)–(2.68) were first developed by G. G. Stokes and are called the Stokes' equations (S1). Using a net radiation method, Siegel (S14) has generalized the Stokes results for layers and surfaces of different optical properties.

Most materials used for optical transmission of solar energy have small values of KL as mentioned above. Therefore, $\tau' \approx 1$ and Eq. (2.67) can be simplified by separating the absorptance and reflectance effects as follows:

$$\tau(\rho',\tau') \approx \tau'[(1 - \rho')/(1 + \rho')]. \tag{2.69}$$

Equation (2.69) is accurate to a few percent for most materials and incidence angles used in solar applications.

Care must be taken in the use of the Stokes' equations to account for polarization properly. The average reflectance $\bar{\rho}$ and the average transmittance $\bar{\tau}$ for both components of polarization, if they are of equal magnitude, are

$$\bar{\rho} = \tfrac{1}{2}[\rho(\rho'_\perp,\tau') + \rho(\rho'_\parallel,\tau')], \tag{2.70}$$

$$\bar{\tau} = \tfrac{1}{2}[\tau(\rho'_\perp,\tau') + \tau(\rho'_\parallel,\tau')]. \tag{2.71}$$

It is incorrect to calculate an average ρ' for both components and use this value in Eqs. (2.64) and (2.65) directly to calculate the average properties. Errors of up to 18% in the calculated transmittance of a layer of glass can result from this improper averaging. The effective optical properties of *multiple transparent layers* are described in (K2). Figure 2.18 shows the calculated transmittance of a layer of glass versus incidence angle i accounting for only reflection losses, i.e., for negligible absorption losses.

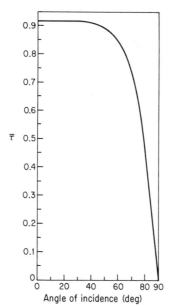

FIGURE 2.18   Transmission versus angle of incidence for glass neglecting absorption ($KL = 0$) with refractive index of 1.53.

B.   Reflection from Opaque Materials

The reflection of radiation from a surface may be *specular* or *diffuse*. In *specular*, or *regular*, *reflection* the angle of incidence of a reflected ray is equal to the angle of reflectance and is in the same plane. Specular reflection, the type encountered in focusing or concentrating solar collectors, is described below. In order for reflection from a surface to be truly specular, all surface microcharacteristics must be smaller than the wavelength of the radiation undergoing reflection. In the case of focusing solar collectors this is usually ensured by using highly polished surfaces or vacuum deposited metals with surface irregularities on the order of nanometers used to reflect light with wavelength of the order of micrometers.

Reflectance of Materials Used in Solar Concentrators   In most advanced radiation heat transfer texts the reflectance of a surface is shown to depend on four angles—the incident, reflected, beam azimuth, and incidence angles. Since few reflectance data of this detailed sort are available, it is a nonessential complication to consider the angular variation of reflectance. In nearly all reflective solar devices it is sufficient (based on comparisons of field test data with prediction) to consider the reflectance as a number dependent at most upon wavelength but independent of incident and azimuth angles.

Usually the reflecting medium used in solar concentrators is rather costly or structurally weak. Therefore, it is generally attached to a substrate of some type. For front-surface mirrors the substrate can be any structural material. For rear- or second-surface mirrors the superstrate, through which the radiation must pass, is transparent and must have high transmittance in the solar spectrum. In addition, the infrared (ir) properties of the transparent medium are important in some concentrator designs since appreciable radiation heat transfer can occur between the hot absorber surface and nearby regions within a few cm of the reflecting surface (see Chapter 4). If the reflecting surface is plastic, it may be necessary to use a high-temperature blend depending upon the radiation properties of the transparent material. The only economic, superstrate transparent material known to be inert and durable is glass.

The most common materials used for solar reflecting surfaces are silver and aluminum. Stainless steel, copper, gold, rhodium, or platinum and other exotic metals could be used but either have low reflectance values or are too costly to be useful in solar applications. Typical specular reflectance values for some materials are given in Table 2.14. The use of specular reflectors is described in Chapter 4.

Dispersion from Specular Reflectors   The idealization that reflection of a beam of radiation occurs about the local surface normal is a satisfactory first-order approximation for most engineering designs. However, reflected radiation is actually distributed in a small cone centered about the nominal reflection direction. For some materials the angular diameter of the cone may be large enough to cause some of the reflected radiation to miss the absorber target.

TABLE 2.14

Specular Reflectance Values for Some Solar Reflector Materials in the Solar Spectrum[a]

Material	Reflectance
Silver (unstable as front surface mirror)	0.94 ± 0.02
Gold	0.76 ± 0.03
Aluminized acrylic, second surface	0.86
Anodized aluminum	0.82 ± 0.05
Various aluminum surfaces—range	0.82–0.92
Copper	0.75

[a] From Pl, © 1977, Pergamon Press, Ltd.

A similar phenomenon occurs in refractive devices used for solar concentration.

Zentner (Z1) has measured the dispersion of light reflected from several materials as shown in Figs. 2.19 to 2.22. Figure 2.19 is for commerical Mylar-S℠, which is seen to be a rather dispersive reflective medium. An experimental Mylar℠ shown in Fig. 2.20 exhibits much less dispersion and a first-surface, aluminized, glass mirror has a nearly perfect specular response (Fig. 2.21). Figure 2.22 shows additional data for other materials at normal incidence (P1).

To conduct mathematical analyses of reflector dispersion it is convenient to have dispersion data expressed in analytical form. Zentner (Z1) suggests the use of a normal distribution about the nominally reflected ray. The reflectance $\rho(\psi)$ at angle ψ measured from the nominal reflected ray direction is then

$$\rho(\psi) = R_1/\sigma_\psi\sqrt{2\pi}\,\exp[-(\psi^2/2\sigma_\psi^2)], \qquad (2.72)$$

where $\sigma_{\psi 1}$ is the standard deviation of the reflected angle distribution ψ and R_1 is the reflectance along the nominal ray direction. Table 2.15 lists

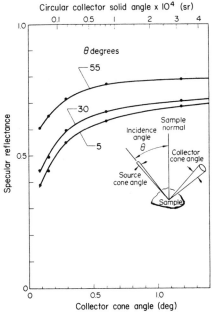

FIGURE 2.19   Dispersion of a Mylar-S℠ reflector for various incidence angles. Sample: first-surface aluminized ⅓-mm Mylar-S film; wavelength 628 nm. [From (Z1), © 1977, Pergamon Press, Ltd.]

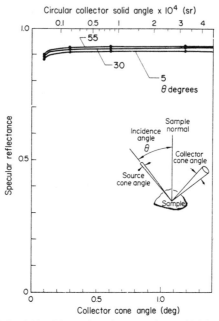

FIGURE 2.20   Dispersion of an experimental Mylar® reflector for various incidence angles. (Z1) Sample: first-surface aluminized 2-mm thick, Dupont experimental film, similar to Mylar type-D; wavelength 628nm. [From (Z1), © 1977, Pergamon Press, Ltd.]

values of R_1 and $\sigma_{\psi 1}$ for glass, plastic, and metals. The standard deviation data show that plastics have the highest dispersion and glass the lowest. For some materials a second normal distribution term is added to Eq. (2.72) to represent the data more accurately.

The data in Table 2.15 are for new, unexposed material samples. Exposure to dust and weathering can decrease specular reflectance by 10–15% owing to increased dispersion. Degradation is particularly rapid for some plastic materials which attract dust electrostatically and are then wiped to clean the surface. Soft material is easily scratched by such cleaning practices. Residual dust or dirt also have a spectral effect on reflectance. Preliminary data on a dirty glass mirror (J3) indicate that a major dropoff of reflectance occurs in the visible range (0.4–0.8 μm) where the solar irradiance (Table 2.2) has its highest value. Therefore, reflectance penalties must be assessed using spectral measurements, not data at a single wavelength. Alternatively, sunlight could be used for reflectance tests, but some experimental difficulties are encountered.

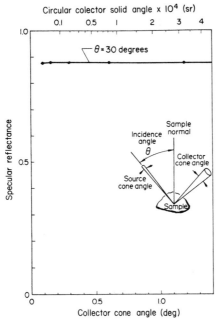

FIGURE 2.21 Dispersion of an aluminized ground glass reflector for various incidence angles. Reference sample: aluminized first-surface mirror on ground glass substrate; wavelength 628 nm. [From (Z1), © 1977, Pergamon Press, Ltd.]

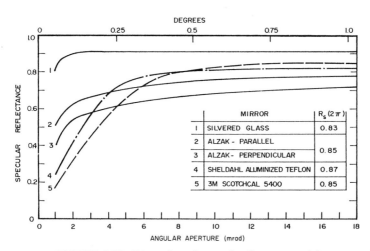

FIGURE 2.22 Dispersion of several reflector materials at normal incidence, R_s is the reflectance integrated over the solar spectrum. [From (P1), © 1977, Pergamon Press, Ltd.]

TABLE 2.15

Reflectance and Dispersion Data for Various Reflective Media[a,b]

Material	Measurement wavelength (nm)	R_1	σ_{r1} (mrad)	R_2	σ_{r2} (mrad)	$R_s(2\pi)$
I. Second surface glass						
(a) Laminated glass—Carolina Mirror Co.	500	0.92	0.15	—	—	0.83
(b) Laminated glass—Gardner Mirror Co.						0.90
Perpendicular to streaks	600	0.92	0.4[c]	—	—	
	500	0.92	0.4[c]	—	—	
Parallel to streaks	800	0.88	<0.05	—	—	
	500	0.92	<0.05	—	—	
(c) Corning microsheet (vacuum chuck)	550	0.77	1.1	0.18	6.2	0.95
II. Metallized plastic films						
(a) 3M Scotchcal 5400	500	0.86	1.9	—	—	0.85
	600	0.86	2.0	—	—	
	700	0.82	2.1	—	—	
	900	0.84	1.9	—	—	
(b) 3M FEK-163	500	0.86	0.90	—	—	0.85
	600	0.86	0.78	—	—	

(c) Sheldahl Aluminized Teflon					
700	0.82	0.86	—	—	
900	0.84	0.86	—	—	0.87
400	0.73	1.4	0.15	12.1	
500	0.80	1.3	0.07	30.9	
700	0.80	1.6	0.04	39.8	
900	0.81	1.4	0.03	31.4	
III. Polished, bulk aluminum					
(a) Alcoa Alzak					
Perpendicular to rolling marks					
670	0.66	0.39	0.21	9.7	0.85
505	0.56	0.42	0.33	10.1	
407.5	0.45	0.53	0.42	9.8	
Parallel to rolling marks					
670	0.70	0.24	0.17	7.7	
505	0.62	0.29	0.27	7.1	
407.5	0.58	0.46	0.29	9.0	
(b) Kingston Ind. Kinglux					
Perpendicular to rolling marks					
498	0.65	0.37	0.23	16.1	0.85
Parallel to rolling marks					
498	0.67	0.43	0.21	18.5	
(c) Metal Fabrications Bright Aluminum	550	0.44	1.4	0.43	10.3	0.84

[a] From (P1).
[b] Alzak is a trade name for Alcoa's anodized aluminum. FEK-163 and Scotchcal 5400 are trade names for Minnesota Mining and Manufacturing's second surface aluminized acrylic.
[c] σ_1 obtained from an area away from a major streak.

SELECTED TOPICS IN THERMODYNAMICS AND HEAT TRANSFER FOR ELEVATED TEMPERATURE SOLAR PROCESSES

3

The sun had long since in the lap
Of Thetis, taken out his nap,
And like a lobster boil'd, the morn
From black to red began to turn

S. Butler

   The analysis and synthesis of solar-thermal systems requires the application of principles of engineering thermodynamics and
heat transfer. This chapter summarizes results of these engineering disci-
plines which are involved in the design of elevated temperature solar
systems. The material in this chapter does not include derivations of governing equations from basic principles. Rather, references to basic texts
are provided for the reader who wishes to review or reexamine assumptions made in arriving at the working equations presented.
   The topics covered in this chapter include descriptions of
first and second law efficiencies of thermal processes, heat engine analysis, and special topics in conduction, convection, and radiation heat
transfer.

I. THERMODYNAMIC PRINCIPLES

Engineering thermodynamics is concerned with various forms of energy—work, heat, chemical energy—and the temperatures at which energetic processes occur. Classical thermodynamics is useful in specifying criteria for the operation of heat engines, the exchange of heat between system components, the partition of entropy, and the efficiency of equilibrium thermal processes. Although the assumption of equilibrium states and quasi-steady processes limits the generality of some results, many conclusions of thermodynamics, when coupled with the nonequilibrium science of heat transfer, are useful in the thermal design of solar systems.

A. Energy Quantity and Quality

Nearly all spoken and written works on the energy question, whether made by technical experts, laymen, or politicians, invariably discuss the conservation of energy. However, any beginning physics student knows that energy must be conserved in any nonrelativistic process. Hence, the conservation of energy is automatically assured by the first law of thermodynamics. Implicit in the erroneous common view is the idea that energy can be saved by not using it. Therefore, no insight is provided into the optimum means of using energy to perform a given task. This econometric view of energy would stipulate that no process be performed at all, to minimize the "use" of energy, a clearly sterile and ineffectual conclusion.

A common measure of energy use efficiency is the first law efficiency η_1. The first law efficiency is defined as the ratio of useful energy output to input energy of a device.

For example, a turbine produces work W from an input of heat Q_i. Therefore, the first law efficiency $\eta_1 = W/Q_i$. For more complex processes, for example, a distillation column with its waste heat used to operate a bottoming cycle turbine, it is difficult to define a first law efficiency. For this example, the outputs of the device are both mass transfer and work, which cannot be easily combined into a single η_1 value.

Another shortcoming of the first law efficiency is its very basis—the ratio of quantities of energy. If an oil-fired boiler operates at a first law efficiency of 80%, little enthusiasm could be expected on the part of a manufacturer in improving its performance. However, use of the same quantity of fuel in an engine-driven heat pump could deliver more heat than the shaft work used to operate the heat pump. Hence, the first law efficiency (sometimes called the coefficient of performance COP) $\eta_1 > 100\%$. Since the first law efficiency is not bounded above by 100%, its utility as an index of possible device performance improvement is very limited.

The second law of thermodynamics—one of the outstanding accomplishments of 19th-century physics (K4)—provides a means of assigning a *quality index* to energy. The concept of *available energy* (G1,K5)—i.e., energy available to do work, the most valuable form of energy—provides a useful measure of energy quality. Using this concept, it is possible to analyze means to minimize the consumption of available energy to perform a given process, thereby insuring the most efficient possible conversion of energy for the required task. Using the concept of availability, it is possible to define a *second law efficiency* η_2 of a process as the ratio of the minimum available energy which *must* be consumed to do a task A_{min} divided by the actual amount of available energy consumed in performing the task.

The second law efficiency is useful in identifying optimal energy conversion processes since it focuses attention on device interactions that transform energy into its two useful types: work (and other ordered forms) and heat. Table 3.1 below shows availabilities and both η_1 and η_2 expressions for several common thermal tasks which may be performed by solar energy or by conventional sources. Table 3.1 is based on values of reversible A_{min} calculated for two energy types, the first involving work W,

$$A_{min} = W, \tag{3.1}$$

the second involving heat Q,

$$A_{min} = Q(1 - T_0/T), \tag{3.2}$$

where $T\ (>T_0)$ is the fixed task end use temperature and T_0 is the environmental sink temperature.

Example Calculate the second law efficiency of a gas space heating system designed to maintain a building at 20°C if the environmental temperature is -10°C. The first law efficiency of the gas furnace is 60%.

From Table 3.1

$$\eta_2 = \eta_1(1 - T_0/T_a)/(1 - T_0/T_r),$$

TABLE 3.1

Availabilities, First, and Second Law Efficiencies for Energy Conversion Processes[a]

	Energy input	
Task	Input shaft work W_i	Q_r from reservoir at T_r
Produce work W_o	$A = W_i$ $A_{min} = W_o$ $\eta_1 = W_o/W_i$ $\eta_2 = \eta_1$ (electric motor)	$A = Q_r[1 - (T_o/T_r)]$ $A_{min} = W_o$ $\eta_1 = W_o/Q_r$ $\eta_2 = \eta_1[1 - (T_o/T_r)]^{-1}$ (solar, Rankine cycle)
Add heat Q_a to reservoir at T_a	$A = W_i$ $A_{min} = Q_a[1 - (T_o/T_a)]$ $\eta_1 = Q_a/W_i$ $\eta_2 = \eta_1[1 - (T_o/T_a)]$ (heat pump)	$A = Q_r[1 - (T_o/T_r)]$ $A_{min} = Q_a[1 - (T_o/T_a)]$ $\eta_1 = Q_a/Q_r$ $\eta_2 = \eta_1[1 - (T_o/T_a)]/[1 - (T_o/T_r)]$ (solar water heater)
Extract heat Q_c from cool reservoir at T_c (below ambient)	$A = W_i$ $A_{min} = Q_c[(T_o/T_c) - 1]$ $\eta_1 = Q_c/W_i$ $\eta_2 = \eta_1[(T_o/T_c) - 1]$ (compression air conditioner)	$A = Q_r[1 - (T_o/T_r)]$ $A_{min} = Q_c[(T_o/T_c) - 1]$ $\eta_1 = Q_c/Q_r$ $\eta_2 = \eta_1[(T_o/T_c) - 1]/[1 - (T_o/T_r)]$ (absorption air conditioner)

[a] The heat reservoirs are considered to be isothermal: T_o is the environmental temperature and processes are reversible: $T_r > T_a > T_o > T_c$.

where T_0, T_a, and T_r are the environment, end use, and source temperatures, respectively.

If the flame temperature is 2300 K,

$$\eta_2 = 0.6(1 - 263/293)/(1 - 263/2300) = 0.07.$$

The second law efficiency is only 7%, indicating an enormous potential for improvement. Note that the flame temperature used in this illustrative example does not actually correspond to an isothermal energy source as required for the use of Table 3.1; see Table 3.2 (below) for an expression for η_2 if the reservoir is not isothermal. The reason for the poor η_2 value in the example is the fundamental mismatch of the low-entropy energy source to a high-entropy task. Show that if a solar heater operating at 30°C were used to perform the 20°C task, the value of η_2 would be 78%.

B. The Thermodynamic Availability of Sunlight

Coherent light from an equilibrium source has zero entropy and is completely available. However, the sun is not an equilibrium source nor is its radiant energy coherent. Therefore, sunlight in near earth space has a finite entropy level. In traversing the atmosphere, the entropy level increases further owing to direction changes (scattering) and frequency and phase shifts.

Parrott (P2) has calculated the availability of beam and diffuse terrestrial solar radiation. For beam radiation the availability A_b is

$$A_b = I[1 - \tfrac{4}{3}(T_0/T_s)(1 - \cos \theta_s/2)^{1/4} + \tfrac{1}{3}(T_0^4/T_s^4)], \tag{3.3}$$

where T_s is the equivalent surface temperature of the sun (~ 5800 K), θ_s is the solar disk angle ($\sim 32'$) and I is the solar flux. For $T_0 = 300$ K, beam radiation is seen to be 99+% available, comparable to or above the availability of the heat of combustion of most fossil fuels. This result elaborates the work of Landsberg (L7,L8).

C. Second Law Efficiencies of Common Thermal Processes

Table 3.1 can be used to calculate the second law efficiencies of many common thermal processes. Table 3.2 summarizes the results of such calculations. It is seen that many low-temperature pro-

TABLE 3.2

U.S. Energy Use Efficiency

Task	Temperatures	η_2	Task	Temperatures	η_2
Space heating			Refrigeration		
Fossil fuels	$T_0 = 275$ K (35°F)	4%	Electric	$T_0 = 294$ K (70°F)	3%
($\eta_1 = 0.6$)	$T_a = 294$ K (70°F)		($\eta_1 = 0.9$)	$T_c = 269$ K (25°F)	
electricity	$T_0 = 275$ K (35°F)	2%	Gas	$T_0 = 294$ K (70°F)	4%
($\eta_1 = 0.9$)	$T_a = 294$ K (70°F)		($\eta_1 = 0.4$)	$T_c = 269$ K (25°F)	
Water heating[a]			Power production		
Fossil fuels	$T_0 = 275$ K (35°F)	10%	$\eta_1 = 0.33$[b]		33%
($\eta_1 = 0.6$)	$T_a = 333$ K (140°F)		Low-temperature power cycles—		
electricity	$T_0 = 275$ K (35°F)	5%	e.g., OTEC-type plant		
($\eta_1 = 0.9$)	$T_a = 333$ K (140°F)		$\eta_1 = 0.02$		2%
Air conditioning			Process steam		
Electric	$T_0 = 308$ K (95°F)	3%	$\eta_1 = 0.85$		$\sim$30%
($\eta_1 = 2.0$)	$T_c = 294$ K (55°F)				
Absorption	$T_0 = 308$ K (95°F)	2%			
($\eta_1 = 0.55 \times 0.8$)	$T_c = 294$ K (55°F)				

[a] The water heating η_2 values assume an approximately isothermal hot water tank. If large temperature excursions are experienced the following expression for η_2 should be used:

$$\eta_2 = \eta_1\{1 - [T_0/(T_a - T_0)] \ln(T_a/T_0)\}.$$

[b] Combined gas turbine, Rankine cycle power plants may achieve $\eta_2 \sim 0.5$ in the future.

cesses are very wasteful of availability. However, higher temperature processes exhibit higher values of η_2. The second law efficiencies of medium- and high-temperature solar process corresponding to those in Table 3.2 are calculated in Chapters 5 and 6.

This brief analysis shows that the second law provides an efficiency index useful in the assessment of energy quality utilization. However, η_2 is not a complete index of process efficacy. The limit $\eta_2 \to 100\%$ can only be achieved with infinitely costly, infinitely slow processes. In addition, η_2 does not contain economic or societal factors which may be equally pertinent in the analysis of solar-thermal systems. Weinberg (W1) has considered both thermodynamic and nonthermodynamic imperatives in the selection of energy sources for the future.

D. Thermodynamics of Heat Engines

Heat engines which produce shaft work by an entropy partition process are used in solar power production, solar pumping, solar cooling, and other applications requiring shaft drive. In this section, three thermodynamic cycles considered for solar applications are described. Also, the ideal Carnot cycle is analyzed.

The Carnot Cycle The Carnot cycle operates at the maximum efficiency permitted by the second law of thermodynamics. Figure 3.1 shows the cycle on a temperature–entropy (TS) diagram with the four cycle steps starting at point 1. Isothermal compression (1–2) is followed by adiabatic compression (2–3). Work is extracted during the isothermal expansion (3–4) and adiabatic expansion (4–1). Since the processes are reversible adiabatic or reversible isothermal, the first law efficiency η_{C1} of the Carnot cycle can be read from Figure 3.1 as

$$\eta_{C1} = 1 - T_c/T_h \qquad (3.4)$$

and the second law efficiency η_{C2} from Table 3.1 is then

$$\eta_{C2} = 1.0. \qquad (3.5)$$

Although the Carnot cycle cannot be realized in practice, it is a useful concept since it represents the upper limit of work production achievable between two heat resevoirs.

The Brayton Cycle The Brayton or gas turbine cycle uses two constant pressure steps instead of the isothermal processes of the Carnot cycle as shown in Fig. 3.2. The working fluid is compressed isen-

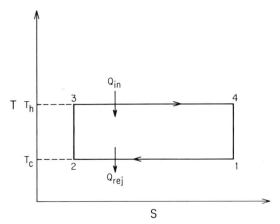

FIGURE 3.1 Ideal Carnot cycle *TS* diagram.

tropically (1–2) to state 2 where heat addition at constant pressure occurs (2–3). Work is produced by an isentropic expansion (3–4) to lower pressure p_L followed by constant pressure expansion (4–1).

The gas turbine cycle is usually operated as an open cycle. That is, the gas at state 4 is exhausted to the atmosphere and fresh air at state 1 is inducted into the turbine. Heat addition (2–3) occurs by means of solar heat addition or fuel combustion in the burner after compression (1–2). Solar-powered Brayton cycles for power production operate the same as gas turbines except that heat addition occurs through a heat exchanger (for which there is an additional second law penalty as described later in the chapter) from the solar collector subsystem.

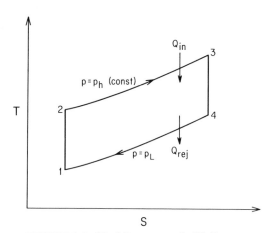

FIGURE 3.2 Ideal Brayton cycle *TS* diagram.

The first law efficiency η_{B1} of the ideal, air standard Brayton cycle is

$$\eta_{B1} = 1 - 1/r_p^{(k-1)/k}, \tag{3.6}$$

where r is the pressure ratio p_h/p_L and k is the specific heat ratio (c_p/c_v).

The calculation of the second law efficiency η_{B2} cannot be done using the equations in Table 3.1 since isothermal resevoirs are not used in a Brayton cycle. The available work used in the cycle ΔA is

$$\Delta A = \int_{T_0}^{T_3} \left(1 - \frac{T_0}{T}\right) c_p \, dT \tag{3.7}$$

and the second law efficiency is (for constant c_p)

$$\eta_{B2} = \eta_{B1} c_p (T_3 - T_2)/c_p[(T_3 - T_0) - T_0 \ln(T_3/T_0)]. \tag{3.8}$$

The Stirling Cycle The Stirling cycle engines utilize isothermal heat addition expansion followed by a constant volume regeneration cooling process. Heat is then rejected to the environment in an isothermal compression followed by regenerative heating. Heat transfer to and from the regenerator is completely within the heat engine and is reversible. Figure 3.3 shows the Stirling cycle schematically.

Since heat exchange within the generator is very slow, Stirling cycle engine speed is slow unless very large heat transfer areas are used. However, the first law efficiency η_{S1} is the same as the Carnot cycle efficiency

$$\eta_{S1} = 1 - T_L/T_h \tag{3.9}$$

and the second law efficiency is unity.

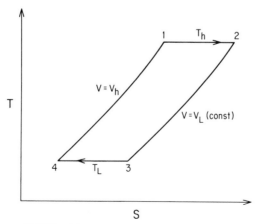

FIGURE 3.3   Ideal Stirling cycle *TS* diagram.

The Rankine Cycle   Steam turbines used for power production operate on the Rankine cycle. The Rankine cycle can be considered as a modified, *vapor-phase* Carnot cycle with the substitution of a liquid-phase pump for the impractical Carnot cycle vapor compressor. For practical reasons, the cycle is rarely operated solely in the vapor phase but the high temperature fluid is usually superheated. Isentropic expansion and work production then occurs solely in the gas phase with condensation (heat rejection) in the vapor phase.

Figure 3.4 shows the ideal Rankine cycle with superheat. Process 1–2 is a heat addition, in order, to liquid, vapor, and gas phases. Isentropic expansion (2–3) provides shaft work and heat rejection occurs by condensing at pressure p_L (3–4). Fluid pressure rise from p_L to p_h occurs in step (4–1) by means of a liquid pump.

Rankine cycle efficiency cannot be written in simple, closed form as was done for the preceding air standard cycles since it is a vapor cycle and phase changes occur. First and second law efficiency can be calculated by using thermodynamic tables of enthalpy and entropy published for many Rankine cycle fluids. The practical design of Rankine cycles is described in Chapter 5 where second law efficiencies of the components are calculated.

II. CONDUCTION HEAT TRANSFER

The principles of conduction heat transfer are described in many texts (H2, K14, O1, F1, J2) and are assumed to be familiar to the reader. However, several topics of special pertinance to high-temperature solar collectors and systems are described in detail in this section including heat transfer in low-pressure enclosures, radial conduction, fin efficiency, and heat removal factors for solar absorbers.

A. Radial Conduction

Radial conduction is of interest in many solar-thermal applications since most concentrating collectors use circular cylindrical pipes for receivers or absorbers. In addition, most line-focus concentrators use glass tubes surrounding the absorber to suppress convection losses. The gap between the glass tube and receiver tube is often evacuated to further reduce or eliminate convection.

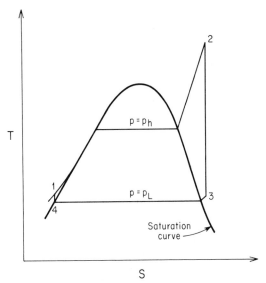

FIGURE 3.4 Ideal Rankine cycle *TS* diagram.

The rate of radial heat conduction q_k in a geometry of the type shown in Fig. 3.5 is given by

$$q_k = [2\pi kL/\ln(r_o/r_i)](T_i - T_o) \qquad (3.10)$$

where L is the cylinder length and end effects are not included. The temperature distribution $T(r)$ is

$$T(r) = T_i - (T_i - T_o) \ln(r_o/r_i)/\ln(r/r_i), \qquad r = [r_o, r_i]. \qquad (3.11)$$

Equations (3.10) and (3.11) apply for circumferentially uniform values of

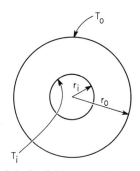

FIGURE 3.5 Radial heat conduction geometry.

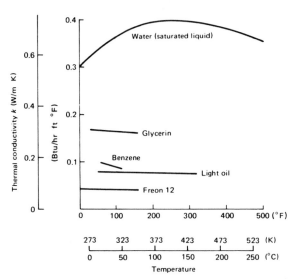

FIGURE 3.6   Thermal conductivity of several common liquids as a function of temperature. [From (K2).]

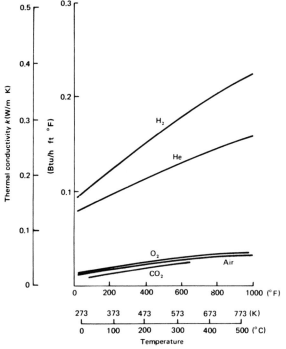

FIGURE 3.7   Thermal conductivity of common gases as a function of temperature. [From (K2).]

T_i and T_o, a situation usually encountered in metallic absorbers in solar concentrators.

For assemblies of several annular conducting zones the thermal resistance R_j of each is given by

$$R_j = \ln(r_{o,j}/r_{i,j})/2\pi kL. \tag{3.12}$$

The sum of the several resistances is used to calculate the heat flow q through a composite structure by

$$q = \sum_j \frac{(T_i - T_o)}{R_j}.$$

The thermal conductivities of many materials are tabulated in (R2) and (K5). Conductivities of some common materials are shown in Figs. 3.6, 3.7, 3.8.

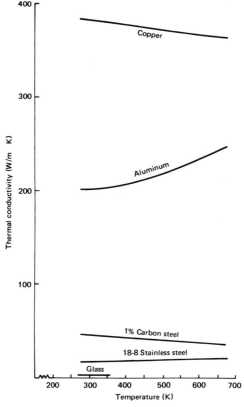

ermal conductivity of some solids.

FIGURE 3.8   Thermal conductivity of metals as a function of temperature. [From (K2).]

B. Conduction in Low Pressure Gases

The removal of gas from the annular gap between a solar concentrator absorber tube and its surrounding glass cover reduces heat transfer by two mechanisms. The first reduction in heat loss is a result of the reduced Grashof number (See the next section). Since the free convection heat transfer coefficient is proportional to the one-fourth power of the Grashof number, heat loss is reduced by the square root of the gas density.

As the pressure is further reduced, convection ceases at a Grashof–Prandtl number product of the order of 10^3 [see Eq. (3.41) in the next section]. At lower pressure, the heat transfer is solely by conduction. The conductivity of most gases is independent of pressure if the molecular mean free path is less than the heat transfer length. Therefore, over several decades of pressure, the rate of heat transfer is independent of pressure. The mean free path λ_m in a gas at rest is given by

$$\lambda_m = (\mu/\rho)(\pi/2RT)^{1/2}. \tag{3.13}$$

For *air* between 100 K and 1900 K,

$$\lambda_m = 2.08 \times 10^{-7}T^2/(T + 198.6)p, \tag{3.14}$$

where λ_m is in inches, T in °R, and p in inches of mercury.

At atmospheric pressure and 310 K, $\lambda_m = 2.6 \times 10^{-6}$ in. or 66 μm. If the density is reduced by a factor of 100, the value of λ_m increases to 6.6 mm and conduction heat transfer is unaffected. However, at a pressure of 10^{-3} Torr (10^{-3} mm Hg) the mean free path is about 5.5 cm, much greater than the heat transfer path length, and conduction is essentially eliminated for practical purposes. (This simplified treatment ignores the wall temperature jump.)

In the region of pressure below free convection, the rate of heat transfer q in planar or cylindrical geometries can be approximated by (D2)

$$q = k \, \Delta T/(g + 2\lambda_m), \tag{3.15}$$

where k is the conductivity at one atmosphere and g is the heat transfer gap width.

C. Fin Efficiency and Heat Removal Factors

Many solar collectors use absorber assemblies consisting of circular fluid conduits in some cases bonded to plates as shown in Fig.

3.9. The plate acts as a fin to conduct absorbed solar heat to the fluid conduit; however some heat losses occur from the fin surfaces as well. This section analyzes the heat loss from absorber assemblies. Elementary heat transfer texts (H2,K14) describe the heat loss from fins by using the term the *fin efficiency*. The fin efficiency F is the ratio of heat transfer to ambient occurring under given conditions to that which would occur if the entire fin were at the base temperature T_b (i.e., if the fin conductivity were infinite). For the rectangular fin of Fig. 3.9a, the fin efficiency is (K14)

$$F = \tanh[m(W + \tfrac{1}{2}t)]/m(W + \tfrac{1}{2}t) \tag{3.16}$$

where

$$m = (2U_c/kt)^{1/2}, \tag{3.17}$$

and U_c is sum of the heat transfer conductance from both the front and

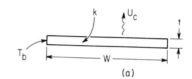

(a)

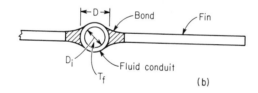

(b)

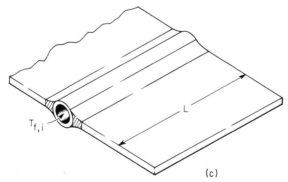

(c)

FIGURE 3.9   (a) Cross section of heat transfer fin of conductivity k, width W, and thickness t; (b) cross section of fin bonded to a fluid conduit (bond shown exaggerated in size); (c) absorber tube and plate assembly of length L and width $2W + D$; U_c is the sum of the heat loss coefficients from both sides of the surface.

rear surfaces of the fin *per unit projected area*. This result and others used in this section are from (K2) and are not redeveloped here.

The heat transfer rate q per unit length from one fin to the environment is

$$q/L = F(WU_c)(T_b - T_a), \tag{3.18}$$

where T_a is the ambient temperature.

Figure 3.9b shows a fin bonded to a fluid conduit of diameter D. Since the heat flow process in this geometry includes heat transfer to the environment from both the tube and fin, a generalized fin efficiency usually called the *plate efficiency F'* is used to calculate the heat transfer rate. The plate efficiency relates fin heat transfer to the fluid temperature T_f instead of the fin base temperature. For the geometry shown for a fin of width W on each side of the tube (Fig. 3.9b) F' is given by (K2)

$$F' = [U_c(D + 2W)]^{-1}$$
$$\times \left[\frac{1}{U_c(D + 2WF)} + \frac{1}{C_b} + R_{\text{pipe}} + \frac{1}{\pi D_i h_f} \right]^{-1}. \tag{3.19}$$

In Eq. (3.19) C_b is the bond conductance (thermal conductivity times bond circumferential length divided by bond thickness), R_{pipe} is the pipe material thermal resistance [Eq. (3.12) per unit length] and h_f is the convection coefficient within the pipe. The plate heat loss q/L per unit length is

$$q/L = F'(D + 2W)U_c(T_f - T_a). \tag{3.20}$$

Note that Eq. (3.20) assumes that the projected area D per unit length is the effective heat loss area for the conduit accounting for bond effects. If a conduit is integral with the plate, $C_b^{-1} = 0$ in Eq. (3.19). If only a tube with small thermal resistance is used with no attached plate,

$$F' = \pi D_i h_{f,i}/(\pi U_c D + \pi D_i h_{f,i}). \tag{3.21}$$

Equation (3.19) applies to the plate cross section shown in Fig. 3.9b. For the full plate assembly, shown in Fig. 3.9c, it is convenient to define a heat removal factor F_R relative to the fluid inlet temperature $T_{f,i}$. F_R is defined as the ratio of the plate assembly heat loss to the heat loss if the entire plate were at the fluid inlet temperature, the maximum available temperature for a heat loss fin. The heat removal factor is given by (K2)

$$F_R = [\dot{m}c_p/U_c(D + 2W)L]\{1 - \exp[-F'U_c(D + 2W)L/\dot{m}c_p]\} \tag{3.22}$$

where $\dot{m}$ is the working fluid flow rate and c_p its specific heat.

Heat loss q_L from the plate assembly can be expressed as

$$q_L = F_R(D + 2W)LU_c(T_{f,i} - T_a). \qquad (3.23)$$

The value of overall heat transfer coefficient U_c used above includes conduction, convection, and radiation losses (see below). It is defined relative to the difference between ambient temperature and surface temperature and per unit frontal aperture area. The value of U_c depends upon surface temperature through the convection and radiation exchanges. Therefore, U_c must be evaluated at the operating temperature and environmental conditions to be experienced in the field. Extrapolation of test data from other temperature ranges should be done only with great caution.

III. CONVECTION HEAT TRANSFER

Convection heat transfer occurs between a fluid in motion and the surface over which it moves. Both free convection, in which body forces cause fluid motion, and forced convection, in which surface, pressure or shear forces are involved, are present in solar collectors and systems. This section summarizes the methods of convection coefficient calculation for flow regimes and geometries present in elevated temperature solar systems.

The convection coefficient $\bar{h}_c$ is defined by the equation

$$\bar{h}_c = |q/A(\overline{T_s - T_f})|, \qquad (3.24)$$

where T_s is the surface temperature, T_f the fluid temperature, and A the heat transfer area. The overbar indicates that $\bar{h}_c$ is the spatial average of h_c over the entire heat transfer surface. Depending on the correlation basis for $\bar{h}_c$, $\overline{T_s - T_f}$ may be the arithmetic or logarithmic mean of inlet and outlet fluid temperature differences.

Dimensional analysis has been a powerful tool in the correlation of many heat transfer measurements into a few relatively simple equations. The dimensionless numbers used in the correlation of these data are listed in Table 3.3 along with their physical meaning. Correlations of $\bar{h}_c$ (i.e., Nusselt number) with Re and Pr are used for forced convection, whereas Nu is correlated with Gr and Pr for free or natural convection. In the remainder of this section various correlations are presented for most of the convection phenomena present in medium- and high-temperature solar-thermal systems. It should be noted that some convection correlations are based on limited data which may not have been col-

TABLE 3.3

Dimensionless Parameter Groups Useful in Correlating Convection Heat Transfer Data[a]

Quantity	Definition	Meaning
Nusselt number Nu	$\bar{h}_c D/k$	Convection/conduction ratio
Reynolds number Re	$\rho \bar{V} D/\mu$ $(\bar{V} D/\nu)$	Inertial/viscous force ratio
Grashof number Gr	$\dfrac{\rho^2 g \beta_T |T_s - T_f| D^3}{\mu^2}$	Body force/viscous force ratio
Prandtl number Pr	$\dfrac{c_p \mu}{k}$	Momentum/thermal diffusivity ratio

[a] The symbols not defined previously have the following definitions: D = characteristic length; g = acceleration due to local body forces, usually gravitational; $\bar{V}$ = mean fluid velocity; ν = fluid kinematic viscosity; β_T = fluid volume expansivity $\equiv (1/V)\partial V/\partial T$; c_p = fluid specific heat.

lected at high temperature. Most equations presented herein should be satisfactory for preliminary design. Some additional tests may be needed to refine the values of convective coefficient prior to a final design for prediction, however.

A. Forced Convection Inside Conduits

Laminar flow exists in a conduit when the Reynolds number is less than 2100 if the fluid stream enters the pipe smoothly. For fully developed laminar flow with constant wall flux the Nusselt number for a conduit of length L is (K14)

$$\text{Nu} = \frac{48}{11} + \frac{0.0668(D_H/L)\ \text{Re}\ \text{Pr}}{1 + 0.04[(D_H/L)\ \text{Re}\ \text{Pr}]^{2/3}}. \tag{3.25}$$

Both the Reynolds and Nusselt numbers are based on the hydraulic diameter D_H defined as

$$D_H = 4 \times \text{flow area/wetted parameter}. \tag{3.26}$$

For a circular pipe D_H is the same as the pipe inside diameter. All fluid properties are evaluated at the average of fluid inlet and outlet temperatures. The heat transfer coefficient is defined relative to the logarithmic mean of the inlet and outlet wall-to-bulk temperature differences. The Nusselt number in fully developed laminar flow is 3.66 for uniform wall temperature.

Turbulent flow exists in conduits if the Reynolds number is above about 7×10^3. For fully developed turbulent flow in a smooth conduit the Nusselt number for uniform wall heat flux is given by the Petukhov equation (K14)

$$Nu = (f/8) \text{ Re Pr}/[1.07 + 12.7\sqrt{f/8}(\text{Pr}^{2/3} - 1)], \qquad (3.27)$$

where Re = $[5 \times 10^3, 1.25 \times 10^5]$ and f is the friction factor given by $f = [1.82 \log_{10} \text{Re} - 1.64]^{-2}$; the heat transfer coefficient is defined as for Eq. (3.24). If the flow is not fully developed, Eq. (3.28) should be used:

$$Nu = 0.036 \text{ Re}^{0.8} \text{ Pr}^{1/3}(D_H/L)^{0.055} \qquad (3.28)$$

in the interval $D_H/L = [10,400]$.

In some solar collector conduits the flow rate may be so small that the mechanisms of free and forced convection coexist. Reference (H2, p. 262) contains Nusselt number expressions for this situation.

In the transition region between laminar and turbulent flow, Eq. (3.29) can be used to estimate the value of $\bar{h}_c$:

$$Nu = 0.166(\text{Re}^{2/3} - 125) \text{ Pr}^{1/3}[1 + (D_H/L)^{2/3}](\mu/\mu_w)^{1/7}, \qquad (3.29)$$

where μ_w is the fluid viscosity at the wall temperature. Transitional flow correlations are not based on extensive testing and should be used with caution.

B. External Forced Convection Over Surfaces

External flows of interest to the solar designer include wind blowing over absorber assemblies, flow of industrial gases across one or more tubes in a heat exchanger, and heat loss from fluid-carrying pipes exposed to the environment.

Laminar flow heat transfer coefficients can be calculated from solutions of the Navier–Stokes and convective energy equations. The Nusselt number based on such a calculation for a planar surface *parallel* to the flow is given by

$$Nu = 0.664 \text{ Re}_L^{1/2} \text{ Pr}^{1/3}, \qquad (3.30)$$

where the Reynolds number is based on the surface length in the direction of flow and is evaluated at the average of the surface and fluid temperatures. Laminar flow exists if $\text{Re}_L < 5 \times 10^5$.

For the more common situation of laminar flow impinging on an isothermal plane at a nonzero angle of attack Sparrow and Tien

(S10) have found that the following correlation fits test data for angles of attack between 25° and 90° (Re = $[2 \times 10^4, 1.5 \times 10^5]$):

$$Nu = 0.931 \, Re^{1/2} \, Pr^{1/3}. \tag{3.31}$$

Equation (3.31) was developed from mass transfer measurements on square plates using the analogy between heat and mass transfer. It was found that the flow yaw angle is also a second-order effect accounting for only a 1% difference over the range 0–90°. The values of Nu and Re are based on a reference length taken to be the length of the side of the square plate. Heat transfer experiments on rectangular plates indicate that Eq. (3.31) is not strictly applicable and that Eq. (3.32) can be used ($\pm$ 10%) instead (S11) for all geometries, both square and nonsquare, and for all yaw and angles of attack:

$$Nu = 0.86 \, Re^{1/2} \, Pr^{1/3}. \tag{3.32}$$

Re for rectangular plates is based on the ratio of the plate area to its perimeter and Re = $[2 \times 10^4, 9 \times 10^4]$.

For turbulent flow over a flat surface (or surface for which the longitudinal radius of curvature is much greater than the boundary layer thickness) the Nusselt number is (K14)

$$Nu = 0.036 \, Pr^{1/3}[Re_L^{0.8} - 23{,}200]. \tag{3.33}$$

If the plate length is much greater than the distance x from the leading edge at which $Re_x = 5 \times 10^5$, the numerical constant 23,200 can be ignored.

Heat transfer from a gas or liquid to the surface of a pipe in the crossflow regime (fluid velocity normal to the pipe centerline) can be calculated from Eq. (3.34):

$$Nu = A \, Re^B \, Pr^{1/3}. \tag{3.34}$$

The constants A and B depend on Reynolds number and are given below (K14):

Re range	A	B
0.4–4	0.989	0.330
4–40	0.911	0.385
40–4,000	0.683	0.466
4,000–40,000	0.193	0.618
40,000–400,000	0.0266	0.805

The Reynolds number is based on the pipe outside diameter and the fluid properties are evaluated at the arithmetic average of the pipe surface and external fluid temperatures.

Crossflow heat exchangers have tubes spaced sufficiently densely that flow interference effects occur. Since the flow is very complex, only empirical correlations are available for predicting Nu. For typical heat exchanger configurations

$$Nu \approx 0.33 \, Re^{0.6} \, Pr^{1/3}, \qquad Re = [10^2, 4 \times 10^4]. \tag{3.35}$$

The Reynolds number is based on the maximum fluid velocity occurring at the plane between two tubes and the tube outer diameter. For further details see (K2).

C. *Free Convection*

Free convection from flat and curved surfaces and within enclosures is present in most solar-thermal systems. Free convection depends upon buoyant forces which in turn are temperature dependent. Therefore, the Nusselt number depends upon temperature much more strongly for free convection than for forced convection.

Natural convection heat transfer from a vertical plane can be calculated from

$$Nu = 0.59(Gr \, Pr)^{1/4} \tag{3.36}$$

if $Gr \, Pr = [10^4, 10^9]$ (laminar flow) where the value of D in the Grashof number (Table 3.3) is taken as the length of the plate in the flow direction. For turbulent flow ($Gr \, Pr > 10^9$)

$$Nu = 0.10(Gr \, Pr)^{1/3}. \tag{3.37}$$

Equations (3.36) and (3.37) apply for both constant heat flux and constant surface temperature regimes if the mean surface temperature is used in the former case.

The free convection Nusselt number for heat loss from isothermal horizontal plates facing upward is given by (L9)

$$Nu = 0.54(Gr \, Pr)^{1/4} \tag{3.38}$$

for the laminar regime, $Gr = [2.2 \times 10^4, 8 \times 10^6]$. For the turbulent range ($Gr = [8 \times 10^6, 1.6 \times 10^9]$),

$$Nu = 0.15(Gr \, Pr)^{1/3}. \tag{3.39}$$

The correlations for free convection [Eqs. (3.38) and (3.39)] are based on mass transfer measurements carried out by Lloyd and Moran (L9) and the well-established analogy between heat and mass transfer. Equations (3.38) and (3.39) apply to many shapes of surfaces including circles, squares, rectangles, and triangles if the reference length D in Nu and Gr is taken as the surface area A divided by the perimeter p, i.e., $D = A/p$.

Natural convection from horizontal cylinders in free air can be correlated by the equation

$$Nu = 0.53(Gr\ Pr)^{1/4} \qquad (3.40)$$

for $Gr = [10^4, 10^9]$. For turbulent flow, Eq. (3.39) can be used.

A common free convection mode in many line-focus solar collectors occurs within the annular gap between a hot absorber tube and a cooler glass tube surrounding the absorber to eliminate forced convection losses to the wind. This heat suppression effect is called the *greenhouse effect* but is usually erroneously attributed solely to a radiation phenomenon (W3). Few data have been collected for this geometry to date. One author (K6) has proposed the following expressions for Nu:

$$Nu = 0.124(Gr\ Pr)^{1/3}/\ln(D_o/D_i) \qquad (3.41a)$$
$$(Gr\ Pr) = [10^7, 10^{10}];$$

$$Nu = 0.44(Gr\ Pr)^{1/4}/\ln(D_o/D_i) \qquad (3.41b)$$
$$(Gr\ Pr) = [10^4, 10^7];$$

$$Nu = 2[\ln(D_o/D_i)]^{-1} \quad \text{(pure conduction)} \qquad (3.41c)$$
$$(Gr\ Pr) < 10^3.$$

The Grashof number is based on the gap $(D_o - D_i)/2$ where D_o and D_i are the outer and inner diameters of the annulus and the Nusselt Number is based on D_i. For values of annulus gas density below which k is pressure independent, the free molecular flow equation (3.15) should be used. Heat transfer in tilted *planar* enclosures is discussed in (K2).

D. Boiling Heat Transfer

Many high-temperature solar processes involve boiling heat transfer within heat exchangers or occasionally within the collector itself. Pool boiling heat transfer coefficients with no superimposed flow can be correlated by Eq. (3.42). The heat transfer rate q per unit area is related to both liquid- and vapor-phase properties as shown (K14):

$$c_{pl}\ \Delta T_x/h_{fg}Pr_l^{1.7} = C[q/\mu_l h_{fg}]^{1/3}[\sigma/g(\rho_l - \rho_v)]^{1/6}. \qquad (3.42)$$

In Eq. (3.42) subscripts l and v denote liquid- and vapor-phase properties at the boiling pressure. (For water the exponent for Pr_l is 1.0.) The surface-to-fluid temperature difference ΔT_x is called the excess temperature and is defined as the difference between the surface temperature and the fluid saturation temperature; h_{fg} is the heat of vaporization, c_{pl} is the liquid specific heat, and σ is the surface tension. The coefficient C depends upon the fluid surface type. Typical values (K14) range from 0.0025 to 0.03 for alcohol/copper. For water, $C = 0.013$ for steel, copper, and platinum surfaces.

If boiling occurs in the presence of an external flow field, the effective convection coefficient is the sum of that for boiling and for forced convection. For example in turbulent flow in a pipe, Eq. (3.27) would be used for the forced convection contribution and Eq. (3.42) for the boiling contribution.

E. Heat Pipes

Heat pipes offer a method of transferring larger amounts of heat with only small temperature differences from the focal area of a high-concentration solar collector to a fluid. The heat pipe (Fig. 3.10) was invented by R. S. Gaugler (Pat. No. 2350348, 1944) and consists of a circular pipe with an annular wick layer situated adjacent to the pipe wall. Solar heat input boils fluid in the evaporator; the vapor migrates to the condenser where heat is transferred to a circulating fluid loop, which carries it to the end use point, for example. The condensate migrates to the boiler by capillary action in the wick or by gravity and the cycle repeats. Gravity return heat pipes omit the wick but cannot be operated horizontally as a result.

Table 3.4 lists properties of heat pipes useful at high temperature. Heat fluxes in excess of 2 MW/m² have been measured. Those very high fluxes are well above those achievable in all but the very highest concentration solar systems.

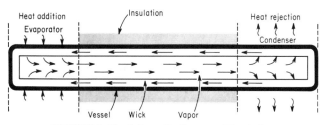

FIGURE 3.10   Schematic diagram of heat pipe.

TABLE 3.4

Properties of Heat Pipes for Medium- and High-Temperature Solar Collectors[a]

Temperature range (°C)	Working fluid	Vessel material	Measured axia[b] heat flux (kW/cm²)	Measured surface[b] heat flux (W/cm²)
−70 to +60	Liquid ammonia	Nickel, aluminum, stainless steel	0.295	2.95
−45 to +120	Methanol	Copper, nickel, stainless steel	0.45 @ 100°C[c]	75.5 @ 100°C
+5 to +230	Water	Copper, nickel	0.67 @ 200°C	146 @ 170°C
+190 to +550	Mercury[e] +0.02% magnesium +0.001%	Stainless steel	25.1 @ 360°C[d]	181 @ 360°C
+400 to +800	Potassium[e]	Nickel, stainless steel	5.6 @ 750°C	181 @ 750°C
+500 to +900	Sodium[e]	Nickel, stainless steel	9.3 @ 850°C	224 @ 760°C
+900 to +1,500	Lithium[e]	Niobium +1% zirconium	2.0 @ 1250°C	207 @ 1250°C
1,500 to +2,000	Silver[e]	Tantalum +5% tungsten	4.1	413

[a] From "Heat Transfer" by J. P. Holman. Copyright © 1976. Used with permission of McGraw-Hill Book Company.
[b] Varies with temperature.
[c] Using threaded artery wick.
[d] Tested at Los Alamos Scientific Laboratory.
[e] Measured value based on reaching the sonic limit of mercury in the heat pipe.

The maximum heat transfer rate of a heat pipe can be calculated from (R3)

$$q = 2A_w g h_{fg} \rho_1^{3/2} [\rho_v \rho_1 / \mu_v \mu_1]^{1/2} (l_m K_1 / L), \qquad (3.43)$$

where A_w is the wick cross-sectional area, h_{fg} is the fluid heat of vaporization, l_m is the wicking height of the fluid, L is the heat pipe length, K_1 is the wick factor; subscripts 1 and v liquid- and vapor-phase properties. The wicking height l_m is given by a simple force balance on a pore as

$$l_m = 2\sigma / r_c \rho_1 g, \qquad (3.44)$$

where σ is the surface tension and r_c is the effective pore radius. A typical wick height for sodium is 40 cm for $r_c \sim 85$ μm.

Elevation effects can decrease the effective value of l_m. For example a heat pipe may be installed at a tilt or with bends in the pipe. If the elevation differential is l_e, then l_m must be replaced with $(l_m - l_e)$ (>0) in Eq. (3.43). If $l_e > l_m$, the heat pipe does not function. Also, if the evaporator temperature falls below the condenser temperature, the heat pipe ceases to function and acts as a heat "diode."

The product of wick area and wicking height $A_w l_m$ is the principal design parameter of a heat pipe of a given length. For a given fluid and wick l_m is determined and A_w can be selected for the desired heat rate. In practice, upper limits on operating and condenser temperatures can constrain the value of A_w since the temperature drop across the wick must increase with wick thickness. If A_w is so constrained, the heat pipe length is also limited by pressure drop and wicking properties, hence limiting the size of solar collector absorber.

One method of alleviating the constraints set by temperature limits is to use a thinner wick at the condenser and evaporator. The performance of such a heat pipe is then given by

$$q = \frac{2gh_{fg}\rho_l^{3/2}[\rho_v\rho_l/\mu_v\mu_l]^{1/2}K_1(l_m - l_e)}{L_e/A_{w,e} + L_a/A_{w,a} + L_c/A_{w,c}}, \tag{3.45}$$

where the subscripts e, a, c refer to the evaporator, adiabatic (insulated), and condenser sections of the heat pipe. Reference (C2) contains extensive design information for heat pipes.

IV. RADIATION HEAT TRANSFER

The principal heat loss mode from high-temperature solar collectors is radiation in the infrared wavelength spectrum. Radiation heat transfer is the subject of several engineering texts [e.g., (H3, S2, S3)] and many of the details are not repeated here. Specific topics—spectral property evaluations, sky radiation, and selective surface properties—are treated in detail.

A. Radiation Fundamentals

Radiative transfer occurs by the emission and absorption of photons whose energy levels $h\nu$ are dictated only by the temperature of the emitter. The spectral distribution of blackbody electromagnetic emission $E_{b\lambda}$ at wavelength λ is given by Planck's law

$$E_{b\lambda}(T) = 2\pi hc_0^2/n^2\lambda^5[\exp(hc_0/n\lambda kT) - 1], \tag{3.46}$$

where T is the blackbody surface temperature, h is Plank's constant $(6.625 \times 10^{-27}$ erg sec), k is Boltzmann's constant $(1.380 \times 10^{-16}$ erg/K), n is the refractive index (assumed independent of frequency—a good approximation for a vacuum and for gases, a fair approximation for solids used in solar collectors), and c_0 is the speed of light in a vacuum. If the various constants in Eq. (3.46) are combined,

$$E_{b\lambda}(T) = C_1/n^2\lambda^5[\exp(C_2/n\lambda T) - 1], \tag{3.47}$$

where $C_1 = 3.740 \times 10^8$ W $\mu m^4/m^2$ $(1.187 \times 10^8$ Btu $\mu m^4/ft^2$ h) and $C_2 = 14{,}387$ μm K $(25{,}896$ μm R).

Equation (3.47) is plotted in Fig. 3.11 as a function of wavelength for refractive index $n = 1.0$. The point of peak spectral intensity can be calculated by differentiating Eq. (3.47) to find that

$$n(\lambda_{max}T)_{max\,E_{b\lambda}} = 2897.8 \quad \mu m\ K \quad (5215.6 \quad \mu m\ R). \tag{3.48}$$

The inverse proportionality of λ_{max} with source temperature T is important in solar applications since solar radiation from a high-temperature source $(\sim 5762$ K) at short wavelength is absorbed by a surface which operates at much lower temperature. Therefore, the emittance ϵ of the receiving surface at longer wavelength may be vastly different from its absorptance α in the solar spectral range.

Table 3.5 contains a tabulation of the blackbody spectral emissive power $E_{b\lambda}(T)$ as a function of $n\lambda T$. In order that $n\lambda T$ be used as

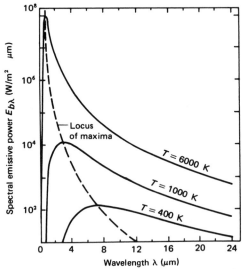

FIGURE 3.11   Variation of the thermal radiation spectrum with wavelength and source temperature for unity refractive index.

TABLE 3.5

Blackbody Radiation Functions[a]

$n\lambda T$	$\dfrac{E_{b\lambda}\times10^5}{\sigma n^3T^5}$	$\dfrac{E_b(0-n\lambda T)}{n^2\sigma T^4}$	$n\lambda T$	$\dfrac{E_{b\lambda}\times10^5}{\sigma n^3T^5}$	$\dfrac{E_b(0-n\lambda T)}{n^2\sigma T^4}$	$n\lambda T$	$\dfrac{E_{b\lambda}\times10^5}{\sigma n^3T^5}$	$\dfrac{E_b(0-n\lambda T)}{n^2\sigma T^4}$
1000	0.0000394	0	7200	10.089	0.4809	13400	2.714	0.8317
1200	0.001184	0	7400	9.723	0.5007	13600	2.605	0.8370
1400	0.01194	0	7600	9.357	0.5199	13800	2.502	0.8421
1600	0.0618	0.0001	7800	8.997	0.5381	14000	2.416	0.8470
1800	0.2070	0.0003	8000	8.642	0.5558	14200	2.309	0.8517
2000	0.5151	0.0009	8200	8.293	0.5727	14400	2.219	0.8563
2200	1.0384	0.0025	8400	7.954	0.5890	14600	2.134	0.8606
2400	1.791	0.0053	8600	7.624	0.6045	14800	2.052	0.8648
2600	2.753	0.0098	8800	7.304	0.6195	15000	1.972	0.8688
2800	3.872	0.0164	9000	6.995	0.6337	16000	1.633	0.8868
3000	5.081	0.0254	9200	6.697	0.6474	17000	1.360	0.9017
3200	6.312	0.0368	9400	6.411	0.6606	18000	1.140	0.9142
3400	7.506	0.0506	9600	6.136	0.6731	19000	0.962	0.9247
3600	8.613	0.0667	9800	5.872	0.6851	20000	0.817	0.9335
3800	9.601	0.0850	10000	5.619	0.6966	21000	0.702	0.9411
4000	10.450	0.1051	10200	5.378	0.7076	22000	0.599	0.9475
4200	11.151	0.1267	10400	5.146	0.7181	23000	0.516	0.9531
4400	11.704	0.1496	10600	4.925	0.7282	24000	0.448	0.9589
4600	12.114	0.1734	10800	4.714	0.7378	25000	0.390	0.9621
4800	12.392	0.1979	11000	4.512	0.7474	26000	0.341	0.9657
5000	12.556	0.2229	11200	4.320	0.7559	27000	0.300	0.9689
5200	12.607	0.2481	11400	4.137	0.7643	28000	0.265	0.9718
5400	12.571	0.2733	11600	3.962	0.7724	29000	0.234	0.9742
5600	12.458	0.2983	11800	3.795	0.7802	30000	0.208	0.9765
5800	12.282	0.3230	12000	3.637	0.7876	40000	0.0741	0.9881
6000	12.053	0.3474	12200	3.485	0.7947	50000	0.0326	0.9941
6200	11.783	0.3712	12400	3.341	0.8015	60000	0.0165	0.9963
6400	11.480	0.3945	12600	3.203	0.8081	70000	0.0092	0.9981
6600	11.152	0.4171	12800	3.071	0.8144	80000	0.0055	0.9987
6800	10.808	0.4391	13000	2.947	0.8204	90000	0.0035	0.9990
7000	10.451	0.4604	13200	2.827	0.8262	100000	0.0023	0.9992
						∞	0	1.0000

[a] To convert the second column from units of $(\mu m\ R)^{-1}$ to $(\mu m\ K)^{-1}$ multiply by 1.80.

the independent variable in Eq. (3.47), both sides of the equation have been divided by σn^3T^5 where σ is the Stefan–Boltzmann constant

$$\sigma = 2\pi^5k^4/15c_0^2h^3. \qquad (3.49)$$

The value of σ is 5.67×10^{-8} W/m² K⁴ (0.1712×10^{-8} Btu/h ft² R⁴).

Also shown in Table 3.5 is the emissive power integrated over the interval $[0,n\lambda T]$. This quantity, denoted by $E_b(0-n\lambda T)$, is made dimensionless by dividing it by $n^2\sigma T^4$, the total wavelength integral over the interval $n\lambda T = [0,\infty]$. This integral, denoted by E_b, is also called the Stefan–Boltzmann law:

$$E_b = \int_0^\infty E_{b\lambda}\,d(n\lambda T) = n^2\sigma T^4. \qquad (3.50)$$

Throughout the remainder of this book $n = 1$ will be assumed in keeping with the definition of a blackbody. If radiation takes place into a medium within which $n \neq 1.0$, however, the appropriate value of n must be used in Eq. (3.49).

B. Radiation Exchange Factors

In general, a fraction F_{12} of the radiation of a diffuse blackbody radiator of area A_1 emitting into its surroundings will be intercepted by a black surface of area A_2. The *exchange of shape factor* F_{12} is a purely geometric quantity which can be calculated from basic trigonometry and calculus. Reference (S2) contains a lengthy tabulation of exchange factors for most geometries of interest in solar system design.

The *net* radiative exchange q_r between two black surfaces can be expressed as

$$q_r = \sigma(A_1 F_{12})(T_1^4 - T_2^4). \tag{3.51}$$

It can also be shown that a reciprocity exists so that

$$A_1 F_{12} = A_2 F_{21} \tag{3.52}$$

and

$$q_r = \sigma(A_2 F_{21})(T_1^4 - T_2^4).$$

For *gray* surfaces, i.e., with uniform emittance less than unity, over a restricted wavelength band a generalized exchange factor $\mathscr{F}_{12}$ is used (K5) in place of F_{12}. For example, within an enclosure consisting of two opaque, gray surfaces (or two gray surfaces and a third surface at 0 K)

$$A_1 \mathscr{F}_{12} = \{\rho_1/\epsilon_1 A_1 + \rho_2/\epsilon_2 A_2 + 1/A_1 F_{12}\}^{-1}, \tag{3.53}$$

where ρ and ϵ are, respectively, the reflectances ($\rho = 1 - \epsilon$) and emittances of gray surfaces 1 and 2.

For solar collectors, two geometries are of special interest. For two concentric tubes

$$\mathscr{F}_{12} = [1/\epsilon_1 + (A_1/A_2)^m(1/\epsilon_2 - 1)]^{-1}, \tag{3.54}$$

where surface 1 is the inner (smaller) tube. For a diffuse outer surface A_2, $m = 1$. If A_2 is a specular reflector, $m = 0$. Equation (3.54) also applies to two large, closely spaced, parallel, plane surfaces for which $A_1 = A_2$.

A second geometry often encountered is an enclosure consisting of three gray surfaces, one of which (surface 3) is well insulated and can be assumed to be adiabatic. For this geometry

$$A_1 \mathscr{F}_{12} = \left\{ \frac{\rho_1}{\epsilon_1 A_1} + \frac{\rho_2}{\epsilon_2 A_2} + \frac{1}{A_1 [F_{12} + (1/F_{13} + A_1/A_2 F_{23})^{-1}]} \right\}^{-1}. \quad (3.55)$$

These types of enclosures are often evacuated in medium-temperature solar collectors (see Chapter 4). In addition, the concept of exchange factors is useful in evaluating the maximum possible concentration of a solar collector (Chapter 4).

C. Sky Radiation

Infrared radiation exchanges occur between components of solar-thermal systems, notably solar collectors, and the sky. Since the temperature of deep space is ~ 2.7 K, the effective sky temperature for radiation is usually below that of the ambient air at temperature T_a. The net exchange q_{sky} between a surface (assumed gray) on earth at temperature T_s and emittance ϵ_1 and the sky at effective temperature T_{sky} can be written from Eq. (3.53) (for sky area $A_2 \rightarrow \infty$)

$$q_{sky} = \epsilon_1 \sigma (T_s^4 - T_{sky}^4). \quad (3.56)$$

The "sky temperature" is a convenient concept for analysis but is actually defined by Eq. (3.56) rather than being an independent variable in it. The flux q_{sky} depends on the radiative properties of the sky (e.g., humidity and cloud cover) and many other secondary phenomena. An equation for q_{sky} including principal climatic effects is given by deJong (D1):

$$q_{sky} = \epsilon_1 \sigma T_a^4 (0.39 - 0.0096e)[1 - a_3(CC)] + 4\epsilon_1 \sigma T_a^3 (T_s - T_a), \quad (3.57)$$

where e is the vapor pressure of atmospheric water vapor in mm Hg and CC is the cloud cover expressed in tenths (see Chapter 2). The empirical constant a_3 is given in Table 3.6 and σ is the Stefan–Boltzmann constant.

The *saturation* vapor pressure of water vapor p_{ws} in atmospheres can be calculated from (A1,A2):

$$\log_{10}(p_{ws}/218.2) = -(\delta/T)[(a + b\delta + c\delta^3)/(1 + d\delta)] \quad (3.58)$$

where $\delta = 647.27 - T$, $T = [K]$, $a = 3.244$, $b = 5.868 \times 10^{-3}$, $c = 1.170 \times 10^{-8}$, and $d = 2.188 \times 10^{-3}$. For more accuracy the steam tables can be used. The partial pressure of water vapor p_w is given by

$$p_w = \phi \times p_{ws} \quad \text{(atm)}, \quad (3.59)$$

TABLE 3.6

Value of Coefficient a_3 for the
Sky Flux Equation (3.57)

Latitude	a_3
5	0.50
10	0.52
15	0.55
20	0.59
30	0.63
40	0.68
50	0.72
60	0.76
70	0.80
80	0.84
85	0.86

where ϕ is the relative humidity. To convert from p_w to e recall that 1 atm = 760.2 mm Hg.

Radiative flux to the sky dome is of particular interest for large solar systems deployed in the desert environment where e and CC are small. Piping systems for large, distributed solar thermal arrays which maintain high fluid temperatures overnight for several possible reasons can experience large heat loss to the sky at night.

D.   Wavelength Selective Surfaces

The spectral separation between the solar spectrum—$\lambda_{max} \approx 0.5$ μm [Eq. (3.48)]—and emission spectra of solar collector surfaces operating at temperatures in the infrared (ir)—$\lambda_{max} \approx$ 2.5–5.0 μm—has been noted earlier. This separation affords a means of reducing radiative losses in the ir by devising a surface with small ir emittance but large emittance (and absorptance by Kirchhoff's law) in the solar spectrum. Surfaces with these properties are called *selective surfaces*. The analysis, fabrication, and testing of selective surfaces is under very rapid development and a summary of all current research is beyond the scope of this book. The basic principles of selective surfaces are described.

Consider the use of a tubular solar collector absorber surrounded by an evacuated glass envelope. The radiative exchange between the absorber and the glass envelope in the absence of convection and conduction can be calculated from Eq. (3.51) where $\mathcal{F}_{ae}$, from Eq. (3.54), is

$$\mathcal{F}_{ae} = [1/\epsilon_a + 1/\epsilon_e - 1]^{-1}, \tag{3.60}$$

where subscripts a and e denote absorber and envelopes, respectively, and ϵ is the total hemispheric emittance. Heat loss can therefore be reduced by reducing the absorber or cover emittance in the ir. Either method would give the same $\mathcal{F}_{ae}$ value since Eq. (3.60) is symmetric in ϵ_a and ϵ_e.

Envelope emittance ϵ_e could be reduced from the usual value for glass, $\epsilon_e \sim 0.9$ to $\epsilon_e \sim 0.1$ or less, by plating the inner surface of the glass with an ir mirror such as stannous or indium oxide. This approach has the advantage that the ir mirror need not have high-temperature stability to achieve a low value of exchange factor $\mathcal{F}_{ae}$. However, the mirrors may have objectionably high values of reflectance in the solar spectrum (5–10% depending on the thickness).

The second method of reducing $\mathcal{F}_{ae}$ is to decrease ϵ_a. This is a somewhat more difficult approach for high-temperature devices for the following reasons:

(1) High absorptance α_s must coexist in the solar spectrum with small ϵ_a in the ir. As operating temperature increases, the overlap of surface and solar spectra increases (see example below). In order to achieve high α_s, many selective surfaces require an antireflectant coating.

(2) The surface must be chemically stable at high temperature in its environment (whether a vacuum or not) for long periods of time. This requirement is particularly severe since chemical reaction rates increase very rapidly with temperature.

(3) The surface must have high-temperature stability. The better the surface, i.e., the lower the emittance, the more severe the problem.

(4) The surface must be strong to accommodate flexure and expansion of the absorber. Lacquer-based surfaces are, therefore, impractical.

(5) Selective surface manufacture requires delicate processes and special equipment. Quality control with low manufacturing cost is a problem.

These five problems indicate that the former process—use of ir mirrors—may be more practical in the long run.

The following example illustrates item (1) above.

Example What is the emittance at 100°C and 800°C of a surface whose spectral reflectance is shown in Fig. 3.12? Also calculate the

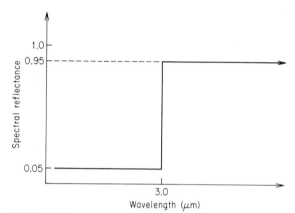

FIGURE 3.12   Model spectral reflectance curve for example.

solar absorptance assuming that 90% of the terrestrial solar spectrum lies below 3 μm and 10% above. Note that Fig. 3.12 represents a model surface useful for illustrative purposes but with a sharp wavelength cutoff not achievable in practice.

Solution   To evaluate the emittances ϵ_{100}, ϵ_{800} at 100°C and 800°C, Table 3.5 can be used. For 100°C the 3-μm spectral cutoff occurs at $\lambda T = 1119$ μm K $= 2014$ μm R. From Table 3.5 a negligible amount of radiation occurs for $\lambda T < 2014$ so

$$\epsilon_{100} = 1 - \rho(\lambda T > 2014), \qquad \epsilon_{100} = \underline{0.05}.$$

For the 800°C surface, $\lambda T = 3219$ μm K $= 5794$ μm R. From Table 3.5, about 32% of the spectrum lies below 5794 μm R; therefore ϵ_{800} is

$$\epsilon_{800} = 0.32(1.0 - 0.05) + 0.68(1.0 - 0.95), \qquad \epsilon_{800} = \underline{0.34}.$$

Therefore, the high-temperature surface has an emittance more than six times that of the same surface at 100°C.

The solar absorptance α_s of the surface can be calculated from the spectral data given relative to the 3-μm breakpoint:

$$\alpha_s = 0.90(1.0 - 0.05) + 0.10(1.0 - 0.95), \qquad \alpha_s = \underline{0.86}. \qquad \blacksquare$$

The earliest surfaces showing significant selectivity were black nickel and copper oxide applied to a polished metal substrate which serves as an infrared reflector. A commercial black nickel surface has measured absorptance $\alpha_s \sim 0.9$ and infrared emittance $\epsilon(100°C) = 0.20–0.25$. Black nickel and copper oxide are both deteriorated by water and do not have adequate selectivity for high-temperature use unless carefully prepared (P4). Black nickel and copper are examples of the *absorber–reflector* tandem selective surfaces. These systems use a polished metal substrate with high infrared reflectance over which a material with high solar spectrum absorptance is deposited. The absorbing material is transparent to ir radiation and as a result the solar absorptance is that of the coating and the ir emittance is that of the polished substrate.

Absorption in the absorptive layer occurs by means of many interreflections between the two interfaces of this layer. Stokes' equations (Chapter 2) can be used to calculate the absorptance if the substrate reflectance and layer extinction coefficient and refractive index are known. It is easy to show (S4) that for α_s to be greater than 90% the refractive index $n < 2.0$ and the extinction coefficient–layer thickness product $KL > 1.0$.

There is an unfortunate contradiction of the $n < 2$ requirement with the desired spectral cutoff for good selectivity in the 2–3-μm range. Most materials with the desired cutoff are semiconductors with $n > 3$ (S4). Therefore, reflective losses are too great. The usual solution to this problem adds cost and complexity by adding an antireflectant layer over the absorption layer.

Black chrome (Cr and Cr_2O_3 with C impurities; 10/1 Cr/Cr_2O_3 ratio at substrate, unity ratio at outer surface) on a nickel substrate has been examined for low-temperature solar-thermal processes and been found to have excellent durability under severe environmental conditions. The spectral reflectance of black chrome is shown in Fig. 3.13; the solar absorptance is ~95% and emittance 12% at 300°C (U1).

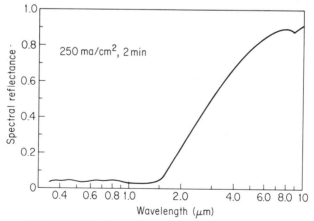

FIGURE 3.13 Spectral reflectance of black chrome on nickel after an 8-day humidity test (MIL STD 810B). [From (M1).]

Although not extensively tested at high temperature, black chrome is expected to survive at temperatures up to 350°C (M1).

Chemical vapor deposition (CVD) is a method of producing optically selective multilayer surfaces for solar applications at high temperature (>650°C). One surface studied at some length (S4) consists of the following layers with thickness as shown:

(1)   stainless steel (bottom)—relatively thick;
(2)   oxide layer diffusion barrier and expansion buffer— 0.03 μm;
(3)   silver reflector—0.10 μm;
(4) Cr_2O_3 diffusion barrier—0.02 μm;
(5)   silicon absorber—1.5–1.6 μm;
(6) Si_3N_4 antireflectant (top)—$\lambda/4$ @ 0.5 μm.

An expansion buffer between the stainless steel and silicon absorber is required owing to different thermal expansion rates. At 500°C the solar absorptance of this surface is about 80% and the emittance 7%. The spectral reflectance is shown in Fig. 3.14. It is expected that a surface of this type could find wide application in medium- and high-temperature solar collectors.

This CVD surface is an example of a *multilayer interference* selective surface. Absorption is achieved by optical wavefront interference effects and multiple reflections in the dielectric layers. Additional work on this CVD surface is required to increase absorptance and high-temperature stability possibly by using amorphous silicon in place of the polycrystalline silicon layer. In addition, the silver layer may be replaced with molybdenum which has good high-temperature stability. Another

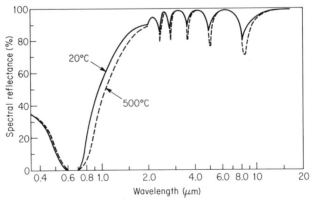

FIGURE 3.14   Spectral reflectance for the CVD surface at 20°C and 500°C. [From (S1).]

area requiring added effort is the antireflectant layer the thickness of which is selected for $\rho = 0$ at only normal incidence. At off-normal incidence, the effective optical path length is greater than $\lambda/4$ and $\rho > 0$. Since the incidence angles on the absorber of some solar concentrators have a broad range, it is important to increase the incidence angle range where $\rho \ll 1$. One method of achieving this is by building a layer of continuously varying refractive index similar to that described for glass in Chapter 2.

Another multilayer, high-temperature selective surface of the interference type is called the AMA surface and has been studied for several years at the University of Minnesota (U1) as part of a solar-thermal power project. The surface consists of 0.06-μm Al_2O_3, 0.02-μm MoO_x, and 0.06-μm of Al_2O_3 on ~0.6-μm Mo on stainless steel (304). It is expected that such a surface could last for 20 yr if operated below 300°C. At temperatures above 450°F extensive interlayer diffusion has been measured.

Another type of selective surface is called the *wavefront discriminant*. This idea first proposed by Tabor makes use of surface rugosities of the order of solar spectral wavelengths for high absorptance resulting from multiple reflections (the cavity effect). The refractive index of such surfaces is also quite low. However, the surface appears smooth to longer wavelengths and has high ir reflectance. These surfaces can be fabricated using CVD technology but the optimum surface morphology and durability in the field are not known. Excellent surveys of selective surfaces are presented by Mattox in (M2) and by Peterson and Ramsey in (P4) Table A.6 in the Appendix lists properties of selective materials.

The absorptance of selective solar absorber surfaces decreases with incidence angle much like the transmittance for transparent media. This decrease must be accounted for in the careful design of solar collector receivers. Reed (R9) has found that the absorptances of many selective surfaces fall on a standard, normalized curve if measured absorptance values are divided by the absorptance value at normal incidence. Hence, for engineering purposes a single curve may be used by knowing only the absorptance at normal incidence. Figure 3.15 shows the standard absorptance $\alpha_s(i)$ curve.

There are presently few materials which meet the five criteria at the beginning of this section for high-temperature applications. Other methods of heat loss control are currently technically proven and are used along with surface treatments and evacuation. Selective surfaces are most important for low-concentration devices operated at high temperature. For high-concentration systems, some selective surfaces may not be appropriate since the requirement for high absorptance is more important than that for low emittance.

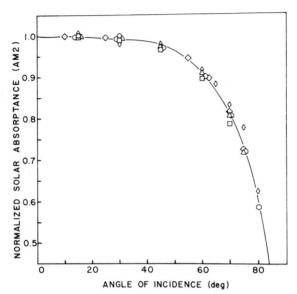

FIGURE 3.15   Standard absorptance vs. incidence angle function for selective surfaces. The various surfaces shown are G.E., Optical Coating Laboratory, and
Owens-Illinois multilayer surfaces, Barry Solar Products black chrome and J. D. Garrison
iron oxide. [From (K9).]

V. COMBINED HEAT TRANSFER
MODES–EXAMPLE CALCULATION

    The three modes of heat transfer described above rarely
exist independently. In most solar-thermal systems all three modes are
present. For example, hot fluid in a pipe loses heat to its environment by
convection to the pipe wall, conduction through the insulation, and radiation and convection from the external surface of the pipe. The design
equations given earlier can be used to calculate the magnitude of the several heat loss terms for various conditions.

    In this section an example calculation is developed to illus-
trate the method. This analysis will be used in Chapter 4 to predict the
performance of line-focus concentrators. The problem to be analyzed is to
calculate the power delivery of the solar concentrator shown in Fig. 3.16.
Solar radiation is reflected onto the tubular absorber assembly consisting
of an *evacuated* glass tube envelope and a heat absorbing pipe. The

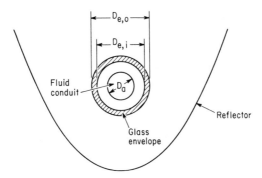

FIGURE 3.16   Cross section of solar concentrator for example calculations.

method of calculating energy delivery uses energy balances on each component as shown below.

Steady-state energy balance for the absorber tube and the envelope can be written directly from the first law of thermodynamics. For the absorber (if free-molecular flow is suppressed the evacuation)

$$q_s = q_{ir,ae} + q_u \qquad (3.61)$$

and for the envelope

$$q_{s,c} + q_{ir,ae} = q_{ir,e} + q_{c,e}, \qquad (3.62)$$

where

q_s = solar radiation absorbed by the absorber (only beam radiation is considered),

$q_{ir,ae}$ = ir radiative exchange between the absorber and its envelope,

q_u = useful energy delivery to the working fluid,

$q_{s,c}$ = solar radiation absorbed by the cover or envelope,

$q_{ir,e}$ = ir radiative exchange between the cover and the environment,

$q_{c,e}$ = convection loss from the envelope to the environment.

Each heat flux term may be calculated from expressions given earlier in this chapter:

$$q_s = \eta_0 I_{b,c} A_c, \qquad (3.63)$$

where η_0 is the collector optical efficiency (see Chapter 4) including absorber absorptance α_a, cover transmittance τ, mirror reflectance ρ_m, and shading, tracking, and misalignment effects. A_c is the collector aperture area and $I_{b,c}$ is the component of beam radiation normal to A_c.

From Eq. (3.51)

$$q_{ir,ae} = A_a \mathscr{F}_{ae} \sigma(T_a^4 - T_e^4), \tag{3.64}$$

where the absorber area $A_a = \pi D_a$, per unit length, and

$$\mathscr{F}_{ae} [(1/\epsilon_a) + (1/\epsilon_e) - 1]^{-1} \tag{3.65}$$

and T_a and T_e are the absorber and envelope temperatures.

The useful energy delivery is given by

$$q_u = \dot{m}(h_o - h_i), \tag{3.66}$$

where $\dot{m}$ is the fluid flow rate and h_o and h_i are exit and entrance values of fluid enthalpy.

Solar radiation absorbed by the envelope is

$$q_{s,e} = I_{b,c} \rho_m \alpha_e A_c, \tag{3.67}$$

where α_e is the envelope absorptance calculated from Eqs. (2.66)–(2.68); α_e usually depends upon the absorber incidence angle but this complication is not considered here. In addition, a small portion of the radiation transmitted through the envelope is reflected from the absorber to the envelope. However, since absorber absorptance α_a is ~95%, the reflected radiation is small and can be ignored in a first-order analysis.

Radiative exchange with the environment $q_{ir,e}$ can be calculated from Eq. (3.57) or from

$$q_{ir,e} = \epsilon_e A_e \sigma(T_e^4 - T_{sky}^4), \tag{3.68}$$

where $A_e = \pi D_{e,o}$ per unit length. For preliminary design it is sufficiently accurate to use the ambient temperature T_{amb} as an approximation for the sky temperature T_{sky}.

Finally, the convective loss to the environment is given by

$$q_{c,e} = \overline{h_c} A_e (T_e - T_{amb}). \tag{3.69}$$

The convection coefficient is for a horizontal cylindrical tube either losing heat by free convection [Eq. (3.40)] or by forced convection [Eq. (3.34)].

In many solar concentrators the resistance to heat flow through the glass envelope and the metal absorber pipe is small and may be neglected. In addition, the resistance offered to heat flow by the boundary layer inside the absorber pipe is generally small and may be neglected. If the flow is slow and the film resistance significant, the heat removal factor F_R [Eq. (3.22)] can be used to account for this effect.

Substituting Eqs. (3.63)–(3.69) into Eqs. (3.61) and (3.62) leads to two simultaneous equations in T_a and T_e:

$$\eta_0 I_{b,c} A_c = A_a \mathscr{F}_{ae} \sigma(T_a^4 - T_e^4) + \dot{m} c_p (T_{f,o} - T_{f,i}), \qquad (3.70)$$

where $q_u = \dot{m} c_p (T_{f,o} - T_{f,i})$ if no phase change occurs, and

$$\rho_m I_{b,c} \alpha_e A_c + A_a \mathscr{F}_{ae} \sigma(T_a^4 - T_e^4)$$
$$= \epsilon_e A_e \sigma(T_e^4 - T_{amb}^4) + h_c A_e (T_e - T_{amb}). \qquad (3.71)$$

The average absorber temperature

$$T_a = \tfrac{1}{2}(T_{f,o} + T_{f,i}). \qquad (3.72)$$

Since T_a and T_e enter as first and fourth powers, a closed form solution is generally not possible. One method of solution is by simple iteration: (a) compute T_e from Eq. (3.71) using given fluid inlet temperature with an estimated $T_{f,o}$ value to estimate T_a; (b) compute $T_{f,o}$ from Eq. (3.70) using T_e from (a); (c) compute T_a from Eq. (3.72). Step (a) can best be done using a Newton–Raphson accelerator method.

Many parameters specific to a given collector must be given for a solution including optical efficiency η_0, collector aperture area A_c, absorber and envelope ir emittances ϵ_a and ϵ_e, fluid flow rate and inlet temperature $\dot{m}$ and $T_{f,i}$, mirror reflectance ρ_m, envelope absorptance and outside diameter α_e and $D_{e,o}$, absorber outside diameter D_a, convection coefficient $\overline{h_c}$, and local insolation and temperature $I_{b,c}$ and T_{amb}. The iterative solution of Eqs. (3.70) and (3.71) can easily be done by a programmable calculator or digital computer.

The parameters present in Eqs. (3.70) and (3.71) are the first-order design parameters for the solar concentrator. The mathematical model developed here can analyze the effectiveness of surface selectivity (ϵ_a) or envelope ir properties (ϵ_e), mirror quality (ρ_m), concentration ratio (A_c/A_a), operating temperature ($T_{f,i}$), and optical efficiency (η_0)—the major design variables for line-focus collectors described in Chapter 4.

The analysis above was carried out on a per unit length basis without specifying the range of sizes. If the absorber is illuminated uniformly along its axis and is less than about 2 m long, the axial distance increment can be simply the collector length to calculate the useful energy delivery for engineering purposes. For collectors several hundred meters long, however, a marching solution must be used with axial calculation increments selected to assure good accuracy. For absorbers not illuminated uniformly, axial increments of a few centimeters may be required depending on the design. In the special case of boiling, for which the absorber temperature is constant, any axial length increment can be used. Also, for boiling the heat removal factor $F_R = F'$, the plate efficiency.

4 | MEDIUM-TEMPERATURE SOLAR COLLECTORS AND ANCILLARY COMPONENTS

I'll tell you how the sun rose—
A ribbon at a time
The Steeples swam in amethyst
The news, like Squirrels, ran

Emily Dickinson

High quality energy from the sun can be used as the energy source for many processes operating at temperatures above 100°C. The efficient conversion of solar radiation to heat at these temperature levels requires the use of concentrating or evacuated solar collectors. Therefore the basic concepts of concentration, the limits imposed by the second law of thermodynamics, and the analysis of several types of high performance collectors comprise this chapter.

I. CONCENTRATION OF SUNLIGHT

A device which increases the flux at the absorber of a solar collector above ambient levels is called a concentrator. The concentration effect is accomplished by use of either reflecting or refracting elements which are located so that solar flux is focused or funneled onto the receiver component. Solar collectors can be of three types. The first are planar, nonconcentrating, flat-plate types which cannot provide heat efficiently at temperatures required by processes considered in this book. Hence, ordinary flat-plate collectors are not considered further.

The second collector type achieves a concentration multiplication in the range 1.5–10. As will be shown shortly, this family of collectors requires neither a sharp focus nor precise solar tracking. Only occasional turning of the collector is needed. These collectors can be used for many intermediate temperature processes considered in this book.

The third generic type of solar collector concentrates solar flux by more than a factor of ten and requires relatively precise optical elements to maintain a sharp focus. Continuous tracking of the sun is also required to maintain the focus.

It is useful to define several terms encountered in the analysis of concentrators. The *concentration ratio* CR is the ratio of the aperture area to the absorber area. The *aperture area* is the area intercepting radiation. The *absorber area* is the area of the component (whether fully illuminated or not) receiving concentrated solar radiation. The absorber is also the component from which the principal heat loss occurs; therefore CR as defined above will be useful in the thermal analysis of concentrators. Other concentration ratio definitions can be used. For example the ratio of absorber flux to aperture flux—the flux concentration—has been used. However, flux or brightness concentrations do not naturally occur in thermal analyses and are not used in this book.

A. The Limits to Concentration

A simple, steady-state energy balance on a concentrator operating at a mean temperature $\bar{T}_c$ can be expressed as

$$q_u = \eta_o I_c A_c - U'_c A_r (\bar{T}_c - T_a), \tag{4.1}$$

where η_0 is the optical efficiency, A_c the collector aperture area, A_r the receiver area, and T_a the ambient temperature; U'_c is based on the receiver area. The instantaneous collector efficiency η_c is defined as

$$\eta_c = q_u / I_c A_c. \tag{4.2}$$

Therefore, from Eq. (4.1)

$$\eta_c = \eta_0 - [U'_c/(\text{CR})][(T\bar{T}_c - T_a)/I_c], \qquad (\eta_c > 0) \tag{4.3}$$

where the concentration ratio is

$$CR \equiv A_c/A_r. \qquad (4.4)$$

It would appear at first glance that the limiting efficiency value η_0 (if η_0 is independent of CR) could be achieved at any temperature $\bar{T}_c$ by making the receiver arbitrarily small, i.e., CR $\to \infty$. However, a concentrator could then be operated at a temperature greater than that of the sun—a violation of the second law of thermodynamics. In this section the thermodynamic limits to concentration are determined for any concentrator (R4).

Figure 4.1 shows a general diagram of any solar concentrator including the radiation source, collector aperture, and receiver. The source represents the line of the sun's trajectory taken to be a moving point source. In Chapter 3 it was shown that radiation shape factors have a reciprocity relation. Referring to Fig. 4.1 we have

$$A_s \mathscr{F}_{sc} = A_c \mathscr{F}_{cs} \qquad (4.5)$$

and

$$A_s \mathscr{F}_{sr} = A_r \mathscr{F}_{rs}, \qquad (4.6)$$

where c denotes the collector aperture, r the receiver, and s the source. The concentration CR can be expressed as

$$CR = A_c/A_r = \mathscr{F}_{sc}\mathscr{F}_{rs}/\mathscr{F}_{cs}\mathscr{F}_{sr}. \qquad (4.7)$$

For the optically ideal concentrator with no losses, all radiation from the aperture enters the receiver zone. Therefore, $\mathscr{F}_{sc} = \mathscr{F}_{sr}$. If the source is a blackbody, $\mathscr{F}_{cs} = F_{cs}$, the view factor of surface s with respect to c. Therefore,

$$CR = \mathscr{F}_{rs}/F_{cs} \qquad (4.8)$$

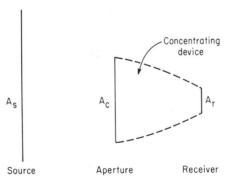

FIGURE 4.1   Diagram of a concentrator with aperture A_c and receiver A_r illuminated by a source A_s.

and since $\mathscr{F}_{rs} \leq 1$ from the second law,

$$CR \leq CR_{max} = 1/F_{cs}. \tag{4.9}$$

That is, the maximum concentration is limited by the second law to the reciprocal of the aperture to source shape factor.

For a single-curvature (trough or linear) concentrator, Fig. 4.2 can be used to find F_{cs}. Over a day a portion of the sun's path can be viewed as a line source of length $2r$. The *acceptance angle* $2\theta_{max}$ is also shown in Fig. 4.2. It is the angle over which solar radiation is to be collected during a period in which no collector tracking need be done. From Hottel's crossed string method (H3) it can be shown that

$$F_{cs} = \sin \theta_{max}. \tag{4.10}$$

From Eq. (4.9) the maximum concentration $CR_{max,2D}$ is

$$CR_{max,2D} = 1/\sin \theta_{max}. \tag{4.11}$$

Values of the half-angle θ_{max} for solar collectors range from $\frac{1}{4}°$ (the sun subtends an angle of $\frac{1}{2}°$) to $90°$ for a flat-plate collector. Equation (4.11) shows why the limit $CR \to \infty$ in Eq. (4.3) cannot be achieved with a sun of finite size. For a compound curvature (dish or three-dimensional) concentrator, the maximum concentration can be shown to be

$$CR_{max,3D} = 1/\sin^2 \theta_{max}. \tag{4.12}$$

Equations (4.11) and (4.12) show that the maximum concentration of a linear concentrator is about 200 and for a dish type about 40,000 assuming the sun to subtend a half-angle of $\sim\frac{1}{4}°$ (4.6 mrad).

Figure 4.3 shows an elevation view of three solar trajectories for the solstices and equinoxes at 40°N latitude. Also shown are the acceptance limits ($2\theta_{max} = 60°$) of an east–west oriented trough collector. The number of hours of acceptance can be estimated from such a figure. An acceptance angle of 60° corresponds to a relatively small maximum concentration ratio of 2.0 according to Eq. (4.11).

Equations (4.1) and (4.12) can be used to find the maximum temperature a concentrator with a given CR can achieve. The amount of

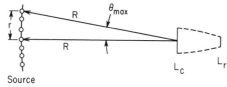

FIGURE 4.2 Diagram used to determine the shape factor F_{cs} of a single-curvature, trough concentrator; the acceptance half-angle θ_{max} is shown along with several positions of the sun relative to the aperture L_c.

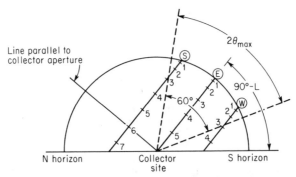

FIGURE 4.3   Elevation view of solstice and equinox sun paths for 40°N with the acceptance limits for a $\theta_{max} = 30°$ solar collector superimposed. The collector aperture is tilted up at the latitude angle. Twelve hours of sunlight are accepted at the equinox and about $5\frac{1}{2}$ h at the solstices. Legend: Ⓢ, summer solstice; Ⓔ, equinox; Ⓦ, winter solstice.

energy absorbed (but not fully delivered to the fluid because of heat losses) by a concentrator's receiver in near-earth space is

$$q_{ab} = \tau \alpha_s A_s F_{sc} \sigma T_s^4 = \tau \alpha_s A_c F_{cs} \sigma T_s^4, \tag{4.13}$$

where T_s is the effective surface temperature of the sun and τ is the transmittance of the concentrator assembly. If the acceptance half-angle θ_{max} is selected equal to the half-angle of the sun's disk ($\theta_s = \frac{1}{4}°$),

$$q_{ab} = \tau \alpha_s A_c \sin^2 \theta_s \, \sigma T_s^4, \tag{4.14}$$

environmental heat losses q_e at very high temperature can be estimated from Eq. (3.56) since radiation is the dominant loss mechanism; therefore,

$$q_e = \epsilon_r A_r \sigma \bar{T}_c^4. \tag{4.15}$$

An energy balance on the absorber gives

$$q_{ab} = q_e + \phi_c q_{ab}, \tag{4.16}$$

where ϕ_c is the fraction of absorber energy delivered to the load (i.e., $\phi_c q_{ab} = q_u$). Solving Eqs. (4.14)–(4.16) for $\bar{T}_c$,

$$\bar{T}_c = T_s[(1 - \phi_c)\tau(\alpha_s/e_r)(CR/CR_{max})]^{1/4}. \tag{4.17}$$

As $\phi_c \to 0$ and $\tau \to 1$, implying vanishingly small energy collection and perfect optics,

$$\bar{T}_c \to T_s(CR/CR_{max})^{1/4}, \tag{4.18}$$

recalling from Chapter 3 that $\epsilon_r \to \alpha_s$ as $\bar{T}_c \to T_s$ because of spectral overlap. Equation (4.18) confirms that the collector temperature cannot exceed the temperature of the sun, as expected.

B. *Acceptance of Diffuse Radiation*

Since diffuse radiation emanates from many directions, part of it will not lie within the acceptance angle $2\theta_{\max}$. The minimum portion which is accepted can be estimated by assuming the diffuse radiation to be isotropic at the concentrator aperture. From shape factor considerations,

$$A_c \mathscr{F}_{cs} = A_r \mathscr{F}_{rc}. \tag{4.19}$$

For most concentrators accepting appreciable diffuse light, $\mathscr{F}_{rc} = 1$ since all radiation leaving the receiver ultimately reaches the aperture. Therefore,

$$\mathscr{F}_{cs} = A_r / A_c = 1/\mathrm{CR}. \tag{4.20}$$

Equation (4.20) is a conservative estimate of the diffuse acceptance since diffuse radiation is not isotropic but normally has higher levels near the sun on sunny days. Therefore the amount of diffuse radiation accepted by a concentrator with concentration CR is greater than or equal to $(\mathrm{CR})^{-1}$.

C. *Relationship of Temperature to Concentration Ratio*

Selection of a concentrator design to perform a task at a specified temperature $\bar{T}_c$ must involve the specification of concentration ratio CR, loss coefficient U_c, and optical efficiency η_0 (ratio of absorbed flux to flux incident on the aperture). Most thermal losses from a concentrator occur by radiation since it is generally cost effective to use an evacuated receiver. An energy balance on the absorber of a concentrator shows the relation of CR and $\bar{T}_c$ for this type of collector to be

$$\bar{T}_c = [(\eta_0 - \eta_c)I_c/\sigma\epsilon]^{1/4}(\mathrm{CR})^{1/4}, \tag{4.21}$$

where I_c is the radiation incident on the collector, η_c is the collection efficiency (ratio of heat delivered to the collector working fluid to aperture flux), and other symbols are as defined previously. The term T_{sky}^4 from Eq. (3.56) is of second order and is not included.

Figure 4.4 based on Eq. (4.21) shows the concentration ratio required to instantaneously operate a collector at 40% efficiency at a specific temperature. The upper curve represents the performance of an evacuated receiver with a nonselective surface ($\epsilon = 0.9$) at 900 W/m² solar flux. An optical efficiency of 65% is assumed. The lower curve shows the CR value needed if a selective surface with $\epsilon = 0.15$ is available. It is a formidable task to achieve this selectivity at high temperature because of the overlap of solar and absorber radiation spectra.

Figure 4.4 should not be used for design since it is only an indication of the minimum CR needed to achieve a given temperature at 40% efficiency in bright sun. A much more detailed analysis is required for rational system design. The curves do show, however, that use of an expensive collector with an excessively high concentration ratio is not necessary to achieve a specific operating temperature at adequate efficiency. The existence of a relationship between CR and $\bar{T}_c$ must, therefore, be recognized to avoid overdesign.

In a subsequent section it will be shown that most practical solar collectors achieve about one-fourth the theoretical maximum concentration CR_{max} given by Eqs. (4.11) and (4.12). For a trough collector $CR_{max,2D} \sim 200$; therefore, practical devices can achieve a concentration of about 50. Referring to Fig. 4.4, CR = 50 corresponds to an operating temperature of 400–800°C at 900 W/m². For a lower flux value of 400 W/m² the same CR = 50 device could operate at 290–600°C. For higher temperature levels CR > 50 is required and compound curvature collectors are necessary.

Based on the preceding analysis it seems appropriate to divide medium- and high-temperature operating ranges at about 300–400°C and CR ∼ 50, corresponding to the use of single-curvature concentrators for the former. For the purpose of organizing the material to be covered in

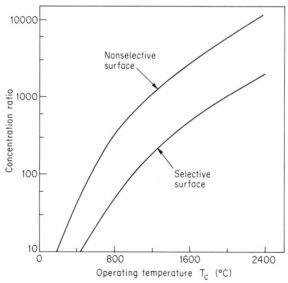

FIGURE 4.4 Minimum concentration ratio required to achieve a given operating temperature at 40% collector efficiency. The upper curve is for an evacuated receiver with a nonselective surface ($\epsilon = 0.90$), the lower for a selective surface ($\epsilon = 0.15$). The optical efficiency is 65% and solar flux $I_c = 900$ W/m².

this chapter and Chapter 6 on high-temperature processes, this division has been made at the indicated temperature and concentration levels.

D. Tracking Requirements

In order to maintain the sun within the acceptance zone of a concentrator, some form of tracking or turning device is needed for CR > 1.5. The amount and frequency of tracking adjustment can be calculated from the equations of the sun's motion in Chapter 2. In general, if a trough collector is oriented east–west, the tracking mechanism and acceptance angle must accommodate a maximum $\pm 30°$/day north–south swing of the sun. For a north–south collector a $\sim 15°$/h motion will be needed. In addition the annual $\pm 23\frac{1}{2}°$/yr solar declination excursion must be accommodated. If the acceptance angle is very large, only infrequent tilt adjustments are needed to maintain acceptance. For very small acceptance angles, nearly continuous tracking is needed.

For intermediate acceptance angles the tilt frequency and number of hours of collection of a single-curvature concentrator can be determined from the equations below for a collector facing the equator. The cutoff time t_c or time at which the sun crosses the acceptance limit θ_{max} can be calculated from (C1)

$$t_c = \cos^{-1}[\tan \delta_s/\tan(\theta_{max}(\delta_s/|\delta_s|) + L - \beta)]/15, \qquad (4.22)$$

where δ_s is the declination, β the aperture tilt, L the latitude, and θ_{max} the acceptance half-angle. The numbers of hours of acceptance is $2t_c$. For maximum t_c with an equator-facing collector, the aperture is tilted so that the sun at noon is at one limit of acceptance. The sun will cross the other limit of acceptance at t_c hours before and after noon (refer to Fig. 4.3). Table 4.1 shows collection times and tilt adjustments for concentration ratios up to about ten. Beyond CR = 10, continuous tracking is usually required.

For continuous tracking concentrators the tracking requirement can be derived from the appropriate incidence angle equations by equating the derivative of cos i with respect to time equal to zero. This operation maximizes the incidence angle as required. For both horizontal east–west and horizontal north–south axes of rotation the rotation angle γ about the axis is given by (S5)

$$\cos \gamma = (\cos \delta_s \cos h_s \cos L + \sin \delta_s \sin L)/\cos i, \qquad (4.23)$$

where the incidence angle is calculated from Eqs. (2.26) and (2.27), respectively.

TABLE 4.1

Number of Hours of Collection per Day and Tilt Frequency for Single-Curvature Concentrators[a,b]

Acceptance half-angle θ_{max} (ideal concentration for perfect mirrors and point sun)	Collection time avg. over year (h/day)	No. of adjustments per year	Shortest period without adjustment (days)	Avg. collection time if tilt adjusted every day (h/day)
19.5° (3.0)	9.22	2	180	10.72
14° (4.13)	8.76	4	35	10.04
11° (5.24)	8.60	6	35	9.52
9° (6.39)	8.38	10	24	9.08
8° (7.19)	8.22	14	16	8.82
7° (8.21)	8.04	20	13	8.54
6.5° (8.83)	7.96	26	9	8.36
6° (9.57)	7.78	80	1	8.18
5.5°[c] (10.43)	7.60	84	1	8.00

[a] From (R4), © 1976, Pergamon Press, Ltd.
[b] CR = [3.10]. Concentrators aligned east–west; at least seven h of collection are achieved per day.
[c] For θ_{max} = 5.5°, minimum collection time is 6.78 h/day.

A polar-mounted concentrator rotates at a uniform rate and at an angle equal to the hour angle, i.e., $\gamma = h_s$.

II. COLLECTORS FOR MEDIUM-TEMPERATURE SOLAR PROCESSES

The analysis of the preceding section has shown that single-curvature, trough-type concentrators can achieve adequate concentration to efficiently convert solar radiation to heat at temperatures below 300–400°C. Therefore, for the medium-temperature processes considered in Chapter 5, the complications of two-axis tracking and the extra cost of manufacture for point-focus collectors can be avoided. This section analyzes the performance of the most common 2*D* concentrator—the parabolic trough—in detail. Several other collector types including intermittent tracking and evacuated tube types are also described. Figure 4.5 shows the common types of tracking concentrators.

A. Parabolic Trough Concentrator (PTC) Performance

The most common commercially available concentrator is the PTC collector. Figure 4.6 shows a photograph of two commercial

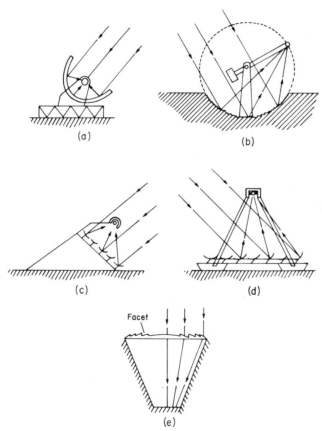

FIGURE 4.5   Line-focus or single-curvature solar concentrators: (a) parabolic trough; (b) fixed circular trough and tracking absorber; (c), (d) Fresnel mirrors; and (e) Fresnel lens. [From (K2).]

FIGURE 4.6   Commercial parabolic trough concentrators (PTC): (a) Solar Kinetics—12.7-m² aperture, CR = 41, 27-cm focal length; and (b) Hexcel—16-m² aperture, CR = 67, 91-cm focal length.

units and Fig. 4.7 describes the major components and identifies several parameters useful in performance analysis.

Parabolic Trough Concentrator collectors usually track the sun with one degree of freedom using one of three orientations—east–west, north–south, or polar (see Chapter 2). The east–west and north–south configurations are the simplest to assemble into large arrays but have higher incidence angle cosine losses than the polar mount. However, the polar mount intercepts more solar radiation per unit area.

The absorber of a PTC collector can either be tubular or planar and is usually housed in a glass tube to reduce convection by the greenhouse effect or to eliminate it by means of a vacuum. The absorber can be rigidly attached to the support structure and act as the center of rotation of the PTC as in Fig. 4.6a, or alternatively, the absorber may be rigidly attached to the reflector as in Fig. 4.6b. By proper design, a PTC can be made free of the gravity couple, usually by counterweighting and rim angle selection.

Optical Analysis of the PTC The *optical efficiency* of a collector is the fraction of sunlight which is intercepted by an ideal collector aperture directly facing the sun to that which is eventually absorbed by the absorber component of a real collector which may not face the sun directly. Eight major collector performance effects can be combined into the optical efficiency denoted by η_o:

(1) mirror reflectance ρ_m (not present in refracting devices);

(2) absorber or receiver cover system transmittance τ_r;

(3) absorber absorptance for solar radiation α_r;

(4) mirror surface slope errors—parameterized by the slope error ψ_1;

(5) solar image spread—parameterized by angular deviation from a beam from the center of the sun ψ_2. The effect of mirror dispersion as described in Chapter 2 (pp. 55 to 61) can be included in ψ_2;

(6) mirror tracking error—parameterized by aperture misalignment ψ_3;

(7) shading of the mirror by the absorber cover tube and supports—parameterized by f_t, the unobstructed fraction of the aperture

(8) incidence angle i. Sunlight passing through the receiver cover tube and not striking the receiver is refracted by the tube and misses the receiver when reflected from the collector mirror.

One expression for the optical efficiency of a PTC is

$$\eta_o = (\rho_m \tau_r \alpha_r f_t)[\delta(\psi_1, \psi_2) F(\psi_3)][(1 - A \tan i) \cos i], \qquad (4.24)$$

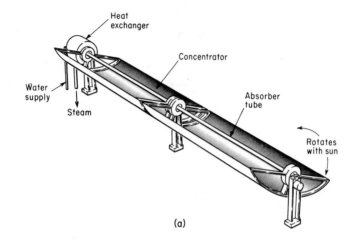

(a)

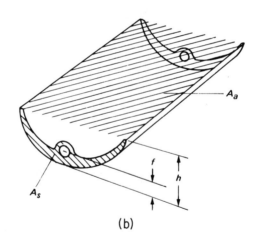

(b)

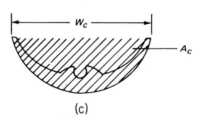

(c)

FIGURE 4.7 (a) Parabolic trough concentration isometric drawing; sketches (b) and (c) identify important parameters for optical design. [From (R6).]

where $\delta(\psi_1,\psi_2)$ is the fraction of rays reflected from a real mirror surface which are intercepted by the absorber for perfect tracking. $F(\psi_3)$ is the fraction of rays intercepted by the absorber for perfect optics and a point sun for a mirror tracking error ψ_3. A is a geometric factor accounting for off-normal incidence effects including blockages, shadows, and loss of radiation reflected from the mirror to beyond the end of the receiver.

For the PTC in Fig. 4.7, A is given by (R6)

$$A = [W_c(f + h) + A_s - A_c]/A_a. \qquad (4.25)$$

A is usually of the order of 0.2–0.3.

Tracking errors, denoted by $F(\psi_3)$ in Eq. (4.24), have been measured on a prototype PTC with CR = 15 by Ramsey (R6) and are shown in Fig. 4.8. The figure shows that a 1° error in tracking reduces the optical capture by 20%. An alternative method of considering tracking errors is to combine the ψ_3 effect into δ in Eq. (4.24). This method is described below.

The order of magnitude of η_0 can be estimated from the foregoing and from data in Chapters 2 and 3. For a well-designed and fabricated PTC, the following values of the various optical constants in Eq. (4.24) should apply: $\rho_m \sim 0.85$, $\tau_r \sim 0.9$, $\alpha_r \sim 0.95$, $f_t \sim 0.95$, $\delta(\psi_1,\psi_2) \sim 0.95$, $F(\psi_3) \sim 0.95$. At normal incidence, therefore, the optical efficiency is the about 62%. If an etched glass absorber cover were used, the optical efficiency could be raised to about 66–67%. Other optical properties are unlikely to improve substantially.

Optical Intercept Factor δ The values of all optical properties other than $\delta(\psi_1,\psi_2)$ and $F(\psi_3)$ can be obtained from standard handbooks (K7), whereas the angular factors must be measured on a collector

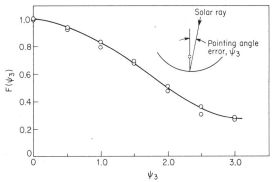

FIGURE 4.8 The angular tracking error factor $F(\psi_3)$ for a prototype PR of concentration ratio 15. [From (R6).]

assembly. During the design process it may be useful to be able to calculate ψ_i effects to determine tolerances on trackers and mirror surfaces. A method of calculating $\delta(\psi_1, \psi_2)$ and, by extension, $F(\psi_3)$ is given below (D3).

Figure 4.9 shows the relationship of a mirror surface segment of a PTC to the tubular absorber. The nominally reflected ray is that ray from the center of the sun which is reflected from the theoretical mirror contour to the center of the absorber. Rays at an angular distance ψ_2 from the center of the sun will intercept the absorber at an angle ψ_2 away from the center of the absorber. Likewise, rays reflected from a mirror surface element whose slope is ψ_1 degrees in error from the theoretical surface slope will intercept the absorber at an angle $2\psi_1$ from the nominal ray.

From Fig. 4.9 it is clear that y, the distance from the receiver center to the ray intersection with a diameter, is

$$y = r \sin(2\psi_1 + \psi_2) \approx r(2\psi_1 + \psi_2). \qquad (4.26)$$

The distance r from the mirror to the absorber tube center for a parabola is

$$r = [L_a(1 + \cos \phi)]/[2 \sin \phi (1 + \cos \theta)], \qquad (4.27)$$

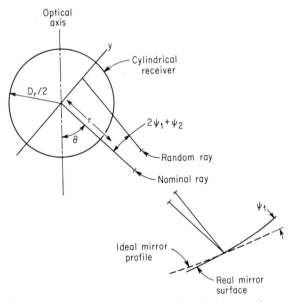

FIGURE 4.9   PTC collector section showing a mirror segment whose slope is ψ_1 degrees away from the theoretical slope. Also shown is a ray from the sun at a distance ψ_2 from the center of the sun and reflected from the real mirror surface.

where L_a is the aperture width, ϕ is the trough rim half-angle, and θ is shown in Fig. 4.9. Using Eq. (4.27), y can be expressed as

$$y = [L_a(1 + \cos \phi)(2\psi_1 + \psi_2)]/[2 \sin \phi (1 + \cos \theta)]. \qquad (4.28)$$

Random rays are seen to intercept the absorber (and cover if present) at off-normal incidence. Hence, reflection losses are larger than for the nominal ray owing to increased reflectance losses (see Chapter 2). In the analysis which follows this effect is not included. However, if unetched glass and selective surfaces without antireflectants are used, the effect may need to be considered.

If the variables ψ_1 and ψ_2 are normally distributed with variances σ_{ψ_1} and σ_{ψ_2} and mean zero, the variance σ_y^2 of the ray intercept point is from Eq. (4.28)

$$\sigma_y^2 = \frac{L_a^2(1 + \cos \phi)^2(4\sigma_{\psi_1}^2 + \sigma_{\psi_2}^2)}{(4 \sin^2 \phi)(2\phi)} \int_{-\phi}^{\phi} \frac{d\theta}{(1 + \cos \theta)^2}; \qquad (4.29)$$

carrying out the integration,

$$\sigma_y^2 = [L_a^2(4\sigma_{\psi_1}^2 + \sigma_{\psi_2}^2)(2 + \cos \phi)]/(12\phi \sin \phi). \qquad (4.30)$$

The assumption of a Gaussian sun with variance $\sigma_{\psi_2}^2$ is not physically correct since the sun has a nonnormal flux distribution. However, the Gaussian model may be used if $\sigma_{\psi_1}^2$ is of the same order as $\sigma_{\psi_2}^2$ and other errors described shortly. If other optical errors are small, however, the Gaussian sun approximation should not be used.

The intercept factor δ is the amount of energy contained in the normal probability flux distribution of variace σ_y^2 which is intercepted by the receiver tube of diameter D_r. That is,

$$\delta(\psi_1, \psi_2) = \frac{1}{\sqrt{2\pi}} \int_{-d_r/2}^{d_r/2} (e^{-Z^2/2}) \, dZ, \qquad (4.31)$$

where the change of variables $Z = y/\sigma_y$ and $d_r = D_r/\sigma_y$ has been made. The right-hand side of Eq. (4.31) is the normal probability integral included in many handbooks and in the Appendix (see Table A.5).*

The method used above can be extended to include other optical effects including tracking and location errors ψ_3 and mirror dispersion effects ψ_4 (see Chapter 3). The effective variance Σ^2 to be used in Eq. (4.30) would be

$$\Sigma^2 = 4\sigma_{\psi_1}^2 + \sigma_{\psi_2}^2 + 4\sigma_{\psi_3}^2 + \sigma_{\psi_4}^2 \qquad (4.32)$$

if the absorber is not rigidly attached to the reflector assembly. If the absorber is attached rigidly to the mirror, the constant 4 in the third term is replaced with unity.

---

* From Table A.5 note that 95% capture requires a receiver diameter $D_r \sim 2\sigma$.

Example   Find the maximum concentration ratio and receiver diameter for a PTC to intercept 95% of the incident radiation striking a perfectly tracking mirror with negligible dispersion. The mirror rim half-angle is 60° and mirror surface accuracy measurements showed that $\sigma_{\psi_1} = 0.2°$. The aperture is 0.5 m wide.

Solution   From Eq. (4.30), σ_y^2 is

$$\sigma_y^2 = \{(0.5)^2(\pi/180)^2[4 \times 0.2^2 + 0.125^2](2 + \cos 60°)\}/$$
$$[12(\pi/180)(60°) \sin 60°]$$
$$= 3.07 \times 10^{-6} \text{ m}^2;$$
$$\sigma_y = 1.75 \text{ mm}.$$

Since the distribution of the sun's rays is more rectangular than normal, the effective standard deviation σ_{ψ_2} for the sun has been taken as $\frac{1}{8}°$ above.

The optical intercept factor is to be 95%; therefore, d_r from the probability integral tables is 1.96. Hence,

$$D_r = d_r \sigma_y = \underline{3.43 \text{ mm}}.$$

The concentration CR is

$$CR = A_c/A_r = L_a/(\pi D_r) = 500/(\pi \times 3.43)$$
$$= \underline{46.4.} \quad \blacksquare$$

The optical intercept factor for a PTC with a *flat* receiver can be computed from Eq. (4.31) and the value of σ_y^2 to be used is (D3)

$$\sigma_y^2 = \frac{L_a^2(4\sigma_{\psi_1}^2 + \sigma_{\psi_2}^2)}{4\phi \tan^2(\phi/2)} \left[\frac{-1}{3 \sin^3 \phi \cos \phi} \right.$$
$$+ \frac{2}{3 \sin^3 \phi} + \frac{2}{\sin \phi} - \frac{\cos \phi}{3 \sin^3 \phi} - \frac{2 \cos \phi}{\sin \phi}$$
$$\left. + \frac{4 \sin \phi}{3 \cos \phi} - \ln \tan((\pi/4) + (\phi/2)) + \ln \tan((\pi/4) - (\phi/2)) \right].$$
$$(4.33)$$

Off-Normal Incidence Effects   The preceding results apply for normal incidence of beam radiation onto the aperture plane. A major off-normal incidence angle effect is the increased shading and blocking of the reflector plus end effects as represented by the last term in brackets in Eq. (4.24). Measurements at off-normal incidence have shown this term accurately represents shading, blocking, and end effects of a PTC (R6). Measured collector optical efficiency data are shown in Fig. 4.10. As expected from Eqs. (2.26) and (2.27), the east–west orientation shows the

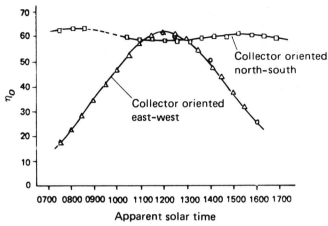

FIGURE 4.10 Measured optical efficiency of a PTC in north–south and east–west orientations. Legend: ○, data from 5/28/1974; △, data from 6/4/1974; □, data from 6/25/1974. [From (R6).]

greatest incidence angle sensitivity. At normal incidence an optical efficiency in the range 0.60–0.65 was achieved as expected from the order of magnitude estimate above.

At off-normal incidence the incidence angle on the receiver cover and absorber surface increases, thereby increasing reflective losses from both. This effect must be considered in the prediction of long-term performance.

A final off-normal effect involves the values of r, y, and σ_y in Eqs. (4.26)–(4.30). At nonnormal incidence r, y and σ_y values for normal incidence must be divided by the cosine of the angle between the aperture normal and the projection of the sun's rays onto the collector optical axis plane. For proper tracking, this angle is the incidence angle. Physically, r increases at off-normal incidence because the beam path from the mirror to the absorber must be longer if the incident beam is out of the transverse plane (i.e., plane of Fig. 4.7c).

Thermal Analysis of a PTC Collector   The thermal performance of a PTC using either a planar or circular cylinder absorber can be analyzed using the heat transfer analyses of Chapter 3. Specifically, Eqs. (3.61)–(3.72) represent a sufficiently detailed model of the absorber to adequately predict performance (R6). If an evacuated receiver is not used, Eq. (3.41) can evaluate the free convection heat transfer coefficient within the envelope enclosure. This free convection term must be added to both Eqs. (3.61) and (3.62). Equation (3.61) shows that the collector efficiency η_c based on beam radiation is given by

$$\eta_c \sim \eta_0 - [\epsilon_r \sigma (\bar{T}_c^4 - T_a^4)]/[I_b CR] \qquad (\bar{T}_c > T_a). \qquad (4.34)$$

For an evacuated tube the approximation has been made that the envelope and ambient temperatures are nearly equal. Ramsey *et al.* (R6) showed that Eq. (4.34) adequately represented the radiation heat loss but found that conduction losses through the absorber assembly end supports could be significant for $\bar{T}_c > 100°C$. Therefore, a conduction term q_k is required in Eq. (4.34) unless the tubes are very long and the end effects negligible:

$$q_k = [(\bar{T}_c - T_a)/R_k](A_k/A_a) \qquad (4.35)$$

where R_k is resistance for conduction heat flow through area A_k. Measured thermal efficiency for a PTC operating at 300°C is shown in Fig. 4.11. Efficiency is calculated relative to direct-normal solar flux. Measured efficiency of commercial PTC collectors is presented in the subsequent section on collector testing.

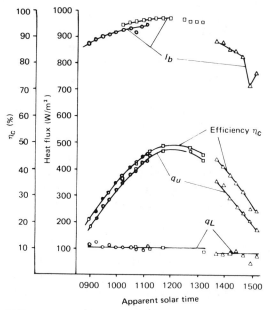

FIGURE 4.11   PTC collector efficiency η_c and useful energy delivery q_u at 300°C measured on several clear days in Arizona. Legend: orientation east–west, absorber 300°C; △, data from 10/24/1974; ⊙, data from 10/25/1974; ☐, data from 11/4/1974. [From (R6).]

B. Performance of Other Single-Curvature Collectors

Several types of focusing collectors are shown in Fig. 4.5 in addition to the PTC concept. The performance of all these may be calculated by using the approach of the previous section. A detailed analysis for each is therefore not repeated here. The optical efficiency is first calculated from values of mirror reflectance, receiver envelope transmittance, receiver absorptance, mirror shading, blocking and end effects, incidence angle, reflector angular uncertainties, and tracking inaccuracies. An equation of the form of Eq. (4.24) can be used for any line-focus collector. The major variations among various trough collectors appear in the f_t, $\delta(\psi_1,\psi_2)$, $F(\psi_3)$, and A terms.

Reflector shading f_t and blocking and end effects embodied in A are a matter of design and can be precisely determined. Tracking errors ψ_3 can be included in the intercept factor $\delta(\Sigma)$ as shown in Eq. (4.32) for reflecting concentrators. References (D3) and (D4) include $\delta(\Sigma)$ expressions for most line-focus collectors.

Fresnel Lens Collectors   Fresnel lens devices (Fig. 4.5e) have been designed and tested in the intermediate concentration ranges. Figure 4.12 shows two realizations of the line-focus Fresnel collector. The first is a large module (16 m²) with an east–west axis (but two-axis tracking) and fabricated by McDonnell-Douglas. The concentrator ratio is about 24 and a second-stage concentrator is used (see below). Test data presented later show an optical efficiency of about 60% for this collector. The second collector shown in Fig. 4.12 and manufactured by Northrup consists of an array of smaller units (~ 1 m²) which can be mounted either east–west or north–south. Tracking is accomplished by a cable drive and sun sensor arrangement. Both collectors use plastic Fresnel lenses.

Several features of the optics of Fresnel lenses differ from PTC optics described above. Since the index of refraction of a material is wavelength dependent, Fresnel lenses exhibit a prism effect called chromatic aberration. That is, sunlight is refracted differently for different wavelengths. The result is a beam spread not present in reflecting concentrators.

The intercept factor for refracting concentrators using lenses such as the Fresnel lens is slightly different than for reflecting devices. This is due in part to the lower sensitivity of refractors to slope errors. Figure 4.13 shows that a slope error ψ_1 in a reflector causes a $2\psi_1$ error in the reflected ray as analyzed earlier. The same error in a refractor surface causes an error $(n_r - 1)\psi_1$, where n_r is the refractive index. Since

FIGURE 4.12   Fresnel lens concentrators: (a) McDonnell-Douglas 16 m²
module; and (b) Northrup 1-m² collectors in an array.

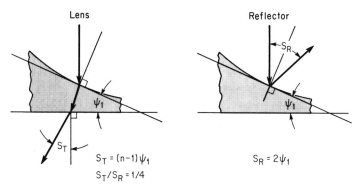

FIGURE 4.13   Comparison of the effect of surface errors ψ_1 on beam displacement for refracting and reflecting elements of a concentrator.

most lens materials have values of $n_r \sim 1.5$, refractors are one-fourth as sensitive to slope errors as are reflectors. An analysis similar to that for the PTC applies for solar ray deviations ψ_2 and tracking errors ψ_3. Mirror dispersion effects ψ_4 [Eq. (4.32)] are zero for Fresnel lens refractors.

Off-normal incidence effects are present in a Fresnel device owing to the change in focal length with incidence angle. This effect is analogous to but not the same as the additional off-normal beam spread noted above for a PTC device.

Figure 4.14 shows the focal point for normal and nonnormal incidence. Note that the effect of the shortened focal length is to cause the sun's image to appear wider at the nominal focal plane during off-normal periods. Figure 4.15 shows the decrease in focal length which occurs for various off-normal conditions (C6, R10). It is seen that a $\pm 60°$ excursion which would be encountered for 8 h of collection at the equinoxes in an east–west axis alignment would diminish the focal length by two-thirds. Hence, the Fresnel line-focus device with CR > 10 is restricted to a north–south orientation ($\pm 35°$ excursion over a year) if the majority of daylight hours are to be collection hours. Collares-Pereira *et al.* have developed a lens–mirror concept (C6) capable of the maximum concentration permitted by Eq. (4.11). A second-stage hyperbolic light funnel (trough) is used to achieve this effect. This concept can also alleviate the off-normal defocusing effect of Fresnel lenses.

Fresnel Mirror Collectors   Two generic types of Fresnel mirror collectors can be identified. The first uses tracking mirror (TM) segments and is exemplified by the collector shown in Fig. 4.16a. The second class uses fixed mirror (FM) segments but a tracking receiver; one example is shown in Fig. 4.16b. The FM design is a fixed aperture device and hence is subject to cosine losses at off-normal incidence. The TM col-

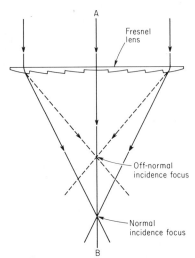

FIGURE 4.14   Line-focus Fresnel lens concentrator section showing nominal focus for normal incidence and shortened focal length at off-normal incidence.

lector can be considered as a series of small rim-angle tracking collectors with a common absorber. However, shading and blocking of one mirror strip by an adjacent one reduces the effective tracking aperture by an amount depending upon aperture tilt, time of day, slat width, slat spacing, and slat thickness.

   The design of the TM collector can be carried out using prin-

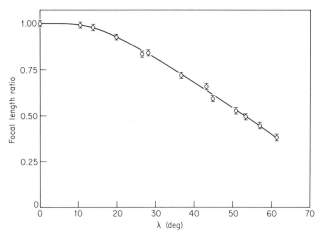

FIGURE 4.15   Effect of off-normal incidence angle λ (measured in plane *AB* out of the plane of Fig. 4.14) on apparent focal length of a line-focus Fresnel lens. The focal length ratio is the focal length at $\lambda \neq 0$ divided by the focal length at $\lambda = 0$ (i.e., normal incidence). [From (C6).]

FIGURE 4.16 Fresnel mirror concentrators: (a) Suntec SLATS™ tracking mirror collector; (b) Scientific Atlanta fixed mirror, moving absorber collector.

ciples of PTC collectors described earlier. However, the FM collector has several unique features which will be summarized herein. Figure 4.17 shows pertinent geometric features of the FM device. The concave array of long mirror strips is located on but not tangent to the surface of a circular cylinder segment. The focal line also moves on the same cylindrical path in response to the sun's virtual motion. An absorber placed at the focal line intercepts the flux reflected from the mirror array. The image width at the absorber is ideally the same as the projected width of a facet; hence the CR is the same as the number of slats, ignoring solar beam spread. Incidence angles away from the transverse plane cause an image spread much like that for a PTC.

FIGURE 4.16 (*Continued*)

Study of Fig. 4.17 shows the method of specifying the angle of each facet. One facet is tangent to the cylindrical surface. The surface angle of any facet, relative to the tangent facet, is simply one-quarter of the angle θ included between the facet location and the tangent facet location measured along the cylindrical surface (R11). Since the aperture of the FM collector is fixed and concave, it will experience some self-shading at very high or very low sun angles. The extent and times of shading are easily calculated from solar geometry principles in Chapter 2.

Measured performance data on FM and TM Fresnel collectors are presented in a subsequent section. The optical intercept factor $\delta(\psi_1, \psi_2)$ for FM and TM collectors developed in (D4) is rather lengthy and is not repeated here.

The CPC Solar Collector A collector concept originated by Winston and others called the compound parabolic concentrator (CPC) is capable of achieving the maximum concentration permitted by the second law [see Eq. (4.11)], i.e.,

$$CR_{CPC} = CR_{max,2D} = 1/\sin \theta_{max}. \qquad (4.36)$$

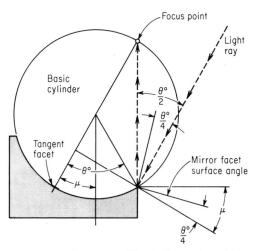

FIGURE 4.17   Geometry of fixed mirror, Fresnel lens concentrator using a moving absorber.

Since the concentration is maximal for given angular acceptance θ_{max}, the CPC can be used in a nontracking mode for concentration ratios up to about 10. This can result in cost saving and simplification or even elimination of the tracking mechanism since no tracking whatever is needed for acceptable performance if an evacuated receiver is used. This section will give a brief overview of CPC performance since it cannot be fully analyzed using the line-focus procedure exemplified by the prior PTC analysis. The reader is referred to (K2, R4, R5) for engineering details of CPC devices and to (W7) by Welford and Winston for the general optical theory of nonimaging concentrators.

Figure 4.18 shows four types of CPC collectors, all drawn to the same scale. Type (a) is the original design using two segments of a parabola. The left curved surface is a reflecting parabola with focus at point Y and axis through Y and parallel to PX (located at angle θ_{max} to the CPC optical axis). The right curved surface is a mirror image of the left. Types (b) and (c) use planar receivers with less exposure of their back surfaces to the environment than for type (a).

CPC type (d) is of the most interest for this book since the absorber is a pipe capable of carrying high-pressure fluids. The reflectors of the collector are not parabolic in shape but the CPC nomenclature will be used to denote the collector type. The shape of the reflector is determined by the requirement that the normal to the reflector surface an any point Q bisect the angle between an incident ray entering the aperture at θ_{max} and striking Q and a line tangent to the absorber through Q. The sec-

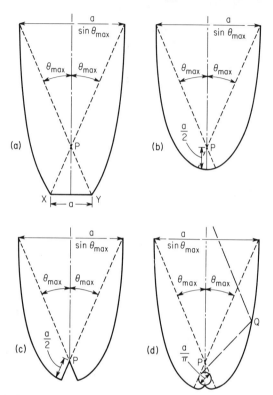

FIGURE 4.18   Four CPC collector concepts drawn to the same scale; (a) basic CPC; (b) fin receiver; (c) bifurcated fin receiver; and (d) tubular receiver with construction line shown.

tion of the reflector below the shadow lines is the involute of the absorber surface. This construction method can be used to design a CPC for any absorber shape. The boundary condition to be met is that the slope at the upper end of the reflector must be parallel to the CPC optical axis. Figure 4.19 shows a commercial CPC module with a tubular receiver and $CR = \frac{4}{3}$.

Optical Efficiency of the CPC Collector   A number of the optical losses present in tracking collectors described above are negligible in CPC collectors because of their broad acceptance band and ability to use imprecise optical elements. Of the seven optical loss mechanisms listed on p. 111, at least four are reduced or nonexistent in CPCs. For example, mirror slope ψ_1 and solar dispersion ψ_2 are of no consequence except very close to the limits of acceptance. Since CPCs would operate near these limits only a few minutes per day, both ψ_1 and ψ_2 are second-order effects. Likewise, tracking errors are not of significance since CPCs need only be

FIGURE 4.19   Commercial CPC collector module using an evacuated tubular receiver with CR = ¾. (Courtesy of Energy Design Corp. Memphis, Tennessee.)

tilted occasionally (frequency depending on concentration) to assure at least 7 h of collection in accordance with Table 4.1. Properly designed CPCs should have no shading loss so $f_t = 1.0$ in Eq. (4.24).

The three remaining optical parameters—mirror reflectance, cover transmittance, and absorber absorptance—are first-order effects for CPCs. Mirror reflectance effects are slightly different for a CPC than for other line-focus reflectors since the CPC does not form a sharp image of the sun at the absorber. Figure 4.20 shows that some solar radiation incident near the aperture edges is reflected more than once on its

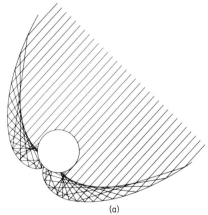

(a)

FIGURE 4.20   Ray trace diagrams of the tubular CPC collector at three values of incidence angle: (a) normal incidence; (b) intermediate; and (c) the limit of acceptance. (Courtesy of W. McIntire, Argonne National Laboratory.)

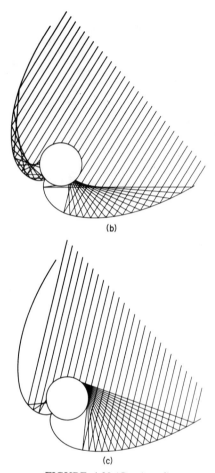

(b)

(c)

FIGURE 4.20 (*Continued*)

way to the absorber. The effect of multiple reflectance can be simply ac-counted for in optical efficiency calculations, to lowest order, by using $\rho_m{}^{\bar{n}}$ where $\bar{n}$ is the number of reflections of all incident rays averaged over the aperture. Some effect of incidence angle on $\bar{n}$ has been noted but Rabl (R5) recommends use of a constant $\bar{n}$ for engineering purposes.

Figure 4.21 can be used to determine the average number of reflections $\bar{n}$ for various concentration ratios CR for the *basic CPC* in Fig. 4.18a. For tubular receiver CPCs (Fig. 4.18d) $\bar{n}$ should be increased by about 0.5 over the Fig. 4.21 values. It is seen that $\bar{n}$ varies as $[1.0 + 0.07 \, \text{CR}]$ $(\pm 0.1) \ln \sqrt{(\text{CR})}$. The dashed line in Fig. 4.21 represents $\bar{n}$ for a fully devel-oped CPC. Since the upper half of the reflector for most CPCs is nearly

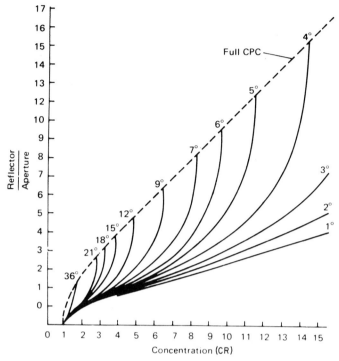

FIGURE 4.21   Effect of truncation on the basic CPC concentration ratio for various values of acceptance angle. [From (R5), © 1976, Pergamon Press, Ltd.]

parallel to the optical axis, it affords little concentration effect. In practice the upper portion (about 40–60% of the mirror) is normally eliminated to reduce CPC size and cost. Figure 4.22 shows the concentration achievable for various truncation amounts. This CR value and the corresponding value of acceptance angle then determine a point on the $\bar{n}$–CR map (Fig. 4.21), and $\bar{n}$ can be evaluated for any truncation. For careful designs a ray trace procedure must be used to find $\bar{n}$ and its dependence on incidence angle.

A final consideration in the optical analysis of the CPC family is the effect of partial acceptance of diffuse radiation. Equation (4.20) showed that at least $(CR)^{-1}$ of the incident diffuse flux reached the absorber. Since CPC collectors operate in the range $CR = [2,10]$ to capitalize on the corresponding reduced tracking requirement, from one-half to one-tenth of the incident diffuse radiation is accepted. This property of CPCs is conveniently included in the CPC optical efficiency by defining the intercept factor δ used for PTC analysis somewhat differently. If δ is

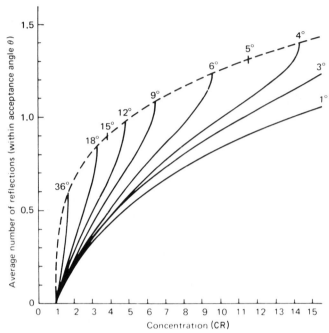

FIGURE 4.22   Average number of reflections for the basic CPC as a function of CR. The dashed line represents a full CPC. The number of reflections for various truncations can be determined by using Fig. 4.21 along with this figure. Values of acceptance angle are shown on the upper curve. [From (R5), © 1976, Pergamon Press, Ltd.]

defined as the fraction of *total* radiation accepted by a CPC, it can be expressed as

$$\delta \equiv [I_{b,c} + I_{d,c}CR]/I_{tot,c} \qquad (4.37)$$

where b, d, and tot refer to beam, diffuse, and total flux incident on the collector aperture.

The optical efficiency of a CPC can then be written as*

$$\eta_o = \rho_m{}^{\bar{n}}\tau_r\alpha_r\delta. \qquad (4.38)$$

Although this expression has the same appearance as Eq. (4.24) it is to be emphasized that δ depends on the characteristics of local solar flux and is not a purely geometric factor such as $\delta(\psi_1,\psi_2)$. For the tubular CPC, $\bar{n} \sim 1.2$, $\rho_m \sim 0.85$, $\alpha_r \sim 0.95$, $\delta \sim 0.95$, and $\tau_r \sim 0.90$. Therefore,

* For the tubular receiver CPC a small gap of width g is required between the mirror cusp and the absorber pipe to accommodate the absorber envelope. Some otherwise collectable rays escape the receiver at this point, reducing the optical efficiency by the factor $(1 - g/p_r)$, where p_r is the absorber perimeter. Other design details are given in Ref (R15).

$\eta_o \sim 0.67$. If an etched glass cover is used, $\eta_o \sim 0.71$. These values of optical efficiency are 7–8% greater than those for a PTC device. Improved optical efficiency partially offsets the concentration limits and associated heat loss effect usually imposed on CPCs (CR < 10) in order to benefit from their reduced tracking requirements.

Thermal Performance of the CPC Collector For a tubular receiver CPC the same heat-loss analysis can be used as for a PTC collector with a tubular receiver. Some uncertainty as to the convection heat loss mode from the tube to the environment [$q_{c,e}$ in Eq. (3.69)] exists since the tube is in a partial enclosure and protected from the environment. Either Eq. (3.34) or (3.40) for external heat loss from a pipe can be used to estimate $q_{c,e}$. The larger value should be used for conservative design.

The thermal efficiency of a CPC (based on *total* beam and diffuse collector plane flux) is given by

$$\eta_c = \rho_m{}^{\bar{n}}\tau_r\alpha_r\delta - [U_c(\bar{T}_c - T_a)^+]/(I_{\text{tot},c}), \tag{4.39}$$

where U_c is based on the aperture area.

By analogy with the PTC analysis developed earlier for an evacuated receiver,

$$U_c \sim [\epsilon_r\sigma(\bar{T}_c{}^2 + T_a{}^2)(\bar{T}_c + T_a)/(\text{CR})] + (A_k/A_a)(1/R_k), \tag{4.40}$$

where R_k is the conduction heat transfer resistance for all conduction paths of total effective area A_k. It has been assumed that the glass tube surface temperature can be closely estimated by the ambient temperature. If a transparent aperture cover has been added to keep dirt from the CPC mirrors, the air within the CPC enclosure acts as an additional thermal resistance between the glass tube and the environment. In that case, tube temperature T_t is generally not well estimated by ambient temperature and must be determined by simultaneous solutions of energy balances on the absorber, tube, enclosed air, and cover.

For the enclosed CPC, the heat transfer analysis of Abdel-Khalik *et al.* (A3) should be used to calculate convection losses. A numerical finite element solution for the basic CPC showed that the onset of convection cells in the CPC enclosure depends on the concentration ratio and truncation amount. These results and Nusselt number values are presented in graphical form in Fig. 4.23.

It is usually preferable to measure rather than calculate U_c if a collector is available for test. A formal test procedure such as ASHRAE 93-77 (modified for concentrator testing as described shortly) can be used to generate an efficiency curve from which U_c can be calculated. At high temperatures the efficiency characteristic will not be linear because of the

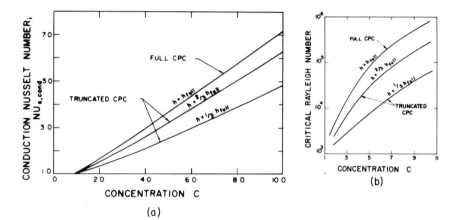

(a)

(b)

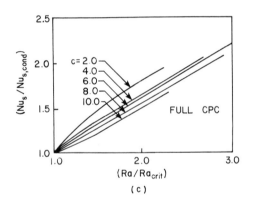

(c)

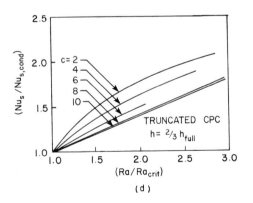

(d)

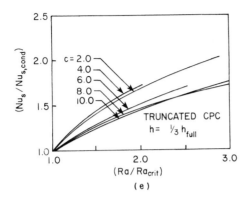

(e)

FIGURE 4.23   Convection heat transfer parameters for the basic CPC enclosure; (a) Nusselt number for conduction (no flow) $Nu_{s,cond} \equiv \bar{h}_c h/k$ where h is the CPC depth and $\bar{h}_c$ is based on the absorber area; (b) critical Rayleigh number for the onset of convection for full, $\frac{1}{3}$-, and $\frac{2}{3}$-truncated CPCs, $Ra \equiv Gr\ Pr$; (c), (d), (e) Nusselt number for convection for various concentrations C and various truncation amounts. [From (A3).]

nonlinearity of radiation heat transfer. Therefore, U_c refers to the heat loss coefficient evaluated at the point $(\bar{T}_c - T_a)/I_{tot,c}$. The numerical value of the loss coefficient is best determined by curve-fitting the data by a polynomial in $(\bar{T}_c - T_a)/I_{tot,c}$ or, as Tabor (T3) suggests, by a law power correlation in $(\bar{T}_c - T_a)^n/I_{tot,c}$—graphical methods of finding U_c from a curvilinear plot are relatively inaccurate. Figure 4.24 shows measured efficiency data for several high performance CPCs.

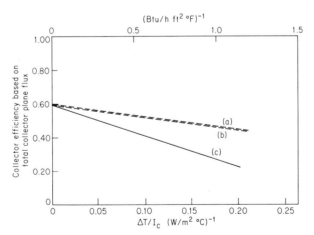

FIGURE 4.24   Plot of measured performance data for several CPC collectors with tubular receivers and aperture covers: (a) CR = 3 [from (S13)]; (b) CR = 1.5 [from (S13)]; (c) CR = 6.5 [from (C7)].

C. Evacuated Tubular Collectors and Concentrator Receivers

Temperatures of the order of 200°C can be achieved at 40% efficiency in bright sun by evacuated tubular collectors without concentration. Figure 4.25 shows sections of several types of commercial tubular collectors. Although receiver geometry varies, all rely on vacuums of the order of $10^{-3}-10^{-4}$ mm Hg to eliminate conduction and convection losses from the absorber surfaces. Typically, U_c values of 0.8–1.2 W/m² °C have been achieved.

The thermal performance of evacuated tubes has been analyzed in Eqs. (3.61)–(3.72). It was shown that radiation is the dominant heat loss mechanism. The optical analysis of tubular collectors varies with the absorber type. For example, the collector shown in Fig. 4.25a has an optical efficiency value relative to total insolation of

$$\eta_o = \tau_r \alpha_r (W/D_t), \tag{4.41}$$

where W is the absorber width and D_t is the tube outside diameter. The effective transmittance τ_r and absorptance α_r vary with incidence angle i according to analysis contained in Chapter 2.

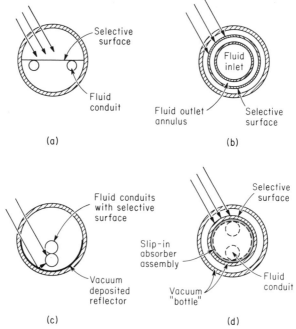

FIGURE 4.25   Various commercial tubular collectors: (a) flat plate; (b) concentric tube; (c) concentrator; (d) slip in heat exchanger for vacuum bottle.

If the receiver is cylindrical (Figs. 4.25b and 4.25d) the optical efficiency is

$$\eta_0 \times (\tau_r \alpha_r)[(I_b \cos i_t + \pi F I_{d,c})/(I_b \cos i + I_{d,c})](D_a/D_t), \quad (4.42)$$

where F is the fraction of the absorber tube circumference which views the sky dome; D_a is the absorber diameter, the tube incidence angle i_t for a north–south tilted tube is calculated from Eq. (2.20), and the collector aperture incidence angle i is calculated from Eq. (2.19). The diffuse radiation on the aperture $I_{d,c}$ is assumed to be isotropic and the collector aperture is taken to be a plane coinciding with the projected area of the tube at noon. Since the incidence angle on a tube i_t is smaller than that on the aperture plane i, the optical efficiency increases at hour angles away from noon. This effect is purely geometric and arises from the definition of efficiency for fixed collectors which is based on the solar flux on a *fixed* aperture *plane*.

If a planar array of tubes is used, shading of one tube by its neighbor will reduce the numerator of Eq. (4.42), the amount depending on the relative spacing and tube size. At least one manufacturer has spaced tubular collectors about one tube diameter apart and has placed a reflector behind the array. The reflector reflects sunlight passing between the tubes onto the rear surface of the absorber, thereby using the full circumference of the absorber. For this geometry and for that in Fig. 4.25c, the calculation of off-normal optical efficiency is not possible in closed form without simplifications and, therefore, is best determined by a formal test.

D. Summary and Generalized Concentrator Performance Representation

The preceding detailed analyses have shown that all line-focus concentrators have a similar instantaneous efficiency representation, namely,

$$\eta_c = \eta_0 - U_c[(\bar{T}_c - T_a)^+/I_c], \quad (4.43)$$

where η_c is the collector efficiency based on solar flux I_c, η_0 the optical efficiency (which may depend on incidence angle and diffuse-to-beam solar flux ratio), U_c the loss coefficient based on aperture area, $\bar{T}_c$ the average collector surface temperature, and T_a the ambient temperature.

Since $\bar{T}_c$ is difficult to measure in practice, the plate efficiency F' and heat removal factor F_R developed in Chapter 3 are usually used to express the efficiency in terms of average fluid temperature $\bar{T}_f$ and inlet fluid temperature $T_{f,i}$, both of which are easily determined. In equation form

$$\eta_c = F'\{\eta_0 - U_c[(\bar{T}_f - T_a)^+/I_c]\} \tag{4.44}$$

where for a tubular receiver

$$F' = [1 + CRU_c\{(D_o/D_ih_i) + [D_o \ln(D_o/D_i)/2k_t]\}]^{-1} \tag{4.45}$$

in which D_o, D_i, k_t, and h_i are the fluid conduit outer diameter, inner diameter, thermal conductivity, and fluid convection heat transfer coefficient, respectively. Since CRU_c is roughly a constant for a given concentrator type to first order, F' is seen to be determined principally by the value of h_i, given a tube size and material. Turbulence promotors are used in the absorber tubes of some concentrators to increase the convection heat transfer coefficient h_i.

In terms of fluid inlet temperature

$$\eta_c = F_R\{\eta_0 - U_c[(T_{f,i} - T_a)^+/I_c]\}. \tag{4.46}$$

For fluid boiling in the collector, $\bar{T}_f = T_{f,i}$ and $F_R = F'$.*

In thermal power applications it is convenient to express collector efficiency in terms of the fluid *outlet* temperature $T_{f,o}$ since this represents the input to the heat engine. The efficiency equation for this representation is

$$\eta_c = F_R e^{(F'U_c/\dot{m}c_p)}\{\eta_0 - U_c[(T_{f,o} - T_a)^+/I_c]\}, \tag{4.47}$$

where $\dot{m}c_p$ is the collector fluid flow rate–specific heat product evaluated at $\bar{T}_f$ (for engineering purposes).

If collector test data η_c are plotted versus $(T_{f,i} - T_a)/I_c$, for example, the intercept of this curve is $F_R\eta_0$ and the slope is F_RU_c if the curve is linear. If the curve is not linear but represented, for example, by

$$\eta_c = [\eta_0 - A(\bar{T}_c - T_a)^n/I_c] \qquad (\bar{T}_c > T_a), \tag{4.48}$$

then the local value of the loss coefficient U_c is $A(\bar{T}_c - T_a)^{n-1}$. Values of n are usually in the range $(1.0, 1.5)$. Note that U_c is *not* the magnitude of the local slope of the efficiency curve which is $(n + 1)A(\Delta T)^{n-1}$.

In summary, the principal collector parameters useful for engineering design are three—the optical efficiency η_0, the heat loss conductance U_c, and the heat removal factor F_R. Each may be calculated from preceding analyses or determined by a formal test. The next section shows the peak instantaneous efficiency of several single-curvature concentrators and tubular collectors. Up to a value of $(\bar{T}_f - T_a)/I_c \sim 0.3°C$ m^2/W the curves are nearly linear and a single value of U_c can be used.

---

* Boiling collectors are subject to the "pinch-point" problem analogues to that described on pp. 192–193.

E. Collector Second Law Efficiency

If a solar collector is regarded as a thermal converter of a quasi-steady source of energy, i.e., the sun, the second law efficiency can be expressed as

$$\eta_{2,\text{coll}} = \eta_c[1 - (T_o/T_{f,o})], \tag{4.49}$$

where $T_{f,o}$ is the collector outlet temperature. Using Eq. (4.47) for η_c we have

$$\eta_{2,\text{coll}} = F_R e^{(F'U_c/\dot{m}c_p)}\{\eta_0 - U_c[(T_{f,o} - T_a)/I_c]\}[1 - (T_o/T_{f,o})], \tag{4.50}$$

where T_a and T_o, the sink temperature, are approximately the same.

F. Results of Collector Testing

In solar collector system feasibility studies and schematic designs the theoretical analyses presented above can be used for satisfactory results. However, in plant design activities more precise performance knowledge is required. It is recommended that formal test data be the basis of final plant design since hardware tests include effects of collector assembly, shipping, and installation problems. These usually tend to reduce performance below the idealized design goals. This section presents typical test data on both commercial tracking and nontracking concentrators.

DOE Sandia Laboratories in Albuquerque undertook a concentrator test program in 1976 as part of its dispersed solar-thermal system program. Most of the test data presented herein were measured at Sandia. The Sandia facility is shown in Fig. 4.26. Collectors are tested up to 300°C using pressurized water or up to 425°C with a heat transfer oil. Both instantaneous and extended efficiency tests can be conducted.

Although no formal test procedure has been adopted for concentrators, the Sandia procedure is similar to that used for flat-plate collectors. An additional thermal loss test is made during daylight to determine thermal losses from the absorber as a function of fluid temperature. It is expected that this loss test, made with the collector defocused but facing the general direction of the sun, best replicates the expected losses during field use. Night loss tests expose the absorber to unrealistic radiative and ambient conditions and are not recommended for commercial testing. Of course, a correction in daytime loss values is required for the $(CR)^{-1}$ acceptance of diffuse radiation.

To date no comprehensive concentrator test has been run.

FIGURE 4.26   Sandia Laboratories Mid-Temperature Solar Systems Test Facility, Albuquerque, New Mexico. (Courtesy J. Leonard, Sandia Laboratories.)

A comprehensive test, containing the information needed for final design would include measurement of

(a)   A thermal efficiency versus operating temperature in several solar environments (i.e., varying ratios of direct, diffuse, and circumsolar diffuse flux),

(b)   optical efficiency as a function of incidence angle,

(c)   tracking accuracy and its effect on optical efficiency,

(d)   mirror reflectance—initial and after lengthy exposure,

(e)   flow rate effects,

(f)   collector time constant.

It is seen that concentrator testing is inherently more complex and costly than flat-plate collector testing owing to additional degrees of freedom in concentrator design.

Figure 4.27 shows efficiency data for five concentrators tested at Sandia Laboratories which have been described previously in this chapter. The two highest efficiency cases correspond to PTC designs and the lowest two to Fresnel mirror types.

In addition to providing the data required to predict performance in a systems context, collector tests can serve as diagnostic tests. For example, the series of tests summarized in Fig. 4.27 indicates that the Hexcel design has good efficiency and could benefit only slightly from redesign or further development. The Solar Kinetics design suffered from a defect in mirror supports, causing much of the reflected flux to miss the absorber. The McDonnell-Douglas Fresnel lens design was penalized by poor lens transmittance. Both Fresnel mirror collectors (Suntec SLATS® and General Atomics) had improperly designed mirrors with incorrect shapes in the former case and poor alignment for the second. In every case, the efficiency was below the collector design goal and was caused by optical problems—either mechanical design or materials prop-

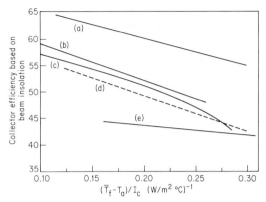

FIGURE 4.27   Measured efficiency for five line-focus concentrators: (a) Hexcel PTC, CR = 67; (b) Solar Kinetics PTC, CR = 41; (c) McDonnell-Douglas Fresnel lens, CR = 24; (d) Suntec SLATS® tracking Fresnel mirror, CR = 35; (e) General Atomics fixed Fresnel mirror, CR = 43. All collectors use a black chrome selective absorber. [From (D6).]

erties. These few tests graphically demonstrate the need for field test data. Of course, many other efficiency tests have been made; the five presented here were selected to illustrate some commercial and prototype designs and field problem types.

G.   Transient Characteristics of Concentrators

The discussions above have related to quasi-steady-state performance prediction and measurement, which is adequate information for engineering design of thermal systems. However, certain rather massive concentrators may have transient characteristics to be considered. The simplified discussion below will provide a first-order estimate of transient effects.

Using a linear model for the collector, the transient energy balance (no useful heat is delivered during warmup) is

$$(mc_p)_r(d\bar{T}_c/dt) = \eta_o I_c A_c - U_c A_c(\bar{T}_c - T_a)^+$$

where $(mc_p)_r$ is the receiver mass specific heat product. The approximation is made that the collector surface temperature $\bar{T}_c$ is an adequate measure of the temperature of the several receiver components, i.e.,

$$(mc_p)_r \bar{T}_c \sim \sum_i (mc_p)_i \bar{T}_i,$$

where $(mc_p)_i$ and $\bar{T}_i$ are the thermal capacitances and temperatures of the absorber components such as the fluid, its conduit, pipe supports, etc. Equation (4.50) can be considered the defining equation for $(mc_p)_r$. Dividing Eq. (4.49) by U_cA_c,

$$\left[\frac{(mc_p)_r}{(U_cA_c)}\right]\frac{d\bar{T}_c}{dt} = \frac{\eta_0 I_c}{U_c} - (\bar{T}_c - T_a).^+ \tag{4.51}$$

The coefficient $(mc_p)_r/(U_cA_c)$ is the time constant of the absorber responding to a step change in insolation. It is seen that the time constant is inversely proportional to concentration ratio if $(mc_p)_r$ is proportional to receiver area and is only a few seconds for CR $\sim$ 100.

III. ANCILLARY COMPONENTS OF SOLAR THERMAL SYSTEMS

The solar collector is the most expensive and most critical component of a solar thermal system. A complete system for using solar heat includes far more than the collector, however. Among the major components of most solar-thermal systems are thermal storage, pumps and piping systems, heat exchangers, and controls. The purpose of this section is to describe the application of these items to solar systems but not to analyze each in detail. Standard process engineering references can be consulted for more information (A5,K15). Controls are discussed later along with specific applications in Chapter 5.

A. Thermal Storage Subsystems

Energy storage is present in most solar systems to act as a buffer between the changeable solar system delivery and relatively more uniform process energy demands. The size of storage to be used is an economic question and must be determined for each application. Therefore, the amount of storage is not addressed in this section. Rather, the types of storage materials and their physical properties are described. Only storage media suitable for use above 100°C are considered. Nonthermal storage systems are considered elsewhere by application type.

Solar-thermal storage is not a prerequisite for economic solar energy use, however. The cost and added system complexity arising from storage may be completely eliminated if solar is used in the so-called fuel-saver mode. With this operational philosophy, solar heat is used when available to displace fuel only during sunlight periods. During periods of solar outage, the nonsolar energy source operates the process. Some investigators have suggested that the fuel-saver mode will be the first economic application of elevated temperature solar heat during the near term when oil is abundant but expensive. This symbiosis seems attractive since solar systems are capital intensive but fuel free, whereas fossil backup systems are not capital intensive but fuel intensive. In addition, in the near term when solar heat will be a small constituent of the national energy mix, the utility grid–solar interface can also be in a fuel-saver mode. Hydro storage in the grid could also serve as an indirect storage subsystem for solar thermal power plants, obviating the need for on-site thermal storage of more than a few hours' duration. Of course, in the far term, as solar energy progresses from its minority status, these fuel-saver operations must be expanded by use of storage so that stored solar heat may be used during solar outages.

The total heat stored Q_s by a material is

$$Q_s = m \left[\int_{T_1}^{T_f} c_{p_s} \, dT + \Delta H_f + \int_{T_f}^{T_2} c_{p_1} \, dT \right], \tag{4.52}$$

where m is the mass of storage, T_1 and T_2 are the lower and upper limits of temperature excursion, c_p is the specific heat defined as $(\partial h/\partial T)_p$, and subscripts s, f, l denote solid, fusion, and liquid phases. A similar expression can be written for liquid-to-vapor transitions but these are rarely used since the evolution of vapor requires a large containment volume for storage.

Three generic types of storage can be identified from Eq. (4.52). In the solar literature they are usually called solid, phase-change, and liquid storage. The remainder of this section describes the three generic storage modes. Any storage material must meet several basic criteria, including high storage energy density (kJ/m^3); good durability under many thermal cycles; high longevity of stored energy; second law match of storage to task; appropriate storage configuration for efficient heat input and extraction; low cost; low toxicity, flammability, vapor pressure, and chemical reactivity; low volume expansivity.

Solid-Phase Storage Sensible heat storage in solid media can be an effective method of storage for high-temperature processes. Most solids have energy density values (kJ/m^3) that are lower than pressurized water but equivalent to many organic liquids. Hence, from an energy density viewpoint, solid media are of interest. Since many solids economically viable for thermal storage have relatively small thermal conductivity values, a large surface-to-volume ratio is needed for efficient heat transfer. In most cases a packed bed of uniformly sized particles is used for this reason.

Particle beds have a very desirable property, as a by-product, in that significant temperature stratification can be achieved. Since the particles touch each other at only a few points, the rate of heat transfer by conduction is quite small. In addition, radiation exchanges are moderate unless extremely high temperatures are encountered. Stratified storage is often desirable because the temperature of fluid leaving the bed is nearly that of the stratum closest to the exit. This zone is usually the hottest zone since it was charged by a counterflow of hot fluid. Hence, energy is extracted at high temperature and high thermodynamic availability. On the contrary, liquid storage is difficult to stratify significantly and heat is removed at roughly the mean storage temperature, not the maximum temperature at which heat was stored during the collection mode. Loss in thermodynamic availability is described shortly.

Table 4.2 lists the densities and specific heats of several useful solid-phase media. Cast iron is seen to have the highest energy den-

TABLE 4.2

Solid in Phase Storage Media Density and Specific Heat[a]

Material	Density[b]		Specific heat[b]	
	kg/m³	lb/ft³	kJ/°C × kg	Btu/lb × °F
Aluminum	2700	168	0.88	0.21
Aluminum sulfate	2710	169	0.75	0.18
Aluminum oxide	3900	240	0.84	0.20
Brick	1698	106	0.84	0.20
Calcium chloride	2510	157	0.67	0.16
Earth–dry	1698	106	0.84	0.20
Magnesium oxide	3570	223	0.96	0.23
Potassium chloride	1980	124	0.67	0.16
Potassium sulfate	2660	166	0.92	0.22
Sodium carbonate	2510	157	1.09	0.26
Sodium chloride	2170	135	0.92	0.22
Sodium sulfate	2700	168	0.92	0.22
Cast iron	7754	484	0.46	0.11
River rocks	2245–2566	140–160	0.71–0.92	0.17–0.22

[a] From (H4).
[b] At 100°C.

sity and has been used for thermal storage in Europe. The materials listed remain solid at relatively high temperature and do not require a pressure vessel for containment as do most liquids.

Heat transfer in fixed beds using a gaseous working fluid has been examined at moderate temperature by chemical engineers. If the particles can be treated as approximately spherical, the Nusselt number is given by (H5)

$$Nu = 0.255 \ Re^{2/3} \ Pr^{1/3}/\epsilon_v, \qquad (4.53)$$

where ϵ_v is the void fraction (usually 0.35–0.5), the equivalent spherical diameter D_s is the length scale for Re and Nu, and the Reynolds number is based on G, the superficial mass velocity based on the total bed cross section. The total surface area of particles A_p for a bed A_b of volume V_b can be calculated from

$$A_p = 6(1 - \epsilon_v)V_b/D_s. \qquad (4.54)$$

Pressure drop Δ_p through a bed of uniform, nearly spherical particles of length L and cross section A_b ($= V_b/L$) is given by (H5)

$$\frac{\Delta p}{\rho V_s^2} = \frac{L}{D_s} \frac{(1 - \epsilon_v)^2}{Re \ \epsilon_v^3} \left[1.24 \ \frac{Re}{1 - \epsilon_v} + 368 \right], \qquad (4.55)$$

where $Re = Gd_s/\mu$ and $V_s = \dot{m}/\rho A_b$; $Re = [100,13000]$. Chapter 5 describes combined solid–liquid storage systems which achieve good stratification but use a liquid working fluid.

Liquid-Phase Storage   The use of liquid-storage media at high temperature usually requires a pressure vessel to avoid boiling. Therefore, the cost for a smaller tank may exceed that for solid media even though these media may have a smaller energy density and require physically larger tanks. Table 4.3 shows the properties of several practical liquid storage substances. In many cases these materials were designed for use as heat transfer fluids, not storage media. Most, other than water, are quite expensive and require an inert gas blanket (N_2) to reduce oxidation. Most petrochemical companies produce high-temperature heat transfer fluids like those shown which are suitable for thermal storage. Organic fluids cannot be used above 350°C however, because of the rapid thermal decomposition and high replacement cost.

Liquid storage materials require leakproof storage vessels usually made from aluminum or steel. If water is used, a glass or rubber liner is required to suppress corrosion. The cost of metal tanks for a given pressure is related to size by a power law expression (G2):

TABLE 4.3

Properties of Liquid Storage Media[a]

Material	Density (kg/m³)	Specific heat (kJ/°C kg)	Boiling point[f] (°C)
Water	~1000	~4.2	100
Therminol 66[™][b]	820 (250°C)	2.4 (250°C)	~330
Dowtherm A[™][c]	859 (250°C)	2.2 (250°C)	257
Dowtherm J[™][c]	706 (200°C)	2.5 (200°C)	180
Hi Tec[™][d]	1890 (250°C)	1.5 (250°C)	d
Caloria HT-43[™][e]	694 (250°C)	3.0 (250°C)	—
Diethylene glycol	1020 (150°C)	2.8 (150°C)	240
Tetraethylene glycol	1110 (66°C)	2.5 (100°C)	280

[a] Data collected from various manufacturers brochures.
[b] Trademark of Monsanto Chemical Co.
[c] Trademark of Dow Chemical Co; Dowtherm is a mixture of diphenyl and diphenyl oxide.
[d] Trademark of E.I. duPont; solidifies below approximately 190°C; does not boil or decompose below 500°C; consists of 40% $NaNo_2$, 7% $NaNo_3$, 53% KNO_3.
[e] Trademark of Exxon Corp.
[f] At 1 atm.

$$\text{cost} = \text{base cost} \times F_m \times (\text{volume/base volume})^n, \qquad (4.56)$$

where F_m is a material multiplier based on carbon steel as 1.0 and shown in Table 4.4. Power law exponents for Eq. (4.56) are given in Table 4.5 (C3). Since the thickness of a tank is proportional to the confined pressure, the cost of a given volume increases with *pressure* approximately linearly.

Phase-change Storage Energies associated with solid–liquid phase changes are in the range 300–2200 MJ/m³. To store a comparable amount of energy in either a liquid or solid medium by sensible heat would require a temperature excursion of 50–500°C for the best materials.

TABLE 4.4

Material Cost Factor F_m for Metal Storage Tanks

Material	Multiplier F_m
Carbon steel	1.00
Aluminum	1.40
Rubber-lined	1.48
Stainless steel	3.20
Glass-lined	4.25

TABLE 4.5

Storage Tank Cost Exponents[a]

Tank type	Scale-up exponent
Atmospheric–steel	0.46
Steel	
100–250 kP$_a$ (15–35 psi)	0.47
250–850 kP$_a$ (35–120 psi)	0.49
Rubber-lined	0.57
Aluminum	0.61
Atmospheric–stainless steel	0.50–0.54
Glass-lined[b]	0.43

[a] From (C3).
[b] Limited to 40 m^3.

Hence, phase-change media are very compact, high energy density materials. They have considerable promise for solarthermal storage at elevated temperature if several criteria are met: (a) high latent heat of fusion, (b) appropriate melting point, (c) high thermal diffusivity and conductivity, (d) low volume expansivity, (e) low cost, (f) negligible chemical reactivity, (g) low vapor pressure and toxicity, (h) minimal maintenance.

Figure 4.28 shows the heat of fusion per unit volume for many materials suitable for use at elevated temperatures (S6). The substances labeled by numbers are metallic fluoride eutectics whose proper-

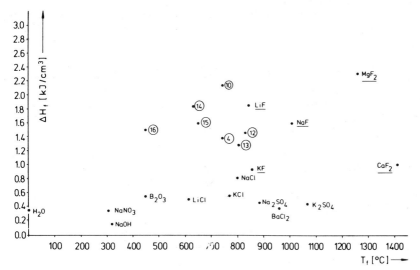

FIGURE 4.28   Volumetric heats of fusion and melting points for various materials. Eutectics shown by numbered circles are listed in Table 4.6. [From (S6).]

ties are given in Table 4.6. It is seen that the metallic fluorides have very high energy densities at a range of temperatures ideally suited for high temperature thermal storage. Figure 4.29 shows the heat of fusion per unit weight for the same materials. On either a volume or weight basis the metallic fluorides appear quite promising. The use of eutectic mixtures is necessary since melting points of the pure fluorides are quite high.

The metallic fluorides also are excellent storage media from a thermodynamic viewpoint. The availability of heat stored at 400–800°C is quite high and is well matched to high performance solar collector requirements and heat-engine conversion to shaft work. Thermal energy recovery from phase-change storage simplifies control system requirements. Schröder (S6) also notes that the materials shown have liquid- and solid-phase densities which differ by less than 20% and no supercooling is noted. In addition, the vapor pressure is below 1 mm Hg up to 900°C and little corrosion is observed. Metallic fluorides have been cycled for more than 5 yr in stainless steel vessels with no observed deterioration in chemical or thermal properties of the materials or containment vessels.

Efficacious removal of heat from a phase-change storage container has been a difficulty in some early research projects. However, if proper account is taken of the factors governing transient heat transfer (K5), namely, surface-to-volume ratio and thermal diffusivity, phase-change storage charging and discharging can be done effectively. Heat pipes using liquid sodium or potassium (Chapter 3) seem particularly promising for this purpose.

In the design of the heat addition/removal subsystem for

TABLE 4.6

Composition, Density, and Melting Point for Various Metal Fluoride Eutectics[a]

No.	Composition (mole %)	MP (°C)	$D_{25°C}$ (g/cm^3)	$D_{MP(solid)}$ (g/cm^3)	$D_{MP(liquid)}$ (g/cm^3)
1	58 KF/35NaF/7MgF$_2$	685			2.090
2	52 LiF/35NaF/13CaF$_2$	615	2.82	2.63[b]	2.225
4	65 NaF/23CaF$_2$/12MgF$_2$	745	2.97	2.77	2.370
10	67 LiF/33MgF$_2$	746	2.88	2.63[b]	2.305
12	75 NaF/25MgF$_2$	832	2.94	2.69	2.190
13	62.5 NaF/22.5MgF$_2$/15KF	809	2.85	2.65	2.110
14	46 LiF/44NaF/MgF$_2$	632	2.81	2.61	2.105
15	60 LiF/40NaF	652	2.72	2.50	1.930
16	12 NaF/40KF/44LiF/4MgF$_2$	449			2.160[b]

[a] From (S6).
[b] Calculated values.

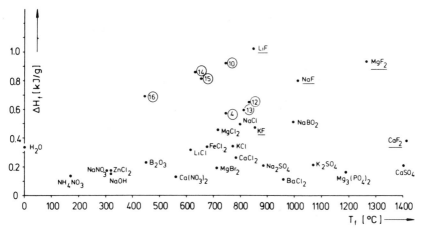

FIGURE 4.29   Gravimetric heats of fusion and melting points for various materials. Eutectics shown by numbered circles are shown in Table 4.6. [From (S6).]

phase-change storage the mode of heat transfer within the melt must be known. It had formerly been assumed that pure radial conduction is the dominant mode of heat transfer. Sparrow *et al.* (S12) have recently shown that the radial conduction assumption is erroneous and that free convection is the dominant mode of heat transfer in the melting process. Therefore, convection coefficient correlations for free convection should be used to predict heat transfer rates. One consequence of this finding is that the heat source elements used to melt storage contained in a horizontal tank should not be placed at the vertical center of the tank since free convection would occur only above the heat element. The lower half of the tank would then not be usable for phase-change storage.

Steam Accumulators   Although liquid–vapor phase changes are generally avoided as a form of thermal storage, the steam–liquid–water phase change can be useful under the proper conditions for short-term storage. Hot, pressurized water can be stored and then later flashed into steam to supply a steam-based process. The storage device in this case is called a steam accumulator and is shown in Fig. 4.30.

A steam accumulator is a pressure vessel about 90% filled with liquid water when fully charged. In the charging mode, steam is bubbled into the vessel through a special nozzle in the center within a vertical circulation pipe. Steam delivered to the pipe rises and draws water with it, thereby maintaining quasi-equilibrium conditions. As steam is condensed

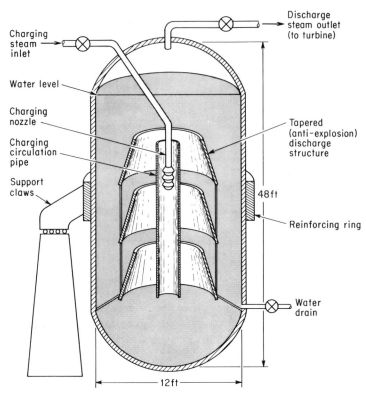

FIGURE 4.30   Cutaway of a vertical steam accumulator.

the water volume increases, the static pressure increases, and the temperature rises.

In the discharge mode, the dome of the tank is vented, boiling results, and steam is evolved. The discharge structure shown maximizes mixing to obviate local explosive boiling. Fully discharged, the accumulator is 60–70% full of water. The maximum discharge rate condition of the accumulator d_{SA} should not be exceeded (G3):

$$d_{SA} = (2.35 + 0.014p)D_w^{0.715} \qquad (4.57)$$

where d_{SA} = (metric tons steam/h/m³ steam volume), p is the pressure (atm), and D_w is the water density (°Baume hydrometric scale, 1.0 = pure water).

Second Law Analysis   The use of storage usually involves the loss of some available energy since the charging fluid is at a higher temperature T_f than the discharged fluid used at a later time, i.e., $T < T_f$. Second law storage efficiency $\eta_{2,s}$ can be expressed as the ratio of availability delivered by storage to that consumed (for constant specific heat and T_f):

$$\eta_{2,s} = \int_{T_i}^{T_f} [1 - (T_0/T)] \, dT \Big/ (T_f - T_i)[1 - (T_0/T_f)], \qquad (4.58)$$

where T_0 is the environmental temperature, T_f the storage fluid charging temperature and the final, maximum storage temperature after a charging cycle, and T_i the initial storage temperature at the start of a cycle. It is easy to show for well-mixed storage that

$$\eta_{2,s} = [T_f - T_i - T_0 \ln(T_f/T_i)]/(T_f - T_i)(1 - T_0/T_f). \qquad (4.59)$$

If storage loses heat, the $\eta_{2,s}$ value above is multiplied by an appropriately defined first law efficiency. Additional information on other storage concepts is contained in Chapter 5 with application types.

B.   Energy Transport Subsystems

Solar heat is produced in a collector and stored in a storage vessel. To transport heat to the collector, to storage, and then to an end use, an energy transport subsystem is used. It consists of pipes, pumps, expansion tanks, valves of various types, and heat exchangers. Heat exchangers are treated in the next section and pipes and pumps in this section. Valves and expansion tanks are standard items and are not described in detail.

Piping Systems In most solar-thermal processes considered in this book, standard circular pipes of steel, copper, aluminum, or special alloys are used to transport heat in the form of internal energy in the pumped fluid. Two parasitic losses occur in pipes—pressure drop and heat loss. Both losses are treated briefly herein. Corrosion problems are described in (K2).

Pressure Drop Viscous and turbulent inertial forces consume mechanical energy and produce heat energy in any flow process. To overcome the dissipative effects of flow resistances pumps are required. The sizing of pumps is a routine engineering exercise and is done by calculating two quantities, the friction factor at a given flow rate and the pipe length. The friction factor f can be read from Fig. 4.31 as a function of the Reynolds number (See Chapter 3) and roughness ratio shown in Fig. 4.32. The Darcy friction factor is defined as

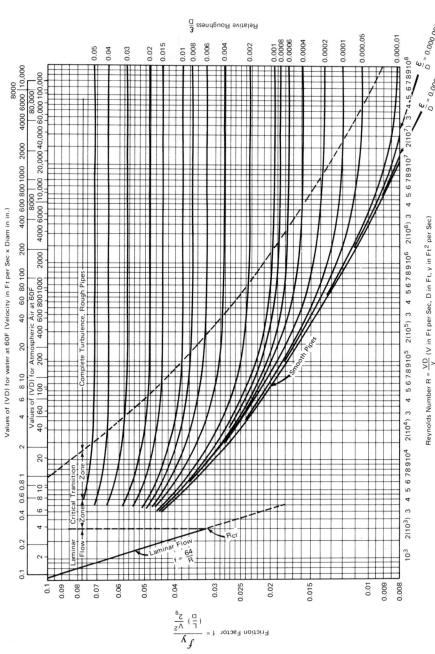

FIGURE 4.31   Friction factor for pipe flow as a function of Reynolds number and roughness ratio ϵ/D. [From (P6).]

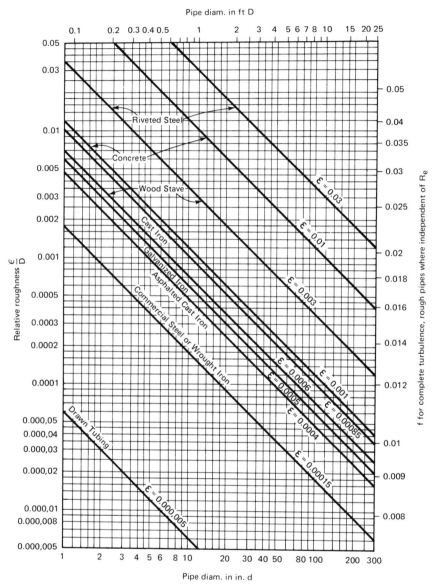

FIGURE 4.32 Roughness ratio ϵ/D for various types of commercial pipe.
[From (P6).]

$$f = \Delta p/(L/D)(\tfrac{1}{2}\rho \bar{V}^2), \tag{4.60}$$

where Δp is the pressure drop, D is the diameter, L is the pipe length including the equivalent length of smooth pipe for all fittings, expansions, and contractions, (K2) and $\bar{V}$ is the space averaged velocity.

Churchill (C4) finds that the curves in Fig. 4.31 can all be represented by one equation for the laminar, transition, and turbulent regimes:

$$f = 8\{(8/\text{Re})^{12} + (A + B)^{-3/2}\}^{1/12}, \tag{4.61}$$

where

$$A = \{-2.457\ \ln[(7/\text{Re})^{0.9} + 0.27\epsilon/D]\}^{16}, \qquad B = (37{,}530/\text{Re})^{16}$$

Pump motor horsepower P_p can be calculated from Eq. (4.60) to be

$$P_p = f(L/D)(\tfrac{1}{2}\rho A \bar{V}^3)/\eta_p\eta_m, \tag{4.62}$$

where A is the flow area, η_p is the pump efficiency, and η_m is the motor efficiency; the volumetric flow $\dot{Q} = A\bar{V}$.

For quick pipe sizing Kent (K10) gives some guidelines which can eliminate the need to use the detailed equations above at the design development level of plant engineering. The typical diameter for liquid pipes is

$$D = 2.607(\dot{m}/1000\rho)^{0.434} \tag{4.63}$$

where $D = [\text{in.}]$, $\dot{m} = [\text{lb/h}]$, and $\rho = [\text{lb/ft}^3]$. If the liquid pipe is a vent or drain, D from Eq. (4.63) is increased by 35%. For gases

$$D = 1.065(\dot{m}/1000)^{0.408}/\rho^{0.343} \tag{4.64}$$

with the same 35% rule applying.

If allowable pressure drop from previous experience is used as the design criterion,

$$D = 1.706 \left\{ \frac{[\mu^{0.16}(\dot{m}/1000)^{1.84}]}{[\rho(\Delta P/100)]} \right\}^{0.207}, \tag{4.65}$$

where $\mu = [\text{cp}]$, $\Delta P/100 = [\text{psi}/100\ \text{ft}]$. Equation (4.65) applies for both liquids and gases. If a control valve is part of the system, the drop should not be less than 30% of the total system drop to ensure good control. Control valve piping should also be capable of handling flows one-fifth larger than the design flow rate.

Although the fluid circuits in most parts of solar-thermal systems are quite simple, the collector field is frequently connected in a complex series–parallel arrangement as dictated by tradeoffs between piping costs, pump size, and fluid temperature rise. The flow in series–parallel arrays can be calculated using iterative techniques such as the Hardy–Cross method. However, for preliminary design a closed form solution has been developed (K2) for flow through arrays of solar collectors connected in parallel. For $j + 1$ identical collectors with the same flow rate connected in parallel with constant diameter D headers or manifolds, the array pressured drop (inlet header inlet port to outlet header outlet port) is given by

$$\Delta P = (K_f \dot{m}^2/6)[j(2j + 1)(j + 1)] + \Delta p_{coll}, \qquad (4.66)$$

where Δp_{coll} is the collector pressure drop, $\dot{m}$ the mass flow rate per collector, and

$$K_f = 8\bar{f}L/(\pi^2 \rho D^5). \qquad (4.67)$$

The average friction factor $\bar{f}$ in the header is evaluated at one-half the total array flow $(j + 1)\dot{m}/2$ and the collectors are all connected into the manifold at a uniform distance L apart. It is to be noted that identical flow $\dot{m}$ through each header does not occur automatically but can be accomplished to within a few percent by requiring 90% or more of the total array pressure drop to occur across the collector, i.e.,

$$\Delta p_{coll} \geq 0.9 \, \Delta P. \qquad (4.68)$$

In the design of piping systems the absolute pressure must not be permitted to drop below the boiling point to avoid vapor lock, boiling, or pump cavitation. The critical design point is where the fluid is hottest and the pressure lowest—often at a circulating pump inlet port. Some common fluids used for high temperature solar systems are shown in Table 4.3. The fluid manufacturer should be consulted for precise viscosity values to be used for system design. Typical values are in the range 0.2 to 5.0 cp. Most are quite temperature sensitive.

Heat Loss Parasitic heat losses from storage tanks, pipes, and pumps can be calculated from equations given in Chapter 3. For cylindrical insulation of length L, inner diameter D, and thickness t for pipes and tanks

$$q_{cyl} = \left[\frac{2\pi kL}{\ln(D + 2t)/D} \right] (T_f - T_a), \qquad (4.69)$$

where T_f is the fluid temperature and T_a the ambient temperature. For spherical storage tanks the heat loss is

$$q_{sph} = [\pi kD(D + 2t)/t](T_f - T_a). \qquad (4.70)$$

In the above it is assumed that the principal resistance to heat loss occurs in the insulation.

Although heat losses can be made arbitrarily small, the law of diminishing returns applies and a point is reached where added insulation costs more than the value of the extra heat retained. An example worked in Chapter 7 derives an equation for the optimum amount of insulation. It can be shown (K2) that the optimum distribution of a given volume of insulation over all the components of a solar-thermal system occurs when the surface heat flux is the same everywhere; note that this does not imply equal insulation thickness t.

Table 4.7 lists the properties of industrial grade insulations useful in elevated temperature solar systems. The principal insulations include calcium silicate, mineral fiber, expanded silica (perlite), and glass. Calcium silicate is a mixture of lime and silica reinforced with fibers and molded into shape. It has good flex and compressive strength. Mineral fiber consists of rock and slag fibers bonded together and useful up to 1200°F (650°C). The compressive strength is less than calcium silicate but is available in both rigid and flexible, shaped segments.

Perlite is expanded volcanic rock consisting of small air cells enclosed by a mineral structure. Additional binders are added to decrease moisture migration and to reduce shrinkage. Cellular glass is available in flexible bats or rigid boards and shaped sections. It has very low moisture absorption.

For very high temperatures, the selection of materials is quite limited. In addition to mineral fiber and calcium silicate noted above, ceramic fibers (to 1400°C), castable ceramic insulation (to 1600°C), Al_2O_3 or ZrO_2 fibers (to 1600°C), and carbon fibers (to 2000°C) are available. It is to be noted that these temperatures correspond to the refractory range of metals where service conditions are very severe.

An excellent summary of industrial insulation practice and design methods and economic analysis is contained in (H6).

C. Heat Exchangers

Heat exchangers are used in many solar-thermal systems when it is necessary to transfer heat from one fluid stream to another if the two fluid streams cannot be mixed. If they can be mixed and are immiscible, a direct-contact heat exchanger can be used. The exchange of heat can be made to occur with nearly unity first law efficiency but some avail-

TABLE 4.7

Properties of Pipe Insulation for Elevated Temperature[a,b]

Insulation type	Temperature range (°F)	Conductivity k [Btu/(h)(ft²)(°F/in.)]	Density (lb/ft³)	Applications
Urethane foam	−300 to 300	0.11 to 0.14	1.6 to 3.0	Hot and cold piping
Cellular glass blocks	−350 to 500	0.20 to 0.75	7.0 to 9.5	Tanks and piping
Fiberglass blanket for wrapping	−120 to 550	0.15 to 0.54	0.60 to 3.0	Piping and pipe fittings
Fiberglass preformed shapes	−60 to 450	0.22 to 0.38	0.60 to 3.0	Hot and cold piping
Fiberglass mats	150 to 700	0.21 to 0.38	0.60 to 3.0	Piping and pipe fittings
Elastomeric preformed shapes and tape	−40 to 220	0.25 to 0.27	4.5 to 6.0	Piping and pipe fittings
Fiberglass with vapor barrier jacket	20 to 150	0.20 to 0.31	0.65 to 2.0	Refrigerant lines, dual-temperature lines, chilled-water lines, fuel-oil piping
Fiberglass without vapor barrier jacket	to 500	0.20 to 0.31	1.5 to 3.0	Hot piping
Cellular glass blocks and boards	70 to 900	0.20 to 0.75	7.0 to 9.5	Hot piping
Urethane foam blocks and boards	200 to 300	0.11 to 0.14	1.5 to 4.0	Hot piping
Mineral–fiber preformed shapes	to 1,200	0.24 to 0.63	8.0 to 10.0	Hot piping
Mineral–fiber blankets	to 1,400	0.26 to 5.6	8.0	Hot piping
Fiberglass field applied jacket for exposed lines	500 to 800	0.21 to 0.55	2.4 to 6.0	Hot piping
Minearl–wool blocks	850 to 1,800	0.36 to 0.90	11.0 to 18.0	Hot piping
Calcium silicate blocks	1,200 to 1,800	0.33 to 0.72	10.0 to 14.0	Hot piping

[a] From (H6).
[b] 1 Btu/(h)(ft²)(°F/in.) = $\frac{1}{12}$ Btu/h ft °F = 0.144 W/m °C.

able energy is lost because of the inevitable thermal driving force necessary to cause heat flow. Typical applications of heat exchangers include liquid-to-liquid heaters, liquid-based air heaters, boilers for heat engines, and condensers or heat rejectors for power plants. A survey of heat exchangers performance calculation procedures is given below using the effectiveness–NTU (ϵ–NTU) method. For a more detailed treatment refer to (K14, P3, R7).

First-order design variables for a heat exchanger are the fluid stream capacitance rates ($\dot{m}c_p$), the overall conductance $\bar{U}A$, and the effectiveness. Heat exchanger effectiveness is defined as the amount of heat transferred q divided by the transfer which could occur if the exchanger were of the counterflow type and infinitely large, i.e., the thermodynamic limit. In equation form

$$\epsilon \equiv q/(\dot{m}c_p)_{\min}(T_{h,1} - T_{c,i}) \qquad (4.71)$$

where $T_{h,i}$ and $T_{c,i}$ are the hot and cold stream inlet temperatures and

$$(\dot{m}c_p)_{\min} = \min[(\dot{m}c_p)_h, (\dot{m}c_p)_c]. \qquad (4.72)$$

The value of $\bar{U}A$ can be calculated from forced convection heat transfer correlations given in Chapter 3 or from manufacturers' data. However, with use, $\bar{U}A$ will generally decrease because of fouling of the heat exchange surface. Asymptotic values of fouling resistances lie in the range 10^{-3}–10^{-4} (W/m² °C)⁻¹ (K2) and should be determined for the specific application.

Kays and London (K8) have calculated effectiveness values for many types of heat exchangers. They found that ϵ can be expressed for a given exchanger type as a function of two variables, the number of transfer units NTU,

$$\text{NTU} \equiv \bar{U}A/(\dot{m}c_p)_{\min}, \qquad (4.73)$$

and the capacitance ratio C,

$$C \equiv (\dot{m}c_p)_{\min}/(\dot{m}c_p)_{\max} = C_{\min}/C_{\max}. \qquad (4.74)$$

For boiling or condensation $C = 0$.

Figures 4.33a–4.33d show effectiveness values for the common heat exchanger types and Table 4.8 contains equations for calculating $\epsilon(\text{NTU}, C)$. For parallel flow and counterflow devices in which boiling or condensation occurs in one stream $\epsilon = 1 - e^{-\text{NTU}}$.

One of the most common uses of a heat exchanger is to isolate a fluid used in a solar collector field from the process fluid which conveys the heat energy to its end use as shown in Fig. 4.34. If the re-

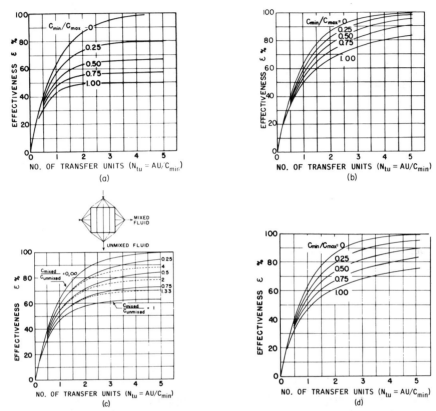

FIGURE 4.33   Heat exchanger effectiveness as a function of NTU and C; (a) parallel flow, (b) counterflow, (c) cross flow, one fluid mixed, and (d) cross flow, no mixing. [From "Compact Heat Exchangers" by W. M. Kays and A. L. London. Copyright © 1964. Used with permission of McGraw-Hill Book Company.]

quired process inlet temperature is $T_{p,i}$, the collector outlet fluid must be at a temperature $T_{c,o}$ above $T_{p,i}$ because of the presence of the heat exchanger. Since collector efficiency decreases with increasing operating temperature, less energy is collected than if no heat exchanger were present. deWinter (D5) has calculated a heat exchanger penalty factor F_{hx} which can be applied to the linear collector model [Eq. (4.46)] developed earlier. Equation (4.46) can be rewritten in terms of the heat exchanger inlet temperature (process return temperature) $T_{p,o}$ as

$$\eta_c = F_{hx}F_R[\eta_0 - U_c(T_{p,o} - T_a)^+/I_c].$$ (4.75)

The heat exchanger factor is given by

TABLE 4.8

Heat-Exchanger Effectiveness Expressions[a]

Flow geometry	Relation
Double pipe	
Parallel flow	$\epsilon = \dfrac{1 - \exp[-N(1 + C)]}{1 + C}$
Counterflow	$\epsilon = \dfrac{1 - \exp[-N(1 - C)]}{1 - C \exp[-N(1 - C)]}$
Cross flow	
Both fluids unmixed	$\epsilon = 1 - \exp\{(C/n)[\exp(-NCn) - 1]\}$ where $n = N^{-0.22}$
Both fluids mixed	$\epsilon = \left[\dfrac{1}{1 - \exp(-N)} + \dfrac{C}{1 - \exp(-NC)} - \dfrac{1}{N}\right]^{-1}$
C_{max} mixed, C_{min} unmixed	$\epsilon = (1/C)\{1 - \exp[C(1 - e^{-N})]\}$
C_{max} unmixed, C_{min} mixed	$\epsilon = 1 - \exp\{(1/C)[1 - \exp(-NC)]\}$
Shell and tube	
One shell pass, 2, 4, 6 tube passes	$\epsilon = 2\left\{1 + C + (1 + C^2)^{1/2}\dfrac{1 + \exp[-N(1 + C^2)^{1/2}]}{1 - \exp[-N(1 + C^2)^{1/2}]}\right\}^{-1}$

[a] $N = \text{NTU} \equiv UA/C_{mm}$, $C = C_{min}/C_{max}$. [From (K8).]

$$F_{hx} = \{1 + [F_R U_c A_c/(\dot{m}c_p)_c][(\dot{m}c_p)_c/\epsilon(\dot{m}c_p)_{min} - 1]\}^{-1} \qquad (4.76)$$

where the subscript c denotes collector properties and ϵ is the effectiveness. The collector loss coefficient U_c is evaluated at the mean collector operating temperature $\bar{T}_c$. Figure 4.35 shows F_{hx} for a range of operating conditions. Unless the collector flow rate is very low, $F_{hx} > 0.95$ for most high-performance collectors. For small $[F_R U_c A_c/(\dot{m}c_p)_c]$, F_{hx} is approximated closely by

$$F_{hx} = 1 - F_R U_c A_c\{[(\dot{m}c_p)_c - \epsilon(\dot{m}c_p)_{min}]/\epsilon(\dot{m}c_p)_{min}(\dot{m}c_p)_c\}. \qquad (4.77)$$

The second law efficiency of a heat exchanger $\eta_{2,hx}$ can be defined as the ratio of availability change in the outlet, cold stream to that in the inlet, hot stream. If no heat losses occur from the exchanger, $\eta_{2,hx}$ is given by

$$\eta_{2,hx} = \left(\frac{(\dot{m}c_p)_c}{(\dot{m}c_p)_h}\right)\frac{T_{co} - T_{ci} - T_o \ln(T_{co}/T_{cic})}{T_{hi} - T_{ho} - T_o \ln(T_{hi}/T_{ho})}, \qquad (4.78)$$

where the numerator is the outlet stream availability and the denominator, the inlet availability relative to an environment at T_o. Irreversibilities

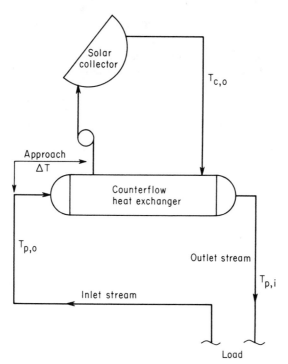

FIGURE 4.34   Use of a heat exchanger to isolate a collector field from a thermal process.

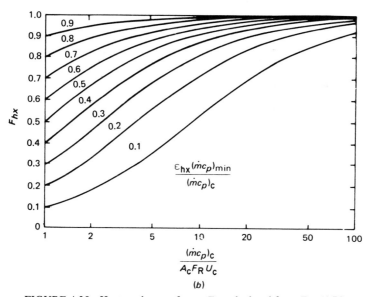

FIGURE 4.35   Heat exchanger factor F_{hx} calculated from Eq. (4.76).

due to viscous losses in the piping are not included. If the heat exchanger is a boiler, i.e., the cold stream is boiling, the efficiency is

$$\eta_{2,hx} = \frac{\dot{m}_c \int_{h_{ci}}^{h_{co}} [1 - (T_0/T_c)] \, dh_c}{(\dot{m}c_p)_h [T_{hi} - T_{ho} - T_0 \ln(T_{hi}/T_{ho})]} \tag{4.79}$$

or, recalling that $\dot{m}_c \, dh_c = -(\dot{m}c_p)_h \, dT_h$ where h is the enthalpy and T_c is constant,

$$\eta_{2,hx} = \frac{\int_{T_{hi}}^{T_{ho}} -[1 - (T_0/T_c)] \, dT_h}{[T_{hi} - T_{ho} - T_0 \ln(T_{hi}/T_{ho})]}. \tag{4.80}$$

MEDIUM-TEMPERATURE SOLAR PROCESSES

5

There is no gamble in solar energy use.
It is sure to work.

F. Daniels

This chapter describes the four principal medium-temperature solar processes which have been engineered and/or built at the prototype level. The four system types are (a) distributed power generation, (b) industrial process heat, (c) shaft work production, (d) solar total energy systems. Each concept is described in this chapter, performance characteristics given, and an example system described. Data from existing facilities are given where available. The emphasis is on systems (b) and (c) for which there is field experience.

In the final sections of this chapter a method of performance estimation is described in detail. The method can be used to predict average annual performance of all concentrator-based systems.

I. DISTRIBUTED SOLAR POWER PRODUCTION SYSTEMS

Solar-thermal power production can be accomplished by two generic types of systems. The first, called the central receiver con-

161

cept, uses large arrays of sun-tracking mirrors which reflect solar flux onto a central boiler atop a tall tower. In the central receiver system, therefore, energy transport from the collector field to the power plant occurs by optical means. Concentration ratios of the order of 10^3 are used along with turbines operating above 550°C. This concept is therefore a high-temperature solar-thermal process for the purposes of this book and will be treated in Chapter 6.

The second concept, using relatively small collectors (a few tens of m^2) spread over a large area, is called the distributed network concept. Energy transport to the power plant turbine occurs through a network of pipes interconnecting the dispersed collector units. These systems use line-focus collectors and operate at temperatures within the scope of this chapter. One distributed system idea uses dispersed compound-curvature (dish) collectors with small turbine generators integral with each collector. Energy transport occurs by buried electrical cables. However, very high collector temperatures are required; therefore, this generation method will be treated in Chapter 6.

There are many types of distributed systems for power production. Factors differentiating among the many generic types include

(1)   point of solar heat input to power system—preheater, boiler, superheater or combination;

(2)   storage size, if any;

(3)   transport fluids—gas, liquid, chemical reactants;

(4)   type of heat engine—Rankine, Brayton, Stirling;

(5)   baseload, intermediate load, peak load;

(6)   size—10-, 100-, 1000-MW_e scales;

(7)   collector type.

It is the purpose of this section to describe some of the more promising system concepts and to outline the major design determinants for such systems. A comprehensive treatment of all proposed systems would be repetitive.

All projected distributed solar-thermal power systems consist of a collector field (0.5–1.0 $mile^2$ per 100 MW_e), a fluid distribution network, a heat engine generator plant, and a control system. Most systems also include some form of short-term storage (see Fig. 5.1). Other than power plant control systems and steam generators [see (W4) for an overview] whose design is not the subject of this book, these major components will be described in this section. Since no solar power plants now exist, detailed costs, design criteria, and test data cannot be presented.

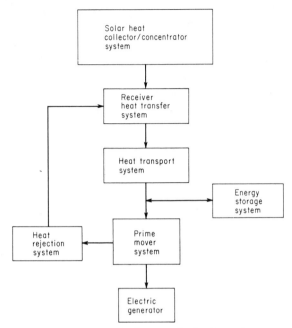

FIGURE 5.1   Conceptual block diagram of solar-thermal power systems.

A.   The Collector Subsystem

Chapter 4 described several solar collectors usable for power production. Among these are parabolic and circular troughs, Fresnel reflectors, and Fresnel lenses. Two generic receiver types can be used—evacuated tubular or nonevacuated cavity—and several tracking modes are possible including north–south, east–west, and polar.

These various line-focus systems operate at temperatures from 350 to 450°C at good efficiency. Parabolic troughs with tubular receivers are near the lower limit and Fresnel mirrors with small rim angles and cavity receivers near the upper. Associated power plant efficiencies of 12–18% are achievable (F2).

Figure 5.2 shows a typical line-focus thermal power system with storage.

Preliminary data indicate that collector efficiencies of 40–45% are achievable with parabolic troughs at 400°C whereas 50–55% can be achieved with Fresnel reflector/cavity types at 450–500°C. These temperatures correspond to the peak system conversion efficiency (not

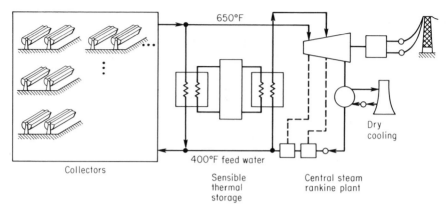

FIGURE 5.2   Distributed solar-thermal system using steam transport and thermal storage.

peak collector efficiency) including the steam Rankine cycle. Although the Fresnel mirror collector shows higher efficiency, the mechanical mirror segment drive device and cavity receiver add about one-third more to the cost.

Winston (W5) has suggested a concept for power production using *fixed* collectors of the CPC type. This concept had formerly been dismissed in solar power feasibility studies and projections without careful study. A state of the art CPC with CR = 1.45, optical effciency of 0.76, and a selective, evacuated receiver was shown to be capable of collecting about 53% of the incident flux in a sunny climate. Collector outlet temperature was specified at 225°C to give an overall power plant efficiency of ~ 10%. Although 40–50% below efficiencies noted above, the less expensive fixed collectors would make this power plant concept viable. Since no moving parts are involved in the collector field, field maintenance costs should be below those for tracking devices. A detailed study of the fixed-field idea is required prior to a final judgment on its economic feasibility. It is likely that the 225°C temperature may find application in solar total energy systems described later rather than in only power production.

Cavity receivers are attractive for several reasons. Evacuation is not required to control convection. Therefore, the reflectance and absorptance losses of a glass envelope are avoided. Similarly, radiation losses are small since the heat loss area is small although the effective emittance of the cavity is approximately unity. Selective surfaces need not be used and absorber surfaces are not critical since the cavity effect assures an absorptance value $\bar{\alpha}_s \approx 1$ even if the actual surface absorptance is only 0.5. Therefore, optical efficiency of these collectors is limited only by mirror reflectance. High absorber heat fluxes need not be a problem since a cavity may use any internal dimensions because the cavity size is quite independent of the cavity aperture.

A number of collector cooling fluids can be used. Pressurized water has excellent heat transfer characteristics but requires high-pressure piping and significant pump work to circulate. Steam is also an excellent transport fluid and seems to have the economic edge over liquid water (F2). Organic fluids and most gases such as helium are also technically feasible. Liquid metals—sodium and potassium—could be used; however, stainless steel absorber conduits and fittings are necessary.

Collector fluid flow rate is an important design variable for several reasons. The collector temperature rise ΔT_c is inversely proportional to flow rate and determines heat engine efficiency for a given condensing temperature. The higher ΔT_c, the higher the heat engine efficiency. For very high flow rates ΔT_c is small but pump power requirements are high—as described later, the fluid loop pipe diameter could be increased to reduce horsepower but increased heat losses would occur from the pipe network. Therefore, an economic analysis is necessary to find the best mix of cycle efficiency, pump horsepower, pipe cost and pipe heat loss (see Chapter 7). In general, however, the cost of energy transport increases with decreasing ΔT_c.

B. Distribution Subsystem

High-temperature fluid heat transfer media described above are all practical methods of heat transport by means of fluid internal energy. A promising alternative transport is by chemical reactants which are dissociated in the collector and recombined at the central plant to release their heat of reaction. The reactants are pumped at near ambient temperature and pressure thereby obviating the need for insulation and high-pressure piping. An example of this concept is given below.

A pipe network for any type of working fluid can be sized using methods described in Chapter 4. Collector field shape and piping layout are variables which can be determined in such a way that pump horsepower is a minimum for given constraints on pipe size. Figure 5.3 shows a typical layout for a square collector field. Numbers shown are in units of single-collector flow rate. It is seen that both diagonal and crosswise headers carry the same exit flow rate but different rates at entry ports along the headers. Multiples of the array shown can be added in an obvious fashion. A trade-off involved in field layout involves collector

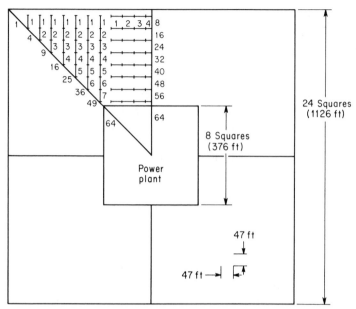

FIGURE 5.3 Piping network for 512 collectors to ensure equal flow at each header exit at the power plant. Numbers shown are in units of single collector flow rate.

shading at low sun angles. If collectors are spaced so that no shading occurs, pipe runs will be longer than if some shading is tolerated for a few weeks a year. Ground cover ratios (ratio of collector aperture area to ground area) of 40–50% seem a good compromise (C5).

Carbon steel is usable below 430°C for noncorrosive fluids whereas more expensive alloys are required for the same fluids at higher temperatures. Liquid metals require stainless steel pipes, pumps, and heat exchangers at substantial cost (12 × carbon steel pipe). Thermal expansion in long pipe runs must be accomodated by expansion loops of the sawtooth or semicircle types. Pipe wall thickness is proportional to internal pressure as shown below in a summary field design calculation procedure.

Pipe insulation to control parasitic heat loss was described in Chapter 4. For high-temperature systems exposed to the weather, calcium silicate with an aluminum jacket can be used to 900°C. Evacuated, multilayer foil insulations can achieve higher R values but are delicate and costly.

Array design can be accomplished in a stepwise process by selecting a pipe size and then calculating first the pump work, then the pipe thickness, and finally the insulation thickness. These parameters all have an associated cost, the minimization of which is the goal of the optimal design. Of course, the pipe material thickness must be greater than that required by the boiler code which applies.

Pump work W is calculated from Eq. (5.1). For various lengths of pipe L_i throughout a collector field with flow rates $\dot{m}_i$ (see Fig. 5.3) the total pump work is

$$W = \sum_i \frac{L_i m_i^3}{\rho_i^2 d_i^5} \frac{8f}{\pi^2 \eta_p}, \tag{5.1}$$

where ρ_i is the fluid density (usually constant unless a chemical reaction takes place) in a pipe element of length L_i, and inside diameter d_i.

Pipe thickness t_p can be calculated from a force balance on a pipe element:

$$t_p = d_i p / 2\sigma, \tag{5.2}$$

where p is the internal gage pressure, d_i is the pipe inside diameter, and σ is the allowable stress depending on the material in accordance with the applicable code, for example the ASME Standard Code for Pressure Piping. In order to account for threading and other effects, in practice Eq. (5.2) is modified to the form used by the piping industry (B2) as suggested by the ASME code:

$$t_p = [d_o p / (2\sigma + \gamma p)] + C, \tag{5.3}$$

where d_o is the outside diameter. Parameters γ and C (the allowance for threads, mechanical strength, and corrosion) are given in (B2). For example, at $T < 525°C$, $\gamma = 0.4$ for steel pipe and C, expressed in inches, is zero for plain end pipe and related to thread depth and pipe size for threaded pipe. Pipe wall thickness is not a major cost determinant (however, diameter is) but must be known to find the insulation thickness.

Insulation thickness t can be calculated by requiring that a given total volume of system insulation V_{ins} be distributed so that the heat loss per unit area over the entire system is uniform (K2). This method is based on an assumed uniform fluid temperature in each part of the network. The total volume of insulation on cylindrical pipes elements of length L_i is

$$V_{ins} = \sum \pi L_i [t_i d_o - t_i^2] \tag{5.4}$$

where t_i is the thickness of insulation on pipe element of length L_i and outside diameter d_o.

Finally the total cost C_n of the three major network components discussed above is found by summing:

$$C_n = W \times (h/yr \times \text{cost of pump power})/\eta_m$$
$$+ \sum L_i C_{p,i}(d_i) + V_{ins} C_{ins} + Q_L C_h \tag{5.5}$$

where η_m is the pump motor efficiency, $C_{p,i}(d_i)$ the present worth (refer to Chapter 7) of pipe cost per unit length of diameter d_i, C_{ins} the present worth insulation cost per unit volume, Q_L the heat lost through the insulation, and C_h its present worth. If fluid costs are significant, the cost of the fluid volume used to fill the pipe network should be added to this cost equation. Network cost C_n is then minimized by selecting the cost–optimal pipe diameter/insulation thickness pairs by an iterative process. Minimum C_n is technically only a system suboptimization for a given fluid temperature, pressure, flow rate, and central heat engine. The complete system optimal design requires exercising of all first-order design variables simultaneously. Typical optimal network designs result in 5–10% parasitic heat loss from the pipe field.

Network Transients   An additional consideration in sizing the distribution network is the morning starting time required to bring the fluid to temperature. A simple energy balance can be used to estimate this effect. During the night the internal energy rate of change $\dot{E}$ is given by

$$\dot{E} = \frac{d}{dt}\left[c_f m_f \bar{T}_f + c_p m_p \bar{T}_p + m_i c_i f(r_i, r_o)\bar{T}_f\right] \tag{5.6}$$

where $m_f c_f$, $m_p c_p$, and $m_i c_i$ are fluid, pipe, and insulation mass-specific heat products, and $f(r_i, r_o)$ is a function relating mean insulation temperature to the fluid temperature in which r_o and r_i are the pipe insulation outer and inner radii. It is given by (B5)

$$f(r_i, r_o) \equiv \frac{r_o^2 - r_i^2 - 2r_i^2 \ln(r_o/r_i)}{2(r_o^2 - r_i^2)\ln(r_o/r_i)}. \tag{5.7}$$

Since most of the fluid-to-ambient temperature gradient occurs across the insulation, $\bar{T}_f \sim \bar{T}_p$; then Eq. (5.6) can be expressed as

$$\dot{E} = [CM](d\bar{T}_f/dt) \tag{5.8}$$

where CM is the quantity in brackets in Eq. (5.6) divided by $\bar{T}_f$, i.e., it is the thermal capacitance of the distribution network. An energy balance on the network of total length L is then

$$\dot{E} = -(\bar{T}_f - T_a)\frac{2\pi k_i L}{\ln(r_o/r_i)} = [CM]\frac{d\bar{T}_f}{dt}. \tag{5.9}$$

It is assumed that the major resistance to heat transfer occurs across the insulation. Solving for $\bar{T}_f$,

$$\frac{\bar{T}_f - T_a}{\bar{T}_f(0) - T_a} = \exp\left[-\frac{2\pi k_i t L}{CM \ln(r_o/r_i)}.\right]$$

The energy E lost overnight for a period of t_n from the pipe network is given by

$$E = \int_0^{t_n} CM \, d\bar{T}_f$$

$$= CM(\bar{T}_f(0) - T_a) \left\{ \exp \left[\frac{-2\pi k_i t_n L}{CM \ln(r_o/r_i)} \right] - 1 \right\}. \qquad (5.10)$$

The heat loss E is the nighttime component of the total loss Q_L in Eq. (5.5). The daytime component is $[2\pi k_i L/\ln(r_o/r_i)]\int(\bar{T}_f - T_a)\,dt$. Generally $(\bar{T}_f - T_a)$ is not a constant so simple integration is not possible. The preceding expression assumes that fluid temperature fluctuations are small and daytime transients after startup, negligible.

Chemical Transport One way of eliminating nearly all parasitic heat leaks from a distribution system is to use ambient temperature reactive fluids which are capable of releasing their heat of reaction at the power plant turbine site. The reactive fluids, usually gaseous, are produced by endothermic dissociation at the collector. All transport lines can then be uninsulated and no expansion loops are needed.

For chemical transport and storage to be successful, the reactants must be (1) free of side products and the reaction completely reversible (most organic reactions are therefore disqualified), (2) compatible with all collector, transport, and reactor materials, and (3) simple to integrate into the solar system.

Many chemical systems could be used for this type of power plant. For example, methane and steam can be dissociated at 700°C to hydrogen and carbon monoxide. Thermal energy left in the outlet stream is exchanged with the collector inlet fluid stream. At the central site the hydrogen and carbon monoxide can be reacted in the presence of a catalyst to release the heat of reaction at about 550°C. The hot gas stream produces steam in a boiler to operate the power plant and is then returned to the collector as condensate and gaseous CH_4 for redissociation. Since energy is transported as chemical energy (26 kcal/mole or 109 kJ/mole) and not sensible heat, the *mass* flow rates required are relatively smaller.

A second reaction proposed for power production is the $2SO_2 + O_2 \rightleftarrows 2SO_3$ oxidation reaction (C8). Dissociation occurs at 800°C or above and recombination in the presence of a catalyst (e.g., platinum or vanadium) can occur at 500–600°C. The chemical energy involved is 23 kcal/mole (96 kJ/mole). A working pressure in the pipe network of only 3–4 atm (300–400 kPa) is required and less than 1% of the electric power produced is needed for pumping (C8).

Gaseous reactions are subject to the constraints of chemical kinetics, which specifies both rate and completion fraction for any reaction. Using the SO_3 reaction as an example, the free energy of reaction $\Delta F(T)$ at temperature T is given by

$$\Delta F(T) = \Delta F(T_0) + RT \ln[P_{SO_2}^2 P_{O_2}/P_{SO_3}^2] \qquad (5.11)$$

where the partial pressures are denoted by P. At equilibrium $\Delta F(T) \equiv 0$. If N moles of SO_3 are initially involved in a reaction and x moles of O_2 are produced, then $(N - 2x)$ moles of SO_3 and $2x$ moles of SO_2 remain after the reaction. Using the ideal gas law to relate x and the partial pressures it is easy to show at equilibrium that

$$\ln[(N + x)(N - 2x)^2/4x^3] = \ln P_{TOT} + [\Delta F(T_0)/RT], \qquad (5.12)$$

where P_{TOT} is the total pressure.

Equation (5.12) shows the effect of pressure and temperature on the extent of dissociation of the reaction as measured by x. For example, at 1 atm 90% of the SO_3 will have dissociated at 1200 K; at 10 atm, about 73% will have dissociated (D7). Further details of reaction kinetics are contained in various chemical engineering texts.

Problems with the chemical transport system include catalyst decomposition and coking, corrosion, suitable reaction rate, low conversion ratios, and the need for large heat exchangers to transport large amounts of sensible heat between inlet and outlet streams at both the collector and boiler. If reactants are stored, the tanks for gaseous components must be quite large. Depending on the reactants, toxicity may be a problem and environmental impacts of an accident releasing CO, SO_2, or SO_3 could be disastrous.

C. Storage Subsystem

Storage media for solar power plants are usually liquid or a combined liquid–solid because of the simplicity and ease of heat transfer using a liquid. Large amounts of storage of the order of days or more for solar power plants are rarely economic (F2,F4). A maximum of about 6 h seems to be the conclusion of several investigators using the standard 1-day outage per 10 yr criterion for the power grid. The physical properties of storage fluids have been given in Chapter 4 along with tank costs. Since storage volume V_s for a given amount of energy storage is inversely proportional to the allowable temperature swing ΔT_s, it is important to relate ΔT_s to the process temperature. Allowable temperature swing is determined by the storage material, temperature differences required for heat transfer, and the operating temperature of the process. Figure 5.4

shows the design storage ΔT_s for a number of solar-thermal processes including power production (B7). It is seen that an approximately linear relation exists between the maximum storage temperature and the allowable swing. A linear expression for the data shown is represented by

$$\Delta T_s = 0.65(T_{max} - 204) \tag{5.13}$$

in °C.

Storage volume for a given amount of stored heat as power plant reserve can be calculated from the expected value of ΔT_s and the cost can be estimated from Eq. (4.56). Since the fluid cost is proportional to volume directly with no economies of scale usually available, the total cost of a tank and its contents C_{TK} will usually be represented by an equation of the form

$$C_{TK} = K_1 F_m V_s^n(K_2 + K_3 p) + K_4 V_s, \tag{5.14}$$

where p is the pressure and the K_i are constants; n (<1) and F_m are from Tables 4.5 and 4.4, respectively. The storage cost per unit energy $c_{TK} \equiv C_{TK}/(V_s \, \Delta T_s \, c_v)$ is

$$c_{TK} = \{[K_1' F_m(K_2 + K_3 p)]/\Delta T_s^n\} + (K_4'/\Delta T_s), \tag{5.15}$$

a decreasing function of ΔT_s. Hence, economies of scale can be expected for the storage subsystem for power plants using liquids confined in metal tanks.

Two-Phase Storage Another effect of decreasing unit costs of storage shown above is the possibility of mixing rock with the working

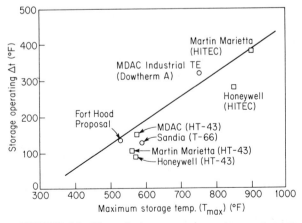

FIGURE 5.4 Thermal storage maximum temperature vs. temperature swing ΔT_s for various high-temperature solar processes. Legend: □, 10-MW$_e$ pilot plant systems; ○, total energy systems. [From (B7).]

fluid to reduce cost. Since rock is cheaper than most heat transfer fluids it can offer cost advantages. The volumetric specific heat of a rock–liquid mix $\bar{c}_v$ is

$$\bar{c}_v = (1 - \epsilon_v)c_{v,f} + \epsilon_v c_{v,r} \qquad (5.16)$$

where ϵ_v is the rock void fraction, $c_{v,f}$ is the fluid heat capacity, and $c_{v,r}$ the rock heat capacity (kJ/m³ °C). Since $c_{v,r}$ and $c_{v,f}$ have roughly the same values for rock and organic fluids, storage volume V_s is unchanged. However, in the cost equation, the second term is reduced since the mixture price is lower than that of the pure fluid. Solid media other than rock can be used. For cast iron $c_{v,CI} > c_{v,f}$ so $\bar{c}_v$ is larger and V_s smaller, thereby reducing both the first and second terms of Eq. (5.14).

In its simplest form the solid–liquid storage subsystem consists of a bed of inexpensive, uniformly sized particles contained in a tank. The high-temperature storage liquid fills the voids and circulates through the bed as heat is added or extracted from the tank. Charging is done by removing fluid from the bottom of the tank—the lowest temperature stratum—and returning it after heating to the top of the tank. Between the high- and low-temperature zones a relatively sharp thermocline is naturally present. As energy is added or subtracted, the thermocline moves downward or upward, respectively. (This is contrary to storage with high throughput rates which destroy the local thermocline and spread the temperature gradient over the full length of the bed.) In addition to the cost advantage of this storage method noted above, the bed approach permits storage of hot and cold fluid in one tank. At all times energy of maximum thermodynamic availability is used as opposed to availability loss in well-mixed tanks.

Tests conducted on biphase storage have shown that a reliable thermocline can be initiated and maintained in a natural steel tank using granite and an organic heat transfer oil (M3). Typical data showed that a 100°C temperature difference was maintained for extended periods. About 20% of the storage volume is consumed by the thermocline. Flow channeling with uniformly sized gravel was not observed. It was also found that the effect of parasitic heat losses from the tank surface penetrated into the tank only a short distance. If washed gravel is used, provision must be made for releasing steam producing during the first few charging cycles. Removal of solid residue must also be done by filtration.

In summary, the solid–liquid storage concept has the same capacity per unit volume as all-liquid storage (if it could be stratified), is less costly, is easily stratified, and is chemically stable. It is expected that this storage mode will be used in the first solar power plant in the U.S.—a 10-MW$_e$ facility to be completed in the mid-1980s. Other variations of two-phase storage showing promise include (1) the use of liquid metals and molten salts for the liquid phase, (2) improved tank aspect ratios, (3) use of two immiscible liquids, and (4) use of metal, ceramic, or metal ores for the solid phase.

Fluid Decomposition Storage cost is a decreasing function of maximum temperature via the ΔT_s dependence on T_{max} shown above if the decomposition limit is not reached. Beyond the decomposition limit, fluid replacement becomes quite costly and will adversely affect system economies (B7). Another high-temperature effect, if water is used, is the rapid increase of pressure p in Eq. (5.15) with T_{max}. In that case, the ΔT_s^{-n} (i.e., T_{max}^{-n}) and ΔT_s^{-1} (i.e., T_{max}^{-1}) unit cost reduction can be overwhelmed by the rapidly increasing p term and cost can begin to increase with T_{max} eliminating any economies of scale. Most organic oils do not require high-pressure confinement and the decomposition effect occurs prior to the high-vapor pressure effect.

Thermal degradation of organic oils has been studied by Morgan *et al.* (M3). It was found that the decomposition rate $\dot{m}_d$ for HT43™ (see Chapter 4) in wt%/h is

$$\dot{m}_d = 5.38 \times 10^{10} \exp[-17650/(T + 273)] \qquad (5.17)$$

and for Therminol 66™ is

$$\dot{m}_d = 1.93 \times 10^{27} \exp[-39580/(T + 273)], \qquad (5.18)$$

where $T = [290,340°C]$. It was also found that viscosity decreased with increasing fluid decomposition rate. Additional data on high-temperature degradation and its effects on economies are given in (B7).

Unit storage costs for all materials listed in Table 4.3 lie in the range of \$3–6/MJ (B7) when fluid, tank pressurization, and fluid replacement are considered if a mixture of 75% rock and 25% fluid is used. Insulation cost is not included. Advanced storage concepts and costs for solar power plants are discussed in (D7).

II. INDUSTRIAL PROCESS HEAT

Industrial process heat is the thermal energy used to prepare goods in manufacturing processes. Process heat can be supplied as steam direct firing as in a kiln, hot air, or hot water (or other hot liquid). Industry in the U.S. consumes about 40% of the national energy budget and

Table 5.1 shows the U.S. use of heat by standard industrial classification (SIC) code. It is seen that nine sectors account for the majority of heat usage—mining, food, textiles, lumber, paper, chemicals, petroleum products, stone–clay–glass, and primary metals. The breakdown of industrial energy usage type is as follows (I1): process steam, 41%; direct process heat, 28%; shaft drive, 19%; feedstock (chemical), 9%; other, 3%.

Although the quantity of energy used by industry has been well known for many years, the thermodynamic quality or temperature

TABLE 5.1

Summary of U.S. Industrial Heat Usage by SIC Category for 1971 with Projections to 1985 and 2000[a]

SIC group	Quantities in 10^{12} Btu		
	1971	1985	2000
10 ⎱	41	59	87
11 ⎟	0.2	0.1	0.05
12 ⎬ Mining	6	10.7	19.8
13 ⎟	971	1,353	1,932
14 ⎰	55	78	113
Subtotal	1,073	1,501	2,152
20 Food and kindred products	738	974	1,310
21 Tobacco products	13	28	68
22 Textile mills	254	440	792
23 Apparel	21	28	37
24 Lumber and wood products	178	204	237
25 Furniture	35	71	152
26 Paper and allied products	1,901	3,119	5,301
27 Printing and publishing	15	15	15
28 Chemicals	2,404	3,764	6,400
29 Petroleum products	2,442	3,148	4,132
30 Rubber	150	220	333
31 Leather	18	18	18
32 Stone, clay and glass	1,461	2,530	4,556
33 Primary metals	3,287	4,972	7,746
34 Fabricated metal products	280	484	871
35 Machinery	268	477	884
36 Electrical equipment	202	347	624
37 Transportation	294	415	601
38 Instruments	50	86	155
39 Miscellaneous	68	127	235
Subtotal	14,079	21,467	34,467
Grand total	15,152	22,968	36,619

[a] From (I1).

level has only been determined recently (Il, B5). Figure 5.5 shows the cumulative distribution of process heat use by temperature. If preheat from 15°C is considered and the process temperature requirement (may differ from current practice in the field) is used to determine heat demand, it is seen that 50% of the process heat usage occurs at or below 260°C (500°F) and 60% below 370°C (700°F). Therefore, between 50% and 60% of all U.S. process heat could be delivered by single-curvature concentrating systems which are the subject of this chapter. (Table 5.3 at the end of this section shows process temperature requirements by SIC code.)

Currently, much heat from fossil sources with high availability levels is used for applications at 300°C or below. The waste of availability is enormous. Table 5.2 below indicates second law efficiencies for the process heat sectors below 260°C (500°F) if fossil fuels are consumed at 80% first law efficiency. Also shown are η_2 values for the

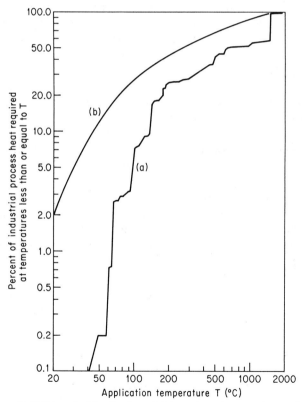

FIGURE 5.5 Distribution of U.S. process heat use by required temperature level: (a) heat requirements; (b) IPH requirements plus preheat from 15°C. [From (I1).]

TABLE 5.2

Second Law Efficiencies for U.S. Industrial Processes below 260°C (500°F)

Temperature	Fraction of U.S. process heat	Fossil fuel (η_2)	Solar (η_2)
85°F	10%	<1%	12%
120°F	5%	6%	52%
150°F	5%	10%	65%
175°F	5%	13%	71%
210°F	5%	16%	72%
250°F	5%	20%	77%
300°F	5%	25%	83%
370°F	5%	30%	85%
460°F	5%	35%	85%
Total/averages	50%	16%	61%

same processes if performed by a solar energy system (without storage) with 10% parasitic losses and a 20°C driving force for heat transfer from collector to load device. Storage and additional heat exchangers degrade this second law efficiency as described in Chapter 4.

      The numbers in Table 5.2 may not be precise in detail, nor need they be to vividly demonstrate the potential for improvement. Solar energy could increase the efficiency of energy use by a factor of more than three by using available technology. Solar applications above 260°C may be somewhat more difficult but the 50% of industrial heat below 260°C represents more than 20% of U.S. energy consumption. The benefits from the use of solar heat to conserve high quality fossil fuels for higher priority uses are obvious.

      Industrial process heat (IPH) systems are quite simple from a hardware viewpoint. They consist mainly of a collector field, fluid conduits, a heat exchanger, and a controller. In some cases storage for an hour or two to buffer short-term insolation outages is used. The size of storage must be evaluated for each process using a cost–benefit analysis. Sun-following IPH systems without storage are basically fuel savers and can be added to existing plants if land is available. In the industrial Northeast of the U.S., land availability may be the critical constraint acting on IPH systems in that area.

      Since IPH systems use components already described for other medium-temperature applications this section will be devoted to the descriptions of *generic types of systems*. The prediction of the long-term performance of these systems is described in the final section of the chapter.

A. Example Process Heat System

Prior to the discussion of generic industrial process heat (IPH) systems, one of the first IPH, steam-based systems in the U.S. will be described to provide the reader with a feel for system sizes and configurations. The Jacobs–Del Solar System Company (JD) collaborated with the Home Laundry Company in 1978–1979 to design and build a solar steam system for laundry and dry cleaning purposes. About 600 m² of PTC collectors are used to produce 170°C steam. The system will furnish 25% of the steam requirements of a laundry in Pasadena, California (E2).

Figure 5.6 shows the major components of the JD system. The solar loop consists of the collector field, a steam generator (heat exchanger), pump, and control valves. Treated city water is used as the loop working fluid and is pressurized by a nitrogen bottle to operating pressure. No boiling takes place in the collector loop. The collector fluid can be used in three modes—steam production, water heating, or short-term storage depending upon the temperature. Steam is produced in unit SG-1 connected in parallel with the fossil-fuel boiler (200 hp). As solar heat is added to SG-1 its tube-side pressure rises to slightly above the boiler set point (105 psig, 0.72 MPa) and solar-produced steam is used in the laundry system. If solar steam pressure falls, the boiler is brought on to pick up the load.

Since this IPH system is of the fuel saver type, the amount of storage is small and amounts to about 15 min heat demand. The role of storage is to buffer solar flux transients during daylight but not to extend solar heat use into nighttime hours. The relatively low, 25% solar load fraction also indicates that storage is not required.

The control system has several functions. First, when fluid leaving the collector is 215°C, the steam generator control valves CV6 and CV8 modulate to produce 170°C steam (110 psig, 0.76 MPa). If collector fluid temperature drops below 182°C, the backup boiler is used to provide the full load. Buffer storage is used for short solar outage periods only. Storage is recharged up to the 215°C level whenever collector outlet temperature is greater than storage temperature. Another control feature inverts the trough collectors for overnight storage or for a loss of coolant episode or for power or pump failure. Freeze control is achieved by circulating storage fluid through the collectors when the ambient temperature is less than 1°C. Collector cleaning with water sprays is also done periodically.

This sample system includes all features of an IPH in a simple and reliable configuration. A solar subsystem, fossil-fuel backup

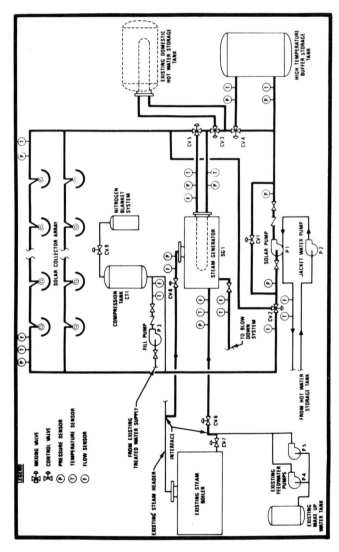

FIGURE 5.6 Schematic diagram of JD IPH solar system used to produce steam for the Home Laundry Company in Pasadena, California. The system began operation in 1979 and is one of the first steam IPH solar systems in the U.S.

and controls are combined together in logical fashion to achieve the design goal. All components other than the collectors (and sun tracker) are off-the-shelf, chemical process industry, standard items.

B. Liquid Heating Systems

Figure 5.7 shows a schematic diagram for a liquid heating IPH system. The liquid to be heated may be passed through the collectors or may be heated indirectly in a heat exchanger (not shown). Collector pump P1 is activated for $T_2 > T_1 + \Delta$ where Δ is selected to provide adequate hysteresis to suppress cycling of P1. Heat is delivered to the process by pump P3 which may use a constant or variable flow rate depending on the relative cost effectiveness of each. If water or other freezing collector liquid is used, a draindown system, activated by operating valves V1 and V2 when pump P1 ceases to operate, is required. If a nonfreezing organic fluid is used, no collector draindown is required but the fluid must be maintained above its pour point in the collectors overnight. If liquid makeup is required to replace fluid consumed in the process, it is provided through pump P2 and valve V3 both operated by a level controller in the storage tank.

The series backup heater is of the rapid response type and can be gas, oil, or steam fired. Fluid stream temperature T_3 is controlled

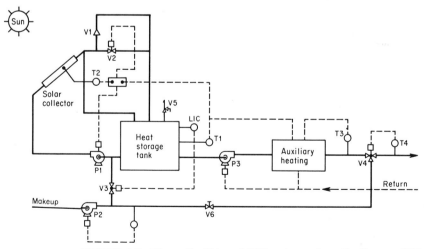

FIGURE 5.7 Direct, liquid-based IPH system schematic diagram. LIC denotes level indicator controller. [From (I1).]

by a feedback controller along with a feedforward signal from the storage tank. The auxiliary heat rate is modulated by temperature signal T_3. If solar heat collection is higher than instantaneous process demand, a tempering or mixing valve—V4, for example—can be used to maintain process inlet temperature at its desired point. Normally, industrial processes have uniform loads and a system controller using a system mathematical model on a minicomputer will give very precise process control.

Hot water is consumed directly in many industries such as food, paper, leather, textiles, and some chemicals. Other systems do not consume hot water directly but use only its internal energy; for example, feed preheat in a petrochemical plant uses only the heat from the hot fluid.

A modification of the direct heating system of Fig. 5.7 is shown in Fig. 5.8. Here the process fluid is heated *indirectly* by means of a heat exchanger placed between the solar collection loop and the process. This type of system would be used if the liquid to be heated is unsuitable for a solar collector working and storage fluid. The control methodology for the indirect system is slightly different. If process return temperature $T_3 > T_1$, valve V5 bypasses storage and the auxiliary source supplies the full process demand. This control feature is needed to prevent auxiliary heat from partially heating a fully depleted storage tank $(T_1 < T_3)$.

During peak insolation periods, T_1 may exceed the desired process temperature T_5. The usual industrial practice in this case is to use process return fluid at T_3 to temper fluid at T_4 if $T_4 > T_5$. Valve V7 is a pressure balancing valve between the two inlet streams to the tempering valve V6. For solar system maintenance, bypass valve V5 can be locked

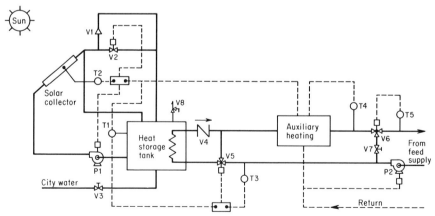

FIGURE 5.8 Indirect, liquid-based IPH system schematic diagram.
[From (I1).]

in the full auxiliary mode and process operation can continue without interruption.

C. Gas Heating Systems

Hot gases are used in industry for several purposes including curing, drying (minerals, food, paint, textiles, lumber), baking, preheating, chemical reaction, and moisture stripping in a stripping column. Figure 5.9 shows a schematic diagram of a solar, indirect gas heater which could be used in many industrial installations.

The solar collection loop and its control are similar to those described above for liquid-based IPH systems. Heat delivery to the gas stream is controlled by the entering gas temperature T_3 which causes the process pump P2 to operate if storage temperature $T_1 > T_3 + \Delta$; the extra temperature increment Δ is determined by the heating coil effectiveness. If the storage temperature is greater than that needed to supply hot gas at the process design temperature, valve V1 tempers the heating coil inlet

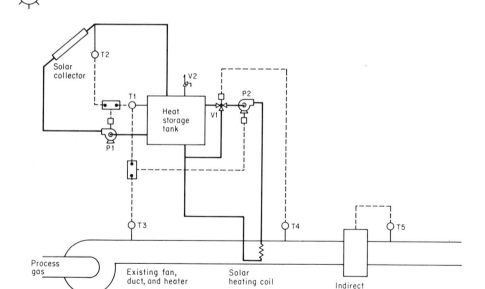

FIGURE 5.9   Indirect gas heating IPH system schematic diagram. [From (I1).]

temperature to provide a proper T_4 value. Temperature T_4 must be measured with an averaging probe to compensate for rather common nonuniform distribution of temperature downstream of an air heating coil.

The backup unit functions in the series mode as is the case for previous liquid heaters. If the backup heater has an appreciable time constant, feedforward control from the storage tank may be needed. Given the usual mass of storage used, it is unlikely, however, that storage temperature would drop at a rate exceeding the warmup rate of the backup device.

D. Steam IPH Systems

Steam is the most common fluid used for process heat end uses in the U.S. In this section two steam systems compatible with solar heat are described—the steam flash chamber and the series solar-backup boilers methods.

Steam Flashing Systems The design of steam accumulators was described earlier in Chapter 4. Figure 5.10 shows one way of using an accumulator in a solar IPH system. Although the flash tank concept is

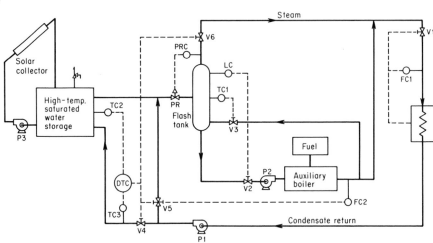

FIGURE 5.10 Steam accumulator or flash chamber IPH system schematic. Legend: FC, flow controller; TC, temperature controller; PRC, pressure regulator controller; DTC, differential temperature controller; PR, pressure regulator. [From (I1).]

simple, control is rather complex since most boilers cannot act as rapid response backup devices. In addition, a minimum flow must be maintained in most boilers to avoid tube damage owing to hot spots.

The rate at which saturated steam is produced at the flash chamber is determined by the temperature of storage and the minimum flow required by the boiler. In principle, if solar storage were sufficiently hot, the boiler could be bypassed. However, minimum boiler flow must be maintained as specified by the manufacturer to avoid tube burnout because of inadequate flow rate and low convection rates in the tubes.

One method of accomplishing this is to use flow controller FC2 to direct process condensate to the flash chamber directly via V5. This fluid bypasses solar storage and reduces the solar heat input to the flash tank. Since condensate return has a lower enthalpy than flash tank output, the boiler injects steam into the chamber, through valve V3, based on a signal from controller TC1.

Auxiliary steam is provided to maintain flash chamber design conditions based on the level control signal. Since chamber temperature correlates with liquid volume in the chamber, the fluid level can be used to determine boiler heat addition above the minimum continuous flow level. If solar storage is completely depleted, i.e., TC3 ≥ TC2, the valves V4 and V6 close and 100% flow occurs through the boiler and valves V5 and V2. The major problem with the flash tank approach is the limited steam rate which can be achieved efficiently.

Solar Boiler–Auxiliary Boiler System Figure 5.11 shows a solar boiler–auxiliary boiler system. The solar boiler is a kettle evap-

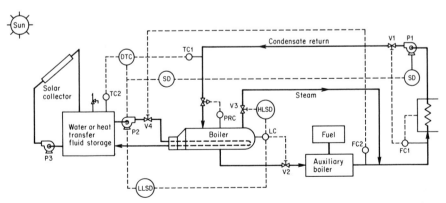

FIGURE 5.11 Solar boiler–auxiliary boiler IPH system schematic diagram. Legend: LC, level controller; LLSD, low-level shutdown; HLSD, high level shut down (other flow symbols defined in Fig. 5.10). [From (I1).]

orator with its solar fluid throughput via pump P2 continuously at a maximum level. Only the boiler minimum flow criterion can override this solar fluid flow rate. If boiler minimum flow is being approached, flow controller FC2 modulates solar flow control valve V4. Since solar firing of the solar boiler is thereby reduced, the process loop liquid level rises in the kettle. Level controller LC then opens valve V2 increasing the boiler flow as required.

In normal operation when the boiler base flow rate criterion does not override maximum solar flow to the solar boiler, it is possible for the solar heat rate to exceed the process demands on a sunny day. As a result, the solar boiler could boil dry. This is to be avoided to prevent scale buildup and can be accomplished by means of a low-level shutdown (LLSD) on the solar delivery pump P2.

As solar storage becomes depleted, the solar boiler ceases to operate and the kettle process fluid level rises. A high-level shutdown (HLSD) switch then closes valve V3 preventing any further fluid buildup in the kettle. Full auxiliary heat is used to operate the process although the solar boiler continues to act as a preheater until TC1 > TC2, at which point solar delivery pump P2 is deactivated.

This system is quite flexible and can even operate a batch process while continuously collecting solar heat since P3 operates independently of the process. The control strategy for this system is quite simple and extracts as much solar heat as possible since solar heat is used for boiling or fluid preheat depending on storage temperature.

The two steam systems described above are only examples of many concepts proposed for solar process steam systems. Each proposal must be examined for controllability, maximum solar usage and flexibility.

E. Possible Field Problems

Although less than a dozen solar IPH systems were built prior to January 1, 1979, several problems have been observed which should be avoided in future systems. Controls require careful design regarding hysteresis or dead band size, mode switch criteria, and sensor location. Controls can make or break a system and therefore must be analyzed for function in all possible modes including normal collection, transient warmup in the morning, daytime solar outage, nighttime storage, loss-of-coolant fail-safe measures, dust storm and hail episodes, high winds, etc. Sensor placement and reliability at high temperature was found to be a particular problem in the early systems. Trackers must accommodate any vibration caused by wind or reciprocating process machinery and sluggishness caused by low temperature of tracker lubricants.

Possible piping problems include leaking rotary joints, fittings, use of plastic pipe for high temperatures and pressures for which it is not suited, and use of control valves which do not fully seal. In addition, dirt and other refuse can be left in lines to foul collectors and damage valves. Manifold design must be done carefully to assure good flow balance in all collectors (See Chapter 4). Although collector roof mounting is usually more expensive than ground mounting (if ground is available) a longer warmup period will be needed in the morning owing to larger fluid volume in extended pipe runs to and from a remote collector field. High viscosity of heat transfer oils when cold is also a potential problem. Retrofit mounting of collectors on a roof may be impossible if the building structure is not designed to carry the extra dead load.

A method for cleaning the collector reflectors or lenses is a necessity in an industrial environment. Airborne pollutants should be identified prior to selection of the collector reflector material and collectors should be sited to minimize particulate pollutant fallout. Most collectors of the reflecting type are shipped disassembled. During field assembly, therefore, a method of aligning and leveling the absorber relative to the mirror must be used. Special jigs seem to be an effective way to accomplish alignment. Long-term average efficiencies of 40–45% are typical for line-focus concentrators, irrespective of specific collector designs, operating in process heat systems below 250°C. The preference of IPH designers of early systems was to avoid the cost and complexity of long-term storage, using only short-term buffer storage.

F. Projected Technical Readiness

The market penetration of solar IPH systems depends upon cost, performance, and reliability of the solar system vis-à-vis the non-solar system. A recent study (B5) has projected the state of technical readiness of solar system for many major IPH sectors. As expected, those systems using simple collectors—flat plates or solar ponds—at low temperature are technically viable today. Those processes operating at higher temperature and requiring concentrating collectors will be technically tested and developed in the more distant future. Table 5.3 lists the major IPH SIC sectors with estimates of technical readiness and the type of solar collector subsystem which seems most suitable. Of course, technical and economic viability are two distinct questions and the latter is considered in Chapter 7.

TABLE 5.3

Estimated Date of Technical Readiness for Various Solar IPH Systems[a]

Industry/process	Energy form[b]	Temperature, (°F)	Estimated time of technological readiness[c]	Shallow ponds or simple air heaters	Flat plates	Fixed compound surfaces	Single-tracking troughs	Central receivers
Aluminum								
Bayer process digestion	Steam	420	1985				X	
Automobile and truck manufacturing								
Heating solutions	Steam(water)	120–180	1980	X	X			
Heating makeup air in paint booths	Air	70–85	1980	X				
Drying and baking	Air	325–425	1985			X	X	
Concrete block and brick								
Curing product	Steam	165–350	1985			X	X	
Gypsum								
Calcining	Air	320	1985			X	X	
Curing plasterboard	Steam(air)	570	1990				X	X
Chemicals								
Borax, dissolving and thickening	Steam	180–210	1980		X	X		
Borax, drying	Air	140–170	1980	X	X			
Bromine, blowing brine/distillation	Steam	225	1985			X		
Chlorine, brine heating	Steam(water)	150–200	1980	X	X			
Chlorine, caustic evaporation	Steam	290–300	1985			X	X	
Phosphoric acid, drying	Air	250	1985			X		
Phosphoric acid, evaporation	Steam	320	1985				X	
Potassium chloride, leaching	Steam	200	1980		X			
Potassium chloride, drying	Air	250	1985			X	X	
Sodium metal, salt purification	Steam	275	1985			X	X	
Sodium metal, drying	Steam(air)	240	1985			X	X	

Process	Medium	Temperature	Year					
Food								
Washing	Water	120–160	1980		X	X	X	
Concentration	Steam(water)	100–200	1980		X	X	X	
Cooking	Steam	250–370	1985				X	X
Drying	Steam(air)	250–450	1985				X	X
Glass								
Washing and rinsing	Water	160–200	1980		X	X	X	
Laminating	Air	212–350	1985		X		X	X
Drying glass fiber	Air	275–285	1985			X	X	X
Decorating	Air	70–200	1980		X	X	X	
Lumber								
Kiln drying	Air	150–210	1980		X	X	X	
Glue preparation/plywood	Steam	210–350	1985	X			X	X
Hot pressing/fiberboard	Steam	390	1985	X				X
Log conditioning	Water	180	1980	X		X		
Mining (Frasch sulfur)								
Extraction	Pressurized Water	320–330	1985		X		X	X
Paper and pulp								
Kraft pulping	Steam	360–370	1985					X
Kraft liquor evaporation	Steam	280–290	1985				X	X
Kraft bleaching	Steam	280–290	1985				X	X
Papermaking (drying)	Steam	350	1985					X
Plastics								
Initiation	Steam	250–295	1985				X	X
Steam distillation	Steam	295	1985				X	X
Flash separation	Steam	420	1985			X		X
Extrusion	Steam	295	1985				X	X
Drying	Steam	370	1985			X		X
Blending	Steam	250	1985				X	X
Synthetic rubber								
Initiation	Steam(water)	250	1985			X	X	X
Monomer recovery	Steam	250	1985			X	X	X
Drying	Steam(air)	250	1985			X	X	X

(Continued)

TABLE 5.3 (Continued)

Industry/process	Energy form[b]	Temperature, (°F)	Estimated time of technological readiness[c]	Shallow ponds or simple air heaters	Flat plates	Fixed compound surfaces	Single-tracking troughs	Central receivers
Steel								
Pickling	Steam	150–220	1980	X	X	X		
Cleaning	Steam	180–200	1980		X	X		
Textiles								
Washing	Water	160–180	1980	X	X			
Preparation	Steam	120–235	1980	X	X	X		
Mercerizing	Steam	70–210	1980	X	X	X		
Drying	Steam	140–275	1980	X	X	X		
Finishing	Steam	140–300	1980		X.	X	X	

[a] From (B5).
[b] Preferred form (secondary form).
[c] Demonstrated thermal efficiency and reliability for 10 yr.

III. SHAFT WORK PRODUCTION

The conversion of solar heat to shaft work in a heat engine was demonstrated more than a century ago. Since that time several dozen experimental systems, most based on the Rankine cycle, have been operated successfully from a technical viewpoint. Those systems were used to produce shaft work for water pumping, air conditioning, industrial shaft drive, and electric power production. This section will describe the design of intermediate sized solar-powered heat engines which operate below 300°C. A simplified schematic diagram of a solar-powered Rankine type heat engine is shown in Fig. 5.12 and an energy flow diagram is shown in Fig. 5.13.

The Rankine engine system shown schematically in

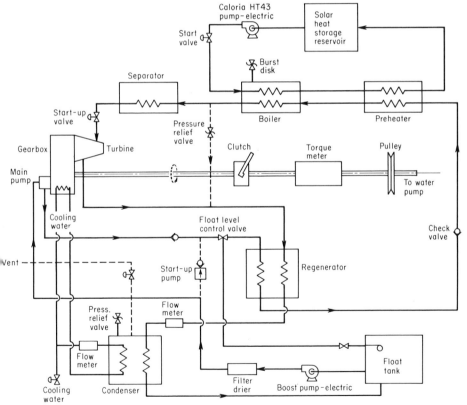

FIGURE 5.12 Schematic diagram of 25-hp example solar-powered Rankine cycle system. (Courtesy of Barber-Nichols Engineering, Arvada, Colorado.)

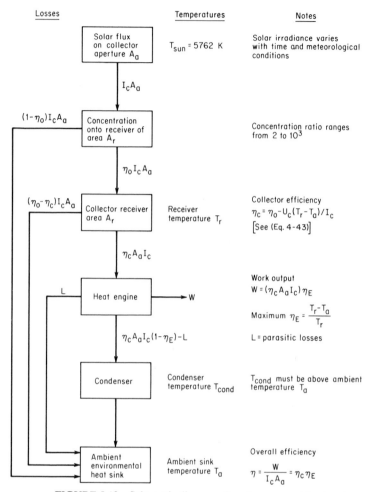

FIGURE 5.13   Schematic diagram of a solar-powered heat engine system. Optical and thermal losses to the environment as well as heat rejection to the environmental sink are shown. Efficiencies of each step are shown to the right.

Fig. 5.12 contains two fluid loops. The solar heat supply loop uses HT-43™ (See Chapter 4) as the working fluid from which heat is extracted in a boiler and preheater. The heat engine fluid passes from the boiler through a liquid–vapor separator to the turbine where shaft work is produced. Sensible heat remaining in the turbine exhaust is transferred to the condensate upstream of the preheater by means of a regenerator. After this step the turbine exhaust gas is condensed in a water-cooled condenser and piped to the fluid tank. Prior to its return to the preheater, the liquid working fluid is passed through a filter. The main fluid pump is driven by the turbine output shaft. The start-up pump shown in Fig. 5.12 is used to initiate fluid circulation prior to turbine start-up. After the main pump begins operation, the start-up pump is shut off.

Two counteracting phenomena are present in solar-powered engine systems. As shown in Chapter 3, the efficiency of heat engines increases as the input fluid temperature increases, whereas the efficiency of a solar collector decreases with outlet temperature as shown in Chapter 4. A system first law efficiency η_{sys} can be defined as the product of collector and engine efficiency as shown in Fig. 5.13. The value of η_{sys} first increases with collector temperature, then decreases as collector heat losses overwhelm thermodynamic gains owing to progressively higher engine inlet temperature. Using relations developed earlier, the efficiency of a solar-powered Carnot engine can be expressed by

$$\eta_{sys} = \{K[\eta_0 - U_c(T_{f,o} - T_a)/I]\} \times [1 - (T_a/T_{f,o})], \qquad (5.19)$$

where T_a is the environmental temperature, $T_{f,o}$ the collector fluid outlet temperature (same as engine inlet temperature if no heat exchanger is used and conduit heat losses are small), and other terms are as defined previously.

Equation (5.19) can be differentiated with respect to fluid temperature to find the temperature corresponding to maximum system efficiency. It is easy to show that the peak efficiency is achieved for a value of $T_{f,o}$ given by

$$T_{f,o:max} = [(\eta_0 I T_a/U_c) + T_a^2]^{1/2}. \qquad (5.20)$$

For example, if $\eta_0 = 0.7$, $U_c = 0.2 \ W/m^2 \ °C$, $T_a = 280 \ K$, and $I = 900 \ W/m^2$, then $T_{f,o:max} = 700°C$ and the maximum system efficiency is 39%. To improve performance it is clear from Eqs. (5.19) and (5.20) that high optical efficiency η_0, very low U_c values requiring high concentration, and high insolation are all required.

Of course, Eq. (5.20) applies only for the idealized Carnot and Stirling cycles with second law efficiency of unity. However, all heat engine systems have an efficiency curve which exhibits a maximum depending upon collector properties, cycle parameters, and weather conditions. Real engines have lower efficiency values than the theoretical limit for several reasons: (1) real fluids must be used with associated thermodynamic penalites as described shortly; (2) turbines and pumps are not perfect, and (3) thermal and mechanical losses occur in piping and all components of the system. These important effects are described in this chapter by describing their impacts on the performance of a common heat engine—the Rankine cycle described in general in Chapter 3.

A. *Working Fluids*

The most common fluid used in Rankine cycles is water since its heat of vaporization is high, its cost low, and its supply plentiful. However, from a thermodynamic viewpoint other fluids may be more desirable, owing to vapor density, phase-change occurrence conditions, or transport properties at intermediate temperatures.

In the ideal Rankine cycle all heat addition occurs at constant temperature equal to the collector outlet temperature. However, the real collector fluid cannot provide heat at a constant temperature since sensible heat removal is associated with a temperature decrease in the collector fluid. Since a heat exchanger is used between the collector and heat engine, some additional thermodynamic availability is lost in accordance with Eq. (4.80). Vaporization of cycle fluid directly in the collector is usually not practical because of flow balance, control, and piping problems.

Three types of cycle fluids can be identified as shown in Fig. 5.14. Fluid type A has a relatively large latent heat addition occurring at the maximum cycle temperature; therefore, cycle efficiency is good. However, the collector outlet temperature must be relatively high to avoid a zero pinch point temperature. (The pinch point temperature is that temperature at which the difference between stream temperatures in a heat exchanger is smallest.) Type A fluids have a high critical point temperature relative to cycle temperature and high collector fluid flow rates are needed.

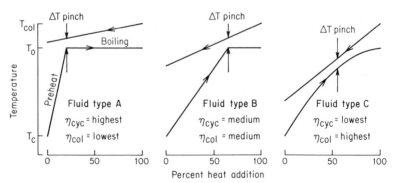

FIGURE 5.14 Temperature characteristics of three types of fluids as they pass through the boiler of a Rankine cycle. Solar collector and engine fluid temperature profiles are shown for the same values of cycle fluid inlet temperature T_c and outlet temperature T_0. The slope of the collector fluid (upper) T_{col} curve is determined by the collector fluid flow rate constrained by the pinch point ΔT.

Type B fluids have a smaller amount of latent heat addition than type A fluids. Therefore, the cycle efficiency is lower but the collector fluid exits the boiler at a lower temperature and as a result collector efficiency is higher than for type A fluids since the average collector temperature is lower. The critical temperature of type B fluids is of the same order as the collector outlet temperature.

Type C fluids are operated in the supercritical range and do not experience constant temperature heat addition. Therefore, cycle efficiency is low. However, the collector fluid experiences a large temperature drop in the boiler and collector efficiency can be quite good for a specific design outlet temperature $T_{f,o}$. Collector fluid flow rates are small relative to those for type A fluids. Each of the three real fluid classes imposes a thermodynamic penalty on the ideal cycle efficiency since heat addition cannot be done isothermally.

Although the thermodynamic characteristics of fluids are important, other properties must be considered. For example, for the same cycle temperature conditions two different fluids may require vastly different pump and expander designs with major cost effects. Operating pressures also effect pressure vessel and pump and turbine seal designs and costs. Various fluids have widely varying cost, durability, flammability, toxicity, and chemical reactivity with other cycle components. Typical practical cycle fluids include the halocarbon refrigerants—R11, R12, R22, R113, R114, R115—pyridine, water, and a number of stable organic fluids with appropriate properties.

B.   The Expander or Turbine

The expander of a heat engine is the component which converts kinetic and internal energy in the vaporized working fluid to shaft work. Piston- or turbine-type expanders are used depending upon the size of the engine and its speed. The performance of expanders has been thoroughly analyzed in the engineering literature and it has been found that similarity parameters can be used to prepare generalized performance maps for the broad range of sizes.

The similarity parameters which are most useful are the Reynolds number, the Mach number, the specific speed N_s, and the specific diameter D_s. The specific speed is defined by

$$N_s \equiv NQ^{1/2}/H_{ad}^{3/4}, \tag{5.21}$$

where N is the rotation rate in rpm, Q is the inlet flow rate in ft³/sec, and

H_{ad} is the enthalpy drop in ft lb/lb across the expander for adiabatic (no heat loss) conditions. It is seen that N_s is a dimensional measure of turbine speed (but with a consistent set of units will be dimensionless). Specific diameter D_s, a dimensional similarity measure of expander size, is defined by

$$D_s \equiv DH_{ad}^{1/4}/Q^{1/2}, \qquad (5.22)$$

where D is the diameter in feet. For most expanders used in solar systems the Mach number is low (<0.7) so that compressibility effects are of second order. The Reynolds number is usually sufficiently high ($>10^6$) so that fluid inertial forces dominate viscous forces and performance effects are independent of Reynolds number.

      Therefore, of the four similarity parameters, only two—D_s and N_s—are of first order. It is possible to plot turbine efficiency as a function of the two parameters for all geometrically similar devices (B3). Turbine efficiency is defined as the work output divided by the fluid enthalpy change in the turbine. Figure 5.15 is a general N_s–D_s map for many common expander types. The figure shows that certain generic expander types are best for certain applications. Selecting an expander type is a major use of the N_s–D_s map. For example, if the specific speed is low, the map indicates that piston engines are the expander of choice. As speed increases piston expanders would be replaced by rotary expanders for smaller sizes and by axial turbines in larger sizes. At very high speed only small diameter axial turbines are useful. For example, if a specific speed of 3.0 is specified, a partial-admission axial turbine with specific diameter of about 10 will have higher efficiency than either a drag turbine or a rotary piston expander.

      Figure 5.16 shows the components of a 250-hp axial flow turbine designed for an intermediate temperature system.

C.   *Pumps*

      The pump used to pressurize the condensed cycle fluid prior to boiling can be of the positive displacement or rotary type. The performance of pumps can be analyzed using the same similarity parameters as are used for expander analysis. An additional parameter, the suction specific speed, is frequently introduced as an indicator of possible cavitation at the pump inlet.

      Figure 5.17 is a performance map for all common types of pumps and compressors. It is seen that specific types of pumps are most suitable for specific subsets of N_s–D_s values.

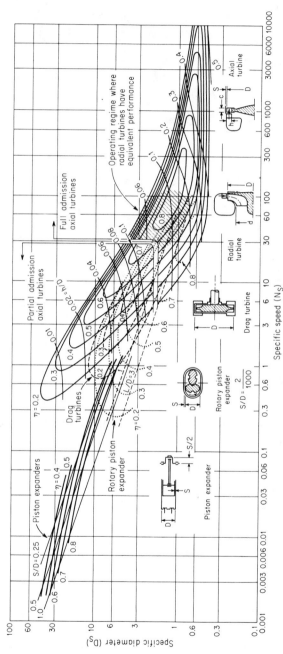

FIGURE 5.15 Specific speed–specific diameter performance maps for all common expander types. η denotes the efficiency related to total inlet pressure and static exhaust pressure. [Adapted from (B3).]

FIGURE 5.16   Components of a 250-hp axial flow turbine for solar or other intermediate temperature applications. The rotor diameter is 5.5 in and the design speed is 66,000 rpm. (Courtesy of Barber-Nichols Engineering, Arvada, Colorado.)

D. Other Components

Heat exchangers are used to transfer heat from the collector or storage fluid to the cycle working fluid. Both preheating and boiling are generally required and separate exchangers are used for each function. The preheater adds sensible heat, the boiler, latent heat. The design of heat exchangers is described earlier in Chapter 4 and follows standard industrial practice. Rankine cycle condensers (either air- or water-cooled) are also designed using conventional methods.

A fourth heat exchanger is used in some Rankine cycles to transfer heat remaining in the turbine exhaust to the boiler (or preheater) inlet liquid stream. This exchanger is called the regenerator and can be used if the turbine exhaust contains appreciable superheat. That is, the exhaust temperature at the condensing pressure is greater than the condensing temperature at the condensing pressure. Fluids which have cycle characteristics of this type are called "drying" fluids and can be identified by the relation, on a pressure–enthalpy diagram, of isentropic lines to the saturated vapor line. If the isentropic lines diverge from the saturation

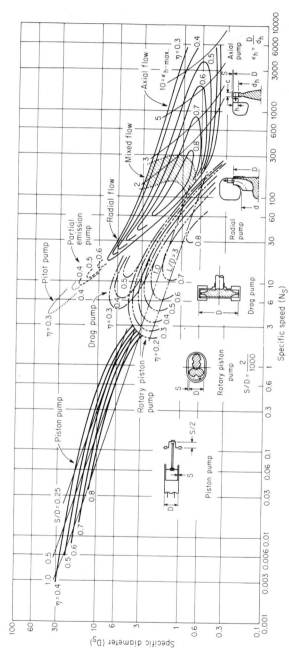

FIGURE 5.17 Specific speed–specific diameter performance maps for all common pump types. η denotes the efficiency related to static exhaust pressure and total inlet pressure. [Adapted from (B3).]

line with decreasing pressure, the fluid is of the drying type. The regenerator is sized using the ϵ–NTU method described in Chapter 4.

Controls for heat engines can be relatively complex and are designed for a specific application. The controller must perform several major tasks including: (1) operation of the solar collection loop, (2) turbine subsystem cold start-up, (3) turbine speed and heat rate control, (4) fail-safe system mode control. The second function is particularly important for systems which use the cycle fluid as the lubricant for expander and pump. Complete lubrication must be established before heat addition to the boiler begins.

Storage for solar shaft drive systems can be either thermal or mechanical. Thermal storage is discussed in detail elsewhere in this chapter and in Chapter 4. Mechanical storage in flywheels or in water reservoirs for solar pump systems is completely available from a second law viewpoint since it is stored as kinetic or potential, i.e., organized energy. The technology of hydropump storage is well developed and water turbine first and second law efficiencies of the order of 80% can be achieved with commercial equipment. Flywheel storage has been used only for small-scale devices and requires additional development. A third type of mechanical storage involves pressurized gas storage in large underground reservoirs; however, compressed gas is not completely available in the second law sense.*

E. System Performance

Long-term performance of intermediate temperature solar systems is analyzed in the final section of this chapter. Some instantaneous performance measurements are presented in this section. Figure 5.18 shows the losses measured in a small Rankine engine (3 hp) designed to operate on R113 heated by a flat-plate collector at 100°C. The

---

* Compressed air storage is not completely available thermodynamically since a heat loss from the warm, compressed gas to the underground storage reservoir occurs. The entropy change of the gas at T_g plus the environment at T_0 during cooling is given by

$$\Delta S = \int_{T_{in}}^{T_{stor} \sim T_0} \left[\frac{1}{T_g} - \frac{1}{T_0} \right] c_v \, dT_g$$

and the loss of availability $\Delta A = T_0 \, \Delta S$. The value of T_{in} depends on the compression pressure ratio r_p, $T_{in}/T_0 = r_p^{(\gamma-1)/\gamma}$, where γ is the specific heat ratio.

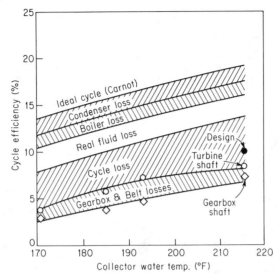

FIGURE 5.18   Measured losses in a Rankine engine owing to thermodynamic and parasitic mechanical inefficiencies in major components. Does not include feed pump (ideal pump work = 0.05 hp). Cooling water temperature = 85°F; condensing temperature = 95°F. [From (B4).]

cycle losses shown include condenser and boiler available energy consumption owing to the requirement of finite temperature differences between fluid streams. The effect of using a real fluid is also shown and has been described above. The cycle losses include expander and pump inefficiencies as well as pressure losses throughout the fluid conduits. The final loss shown is that associated with the output drive subsystem. In larger systems, parasitic mechanical losses would be relatively smaller.

F.   Example System

Figure 5.19 shows the major components in a Rankine-powered irrigation system located near Gila Bend, Arizona. A 5,500-ft² (46-m²) field of PTC collectors heats water to 300°F (150°C) to operate the heat engine rated at 50 hp (37 kW). A halocarbon working fluid is used in the engine as described above. The pump lifts water 14 ft (4.3 m) at a flow rate of 10,000 gpm (630 1/s). Control valves, heat exchangers, pumps are shown in the foreground of Figure 5.19. The turbine and gearbox are located behind the upper fluid tank. A global and shadow-band diffuse pyranometer are shown at the upper right.

FIGURE 5.19　Photograph of the 50-hp Rankine engine used for the Gila Bend pumping system. (Courtesy of Northwestern Mutual Life Insurance Co. and Battelle Memorial Institute.)

G. *Second Law Analysis*

The second law provides a method of calculating the thermodynamic losses in power cycles and gives insight into their possible reduction. The availability ΔA contained in a constant pressure fluid stream at temperature T_f can be calculated from the definition

$$\Delta A = \Delta H - T_0 \, \Delta S. \tag{5.23}$$

In the case of a liquid stream

$$\Delta A = c_p[(T_f - T_0) - T_0 \ln(T_f/T_0)] \tag{5.24a}$$

and for a saturated vapor stream

$$\Delta A = h_{fg}[1 - (T_0/T_f)] + c_p[(T_f - T_0) - T_0 \ln(T_f/T_0)], \tag{5.24b}$$

where T_0 is the environmental temperature and h_{fg} is the latent heat. Physically, Eq. (5.24a) states that an ideal heat engine can extract $c_p(T_f - T_0)$ heat from the liquid stream. Part is converted to work and $c_p T_0 \ln(T_f/T_0)$ is rejected as waste heat. Equation (5.24) shows that a reduction of T_0 will increase the available work more than an equal increase in T_f.

The second law efficiency can be written for the fluid stream as

$$\eta_2 = \frac{W}{\dot{m}_f c_p[(T_f - T_0) - T_0 \ln(T_f/T_0)]}, \qquad (5.25)$$

where W is the useful work produced at a flow rate $\dot{m}_f$. The first law efficiency is

$$\eta_1 = W/(\dot{m}_f c_p \, \Delta T_f), \qquad (5.26)$$

where ΔT_f is the fluid temperature drop required to produce work W. Usually $\Delta T_f < (T_f - T_0)$. Combining (5.25) and (5.26),

$$\eta_2 = \eta_1 \frac{\Delta T_f}{(T_f - T_0) - T_0 \ln(T_f/T_0)}. \qquad (5.27)$$

It is seen that high values of η_2 require *both* high cycle efficiencies η_1 and high fluid temperature drops.

In a real cycle such as a Rankine cycle additional losses are incurred above those noted above for an idealized extraction of heat from a liquid stream. Each component of the power cycle–fluid pump, heat exchanger, turbine, and condenser—has an associated thermodynamic irreversibility I_t analogous to the last term of Eq. (5.23). For the fluid pump

$$I_{t,p} = [(1 - \eta_p)/\eta_p]v_l(T_0)[\Delta p], \qquad (5.28)$$

where η_p is the pump efficiency, $v_l(T_0)$ the liquid specific volume at T_0, and Δp the pump pressure rise. (The small effect of $\int p\,dv$ work is ignored in the liquid phase.) For small values of $I_{t,p}$ small pressure rises and high pump efficiencies are needed.

The boiler thermodynamic irreversibility $I_{t,b}$ is given by

$$\dot{I}_{t,b} = \dot{I}_{t,s} + \dot{I}_{t,tf}, \qquad (5.29)$$

where the second subscripts s and tf refer to the solar-heated and turbine fluid streams, respectively. The values are

$$\dot{I}_{t,s} = - \int_{\text{boiler}} \left[\dot{m}_s \frac{c_p \, dT_s}{T_s} \right] T_0, \qquad (5.30)$$

$$\dot{I}_{t,tf} = - \int_{\text{boiler}} \left[\dot{m}_{tf} \frac{dh_{tf}}{T_{tf}} \right] T_0. \qquad (5.31)$$

Recalling that $\dot{m}_s c_p \, dT_s = -\dot{m}_{tf} \, dh_{tf}$ from the first law, the total irreversibility is

$$\dot{I}_{t,b} = \dot{m}_s c_p T_0 \int_{\text{boiler}} \left[\frac{1}{T_{tf}} - \frac{1}{T_s} \right] dT_s, \qquad (5.32)$$

where T_{tf} is usually constant in the boiler. On a per unit working fluid mass basis (using the overall energy balance $\dot{m}_s c_p / \dot{m}_{tf} = \Delta H_{tf} / \Delta T_s$)

$$I_{t,b} = (T_0 \, \Delta H_{tf} / \Delta T_s) \int_{\text{boiler}} \frac{(T_s - T_{tf})}{T_{tf} T_s} \, dT_s, \qquad (5.33)$$

where ΔH_{tf} is the turbine fluid enthalpy increase in the boiler and ΔT_s is the temperature drop in the solar-heated fluid. To minimize the boiler irreversibility the solar-heated fluid-to-turbine fluid temperature difference must be minimized.

Turbine irreversibility $I_{t,t}$ arises from the nonisentropic expansion of fluid in the blading. The turbine irreversibility $I_{t,t}$ is

$$I_{t,t} = T_0 \int_{T_{te(ideal)}}^{T_{te(real)}} \frac{dH_{tf}}{T_{tf}}, \qquad (5.34)$$

where T_{te} is the turbine exhaust temperature. Since the real and ideal turbine exhaust temperatures are about the same for a well-designed turbine, $T_{te(real)} = T_{te(ideal)}$, denoted as T_{te} hereafter. The turbine irreversibility is then

$$I_{t,t} = (T_0 / T_{te})[(1 - \eta_t)/\eta_t] \, \Delta H_t \qquad (5.35)$$

where ΔH_t is the turbine enthalpy drop. Small values of $I_{t,t}$ can be achieved with large turbine efficiencies. If the turbine exhaust contains superheat which is rejected in the desuperheater region of the condenser, some availability is lost unless a regenerator is used and this additional irreversibility is given by

$$I_{t,c} = \Delta H_{tf_{desuperheater}} - T_0 \int_{\text{desuperheater}} \frac{dH_{tf}}{T_{tf}}. \qquad (5.36)$$

To minimize $I_{t,c}$ small values of sensible heat rejected to the desuperheater ΔH_{tf} are needed along with low values of turbine exhaust temperature.

The final reversibility source is the finite temperature difference across the condenser. An expression similar to that for the boiler, Eq. (5.32), is used to calculate the condenser irreversibility.

Evaluation of the several irreversibilities requires the use of an equation of state for the turbine working fluid. Values of enthalpy and specific volume are calculated from the equation of state. Milora and Tester (M5) have tabulated equations of state for fluids useful up to 300°C. For subcritical cycles (fluids type A and B, Fig. 5.14) the heat exchanger, desuperheater, and condenser irreversibilities are the largest of the five terms described above. As cycle pressure is increased for a supercritical cycle, the pump and turbine shaft work levels increase for most fluids. However, the boiler irreversibility is smaller since the solar and cycle fluid temperature profiles are nearly parallel (type C fluid).

Milora and Tester, using the principle of corresponding states, have developed an accurate but simple method for screening turbine fluids. They showed that for maximum η_2 (minimum cycle irreversibility), the critical temperature of the fluid T_{crit} should be related to the solar collector (or storage) outlet temperature $T_{f,o}$ (°C):

$$T_{f,o} = T_{crit} + 790[(\gamma - 1)/\gamma], \tag{5.37}$$

where γ is the specific heat ratio c_p/c_v for the cycle fluid. A survey of fluids gives a value of $T_{f,o}$ which can then be matched to a solar concentrator design. For example, R113 gives a peak cycle η_2 value at 299°C whereas isobutane (R600a) is best for near 200°C operation. Further data are given in (M5).

IV. SOLAR TOTAL ENERGY SYSTEMS (STESs)

Solar total energy systems are designed to provide electric power (or shaft work for other purposes) with associated use of turbine exhaust heat to provide hot water, space heating, and/or cooling and other thermal loads. The principle reason for considering an STES is to use the thermodynamic availability present in solar heat produced by concentrators in the most efficient way by insuring a close second law match to the various tasks to be performed. The several processes are usually cascaded in order of increasing entropy with work production occuring first and low-temperature heating last. Figure 5.20 shows a typical STES, this example being a proposal for the Solar Energy Research Institute in Golden, Colorado.

This section will describe the first-order variables in the design of an STES and give the results of performance estimates for several locations in the U.S.

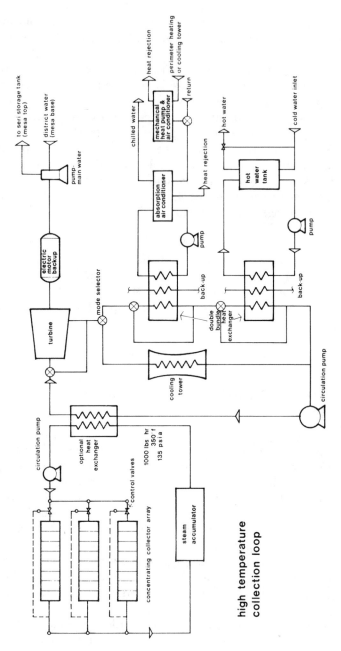

FIGURE 5.20   STES proposed for use at the Solar Energy Research Institute, Golden, Colorado (low-temperature storage not shown).

A. Load Type

An STES exists between two application limits—an electric power plant whose sole output is electricity (or other shaftwork) and a thermal plant whose sole output is heat. A convenient parameter to describe the nature of the STES load is the thermal to electrical load ratio T/E:

$$T/E \equiv \text{end use thermal demand/end use electrical demand} \quad (5.38)$$

where the total STES load $L = T + E$. T and E values for nonsolar systems include conversion efficiencies at the boiler or combustor and power plant, respectively.

A second important design characteristic of an STES is the load phasing for a given T/E ratio. Figure 5.21 shows three generic types of phasing patterns which can be encountered in an STES. Idealized pattern I is found in office buildings, schools or other institutions, and single shift industrial plants. This pattern has a uniform T/E ratio for the hours of occupancy and $T = E \approx 0$ otherwise.

Generic pattern II represents the idealized load phasing which might be found in residences, some industries, shopping centers, and hospitals. Pattern III is found in larger industries continuously operating on three shifts and is characterized by a fixed T/E ratio for 24 h.

The third major characteristic of an STES is its thermal performance as measured by its energy delivery, which is some fraction of the total load L depending on the solar system size relative to the value of L. Since few such systems have been built, most performance estimates are produced from short-time scale computer simulations carried out for periods of the order of years. A convenient dimensionless ratio for ex-

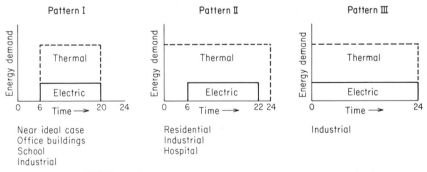

FIGURE 5.21 Three idealized thermal and electric load phasing patterns for an STES.

pressing thermal performance is $\bar{f}_s$, the solar load fraction or percent of annual demand carried by solar energy. The solar load fraction depends upon the T/E ratio and load phasing for a given system configuration. In addition, the local solar climate determines the response of an STES to various load profiles. The quantity $\bar{f}_s$ refers to end uses of energy and includes both power and heat requirements and deliveries by the STES or utility. The fraction of thermal load carried is $\bar{f}_{s,t}$ and of electrical load $\bar{f}_{s,e}$. Typical T/E ratios are shown in Table 5.4. for major U.S. industries.

    Since the incentive to use an STES is primarily economic in industrial economies, a cost–benefit ratio C/B is an alternative way of measuring the performance of an STES. The details of C/B calculation

TABLE 5.4

T/E Ratios for Some U.S. Standard Industrial Classification (SIC) Code[a]

SIC		Name	Purchased fuels and electric energy		$ per 1000 kWh	Purchased fuels	
			(kWh eq × 10⁹)	(cost × $10⁶)		(kWh eq × 10⁹)	(cost × $10⁶)
21		Tobacco products	5.9	38.6	6.54	4.8	20.2
	2111	Cigarettes	3.8	22.8	6.00	3.2	12.1
	2141	Tobacco stemming/drying	1.5	11.7	7.80	1.2	6.5
22		Textile mill products	94.5	692.1	7.32	67.6	286.1
225		Knitting mills	16.5	121.6	7.37	12.6	59.0
226		Textile finishing, except wool	21.8	117.3	5.38	19.9	85.1
227		Floor covering mills	8.8	45.6	5.18	7.8	29.2
228		Yarn and thread mills	12.0	116.3	9.69	5.5	21.2
229		Misc. textile goods	8.0	62.3	7.79	5.9	26.2
23		Apparel, other textile prod.	19.0	171.0	9.00	12.6	47.3
232		Men's, boys' furnishings	4.2	37.0	8.81	7.7	9.6
233		Women's, misses' outerwear	4.2	45.9	10.93	S	S
234		Women's, children underwear	1.1	11.7	10.64	0.6	2.1
235		Hats, caps, millinery	0.3	2.9	9.87	0.2	1.1
236		Children's outerwear	0.6	6.5	10.83	0.3	1.3
238		Misc. apparel	0.8	7.6	9.50	0.5	2.1
239		Misc. fabricated text. prod.	5.8	40.4	6.97	4.4	16.6
24		Lumber and wood products	79.6	482.4	6.06	64.7	274.7
242		Sawmills, planing mills	26.9	180.0	6.69	17.5	85.8
	2421	Sawmills, planing gen.	24.0	160.1	6.67	17.2	76.9
243		Millwork, plywood	15.9	94.9	5.97	12.4	45.2
	2435	Hardwood veneer	2.3	15.0	6.52	1.8	6.6
	2436	Softwood veneer	9.4	48.1	5.12	7.6	26.8
244		Wood containers	1.7	13.3	7.82	1.3	6.2
245		Wood bldgs. and mobile homes	2.6	16.5	6.35	2.2	8.4
249		Misc. wood products	16.1	94.4	5.86	13.2	50.8

[a] D = data withheld to avoid disclosure. S = data inconsistent or large standard error. [From (B7).]

are given in Chapter 7 but, stated simply, C/B is the ratio of annual STES costs—capital, interest, maintenance, insurance, and operating—to the dollar value of the annual fuel savings. In equation form,

$$C/B = \text{STES cost}/(\bar{f}_{s,t}TC_t + \bar{f}_{s,e}EC_e), \qquad (5.39)$$

where C_t and C_e are the annual worth of unit fuel prices ($/GJ, for example) over the system useful life including taxes and inflation in fuel prices. For $C/B < 1$ there is an economic incentive to use an STES; if $C/B > 1$, an STES is not economically viable. A convenient method of finding the lower bound of C/B at a given site for a given process is to determine the performance of an STES for the *best possible set of T/E and phasing values*. If $C/B > 1$ for this best case, no possible STES configuration will be viable.

$ per 1000 kWh	Purchased electric energy		$ per 1000 kWh	Electric energy generated less sold (kWh × 10⁶)	Total elec energy used E_T (kWh × 10⁹)	Net therm energy used E_{TH} (kWh × 10⁹)	Ratio (E_{TH}/E_T)
	(kWh × 10⁶)	(cost × $10⁶)					
4.21	1027.9	18.4	17.92	D	1.028	4.8	4.67
3.78	624.8	10.7	17.13	D	0.625	3.2	5.12
5.42	265.6	5.2	19.58	S	0.266	1.2	4.51
4.23	26908.4	406.0	15.09	374.6	27.283	66.776	2.44
4.68	3923.0	62.6	15.96	S	3.923	12.6	3.21
4.28	1943.7	32.2	16.57	150.7	2.094	19.448	9.29
3.74	1013.5	16.4	16.18	D	1.014	7.8	7.7
2.98	6458.8	94.4	14.62	D	6.459	5.5	0.85
4.53	2092.0	35.6	17.02	D	2.092	5.9	2.82
3.75	6357.0	123.7	19.46	S	4.357	12.6	1.98
3.56	1490.2	27.4	18.39	1.9	1.492	2.694	1.81
—	S	S	—	S	—	—	—
3.50	544.6	9.6	17.63	D	0.545	0.6	1.10
5.50	74.3	1.8	24.23	—	0.0743	0.2	2.63
4.33	268.6	5.2	19.36	S	0.269	0.3	1.12
4.20	271.5	5.5	20.26	S	0.272	0.5	1.84
3.77	1441.9	23.8	16.51	S	1.442	4.4	3.06
4.25	14790.7	207.7	14.04	500.5	15.291	63.199	4.13
4.40	7345.4	94.2	12.82	316.9	7.662	18.609	2.42
4.47	6777.7	83.2	12.28	308.5	7.026	16.274	2.30
3.65	3452.6	49.7	14.39	123.3	3.576	12.030	3.36
3.67	512.9	8.4	16.38	D	0.513	1.8	3.51
3.53	1827.1	21.3	11.66	D	1.827	7.6	4.16
4.77	370.8	7.1	19.14	D	0.371	1.3	3.50
3.82	408.9	8.1	19.81	S	0.409	2.2	5.38
3.85	2902.3	43.6	15.02	D	2.902	13.2	4.55

The Aerospace Corporation (B7–B9) has conducted a study of STESs in several U.S. locations to determine whether $C/B < 1$ for the optimum demand profile. A central receiver collector was assumed (see Chapter 6 for a description of the collector) but the results are similar for any high performance concentrator. Some of the reported results of this study will be summarized below. It was found that the ideal load phasing occurred if the electrical load E occurred for a time period slightly longer than daylight and the ideal T/E ratio (3.64) was such that all turbine exhaust heat could be used, subject to limits imposed by finite heat exchangers, and no turbine throttling required. The ideal T/E ratio is calculated from $(1 - \eta_t)\eta_{th}/\eta_t$, where η_t is the turbine efficiency and η_{th} is the heat use efficiency of the thermal process. For a line-focus powered heat engine $\eta_t \sim 0.18$ at 400°C (see previous section on power production) and for commercial thermal processes $\eta_{th} \sim 0.8$. For a 400°C turbine inlet, most thermal applications (i.e., those operated by the turbine exhaust) would be limited to 260°C and below. The *cost–benefit ratio for the ideal case* is a minimum for all possible cases and is denoted by C/B_{min} in the discussion below.

The major variables in a given solar conversion subsystem of an STES are collector area and storage size. These variables are important in all solarthermal applications and are also first-order variables for an STES as expected.

B. General Performance Characteristics of STESs

Using the results of computer simulations (B7), it is possible to draw several general conclusions about the performance of STESs in various parts of the U.S. For this purpose it is useful to define a set of standard economic assumptions, each of which can be later varied for sensitivity studies. To assign a cost to the results of this section, solar collectors at \$150/m², high-temperature storage (up to 400°C) at \$4265/GJ and low-temperature storage at \$950/GJ are used along with a turbine cost of \$400/kW$_e$. The factor used to annualize all costs (see Chapter 7 for details) is 15% of the initial cost per year. Competitive energy is assumed to cost 4.4¢/kW h for electricity and \$6.90/GJ for thermal energy. These costs include a small inflation factor (1.5–3% per year, real rate) over a 30-yr period.

Phasing, T/E, and Solar Flux Level Effects   Figure 5.22 shows the effect of load T/E ratio on the cost–benefit ratio C/B. It is seen that for $T/E > 2$, the cost–benefit ratio is near its asymptotic and min-

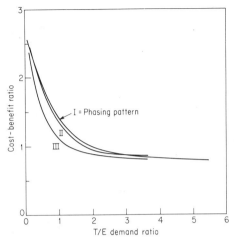

FIGURE 5.22   Effect of load phasing and T/E ratio on the cost–benefit ratio C/B for an STES in Albuquerque. [From (B7).]

imum value. Therefore the conclusion is that relatively high thermal loads result in the best economic performance of an STES. Table 5.4 shows that most industrial users of consequence have $T/E > 2$.

Figure 5.22 also shows that the load phasing (see Fig. 5.21) is important only for systems where the electrical load is relatively large and even then the maximum effect is only 20%. For larger T/E ratios the phasing effect can be ignored. It is seen that the ideal STES described above ($T/E = 3.64$) is on the asymptotic portion of the curves. Figure 5.22 was prepared by choosing a phasing pattern and T/E ratio followed by performance simulations of several dozen system configurations capable of providing the specified T/E. The lowest C/B configuration of all considered is plotted in Fig. 5.22.

Two systems exist for the production of a cooling effect—absorption and vapor compression—the former using a heat input, the latter, shaft work. If cooling loads are high it may be efficacious to shift to absorption to increase the T/E ratio, hence improve the cost–benefit ratio C/B. However, the COP of absorption units is in the range of 0.6 ~ 0.8 and for vapor compression units 2.0 or above. This COP disadvantage tends to offset possible higher T/E ratio benefits. The best strategy for cooling in an STES is not yet understood.

STES performance for several locations is shown in Fig. 5.23 for various T/E ratios. The same asymptotic behavior is seen for all sites with the poorest solar radiation sites showing the highest values of C/B. In generating these results, high temperature storage (used for

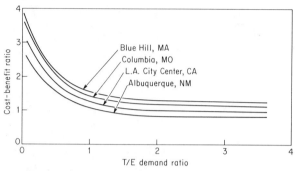

FIGURE 5.23   Cost–benefit vs. T/E ratio for various locations. Blue Hill has the lowest solar flux and Albuquerque the highest on the average. [From (B7).]

power production) was increased with decreasing T/E and low-temperature storage (for thermal end uses) has been decreased proportionately. The results of STES studies for 34 locations show a strong correlation between C/B ratio and the yearly averaged, daily beam radiation $\bar{I}_b$. It is therefore expected that STES feasibility at any site could be analyzed using this information. For the ideal STES with costs as defined above, the relationship between C/B and $\bar{I}_b$ is (B9)

$$C/B = \begin{cases} 2.5 - 0.28\bar{I}_b + 0.009\bar{I}_b{}^2 & \text{(phasing pattern I)} \\ 2.4 - 0.25\bar{I}_b + 0.007\bar{I}_b{}^2 & \text{(phasing pattern II)}, \\ 2.6 - 0.30\bar{I}_b + 0.011\bar{I}_b{}^2 & \text{(phasing pattern III)} \end{cases} \quad (5.40)$$

where $\bar{I}_b$ is in units of kW h/m² day. Likewise, the solar load fraction $\bar{f}_s$ in percent can be estimated from (B9)

$$\bar{f}_s = \begin{cases} 32.5 + 0.98\bar{I}_b + 0.46\bar{I}_b{}^2 & \text{(phasing pattern I)} \\ 57.6 - 9.2\bar{I}_b + 1.27\bar{I}_b{}^2 & \text{(phasing pattern II)}. \\ 58.4 - 10.9\bar{I}_b + 1.32\bar{I}_b{}^2 & \text{(phasing pattern III)} \end{cases} \quad (5.41)$$

If fuel costs differ from those assumed or $T/E \neq 3.64$, adjustments in C/B and $\bar{f}_s$ can be made. Equations (5.40) and (5.41) are based on an electrical load E of 500 kW$_e$.

Competing Energy Cost Effects   Previous results based on 4.4¢/kW h power and \$6.90/GJ fuel can be generalized to any cost structure and T/E ratio very simply. Equation (5.39) defined the C/B ratio. The denominator of the equation contains cost and T/E effects, therefore:

$$\frac{C/B}{(C/B)_{\min}} = \frac{[\bar{f}_{s,t}TC_t + \bar{f}_{s,e}EC_e]_{\text{ideal}}}{\bar{f}_{s,t}(T/E)TC_t + \bar{f}_{s,e}(T/E)EC_e}. \quad (5.42)$$

Denominator values for solar load fractions $\bar{f}_{s,t}(T/E)$ and $\bar{f}_{s,e}(T/E)$ are calculated from the performance model. Fuel costs C_t and C_e are local costs including the effect of real inflation in fuel price over a 30-yr period.

Figure 5.24 shows minimum achievable C/B ratios for the ideal STESs in various parts of the U.S. Fuel costs are based on data collected in 1977 with net price escalation of 3% for fuel and 1.5% for electricity. Although the ideal STES has a T/E ratio of 3.64, most industrial applications require $T/E > 2$ for which the C/B versus T/E curve is near its asymptotic minimum value. Hence, the C/B ratios for ideal STESs should be approximately correct for the range of real T/E ratios encountered in industry. If residential and commercial STESs have relatively high T/E ratios, the map should apply as well. However, some commercial systems have high electrical demands ($E > T$) and the values given on the map would not apply.

Data overlaid on the map show the dramatic effect of local fuel prices. It will be recalled that the highest insolation areas of the country had low C/B ratios according to Fig. 5.23 and Eq. (5.40) if constant prices were used. However, the effect of locally high energy prices in the Northeast and South are seen to give STES C/B ratios less than one and equivalent or less than those in the sunny Southwest. These conclusions are based on solar system prices given earlier which were estimated for mass production. Current prices of solar components are somewhat higher and may shift C/B ratios calculated as slightly below unity to the infeasible range $C/B > 1$.

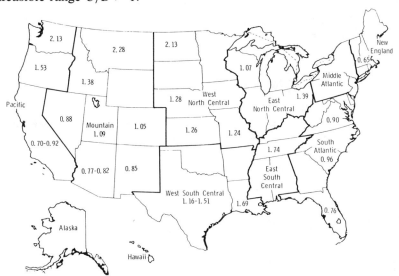

FIGURE 5.24   Minimum cost–benefit ratios C/B for STESs in the U.S. based on local fuel prices. $C/B < 1$ implies economic feasibility. [From (B7).]

The earliest date for any appreciable penetration of STESs into the U.S. energy mix is ~ 1995 (B9) assuming nominal energy inflation rates. After 1995 the largest impact is speculated to be in California, Hawaii, and Texas. By the year 2000 some 500 applications in 44 states could displace $4.7Q$ ($1Q = 10^{15}$ Btu) of energy. Any projection of this type is subject to economic decisions by industry, political decisions by government, and the prevailing costs and efficiencies of solar system and fossil fuels.

C. Second Law Efficiencies of STES

The fundamental thermodynamic reason for the use of an STES is to match the decreasing quality of energy in collector fluid stream, as available energy is extracted, to a set of tasks which can use a progressively lower temperature source of heat. If the matching is done properly, the second law efficiency can be increased over that which could be achieved if the processes were not cascaded but operated in parallel from the same heat source. The outputs of the process are obviously the same for either approach. An example illustrates this phenomenon.

Example Find the second law efficiencies of an STES consisting of a turbine and space heating system operating from a solar collector field at $T_{c0} = 400°C$ and in a STES operating for the same field. The turbine exhaust vapor is supplied at $T_{t,0} = 50°C$ to the condensing space heater which delivers heat at $T_h = 30°C$. The T/E ratio is 3.0, turbine efficiency $\eta_t = 18\%$, and the environmental temperature $T_0 = 0°C$. Ignore heat exchanger and pump irreversibilities.

Solution Second law efficiencies can be calculated from expressions given in Table 3.1. For the turbine superheated vapor assume that an effective constant specific heat applies. For a variable temperature heat source such as the turbine vapor, the availability consumed is given by the fluid enthalpy change multiplied by

$$1 - [T_0 \ln(T_{c0}/T_{t0})/(T_{c0} - T_{t0})]$$

[see Eq. (5.24a)]. The various process efficiencies are calculated below.

Turbine

$$\eta_2 = \eta_t \left[1 - \frac{T_0 \ln(T_{c0}/T_{t0})}{T_{c0} - T_{t0}}\right]^{-1} = 0.42$$

Space Heater [Eq. (3.2)]

$$\eta_2 = \eta_{1,h}[1 - (T_0/T_h)]/[1 - (T_0/T_{c0})] = 0.17$$

for $\eta_{1,h} = 1.0$ for simplicity.

STES

$$\eta_2 = \frac{1 + (T/E)[1 - (T_0/T_h)]}{(1/\eta_t)\{1 - [T_0 \ln(T_{c0}/T_{t0})/(T_{c0} - T_{t0})]\} + T/E(1 - T_0/T_{t0})(1/\eta_{1,h})}$$

by analogy with the preceding two expressions. For $T/E = 3$,

$$\eta_2 = 0.46$$

greater than the second law efficiency of either component taken alone. Note that the STES does not use the entire heat rate provided by the collector field since $T/E = 3.0$. To do so, the T/E ratio would need to be $(1 - \eta_t)/\eta_t = 4.56$.

D. Example System

Most STESs are quite complex and are specific to the process and T/E ratio to be met. Hence, no example system is described herein. References (E3), (G7), and (W8) contain detailed engineering designs of STESs for federal demonstration projects in the South.

V. LONG-TERM PERFORMANCE OF MEDIUM-TEMPERATURE SOLAR PROCESSES

In order to assess the economic viability of any solar process, its cumulative energy delivery over its economic life of the order of years or decades, must be known. It is very difficult to calculate this number in detail since (1) solar systems and their energy delivery are subject to the vagaries of local microclimate which can change on a time scale on the order of hours, and (2) future weather cannot be predicted at this level of detail. The standard approach used to estimate future performance of a solar system is to use a typical year of past weather data and assume that it will represent the future on the average, to engineering accuracy. The first difficulty above can be avoided by using long-time scale calculations instead of hourly or smaller scales. In order to use long-term means of solar and weather data, the statistical distribution of these data must be known. A long-term calculation method based on these ideas is the subject of this section.

A. Critical Solar Intensity Ratio X

The instantaneous efficiency equation for many solar collectors has been shown to be of the form

$$\eta_c = F(\eta_0 - U_c \, \Delta T^+/I_c), \qquad (\eta_c > 0) \qquad (5.43)$$

where ΔT^+ is the value of a collector to ambient temperature difference if positive and η_c is zero otherwise and F is a heat exchanger factor the expression for which depends on the definition of ΔT^+ [see Eqs. (4.43)–(4.47)]. It is technically correct but not always economical to operate the solar collector system if $\eta_c > 0$. In practice $\eta_c \geq \eta_{min} > 0$ is usually the system turn-on criterion since it is not worthwhile to operate collector loops for cases where η_c is very small.

Equation (5.43) can be used to determine the solar intensity level above which useful energy collection can take place. Solving Eq. (5.43) for I_c,

$$I_c \geq U_c \, \Delta T^+/(\eta_0 - \eta_{min}/F). \qquad (5.44)$$

A dimensionless critical intensity ratio X is generally used and since $\eta_{min} \ll 1$, for convenience the second term in the denominator above is dropped:

$$X \equiv U_c \, \Delta T^+/\eta_0 I_c \leq 1.0. \qquad (5.45)$$

X is seen to be the ratio of collector heat loss to absorbed solar flux at $\eta_c \equiv 0$, i.e., at the no-net-energy-delivery condition. In many cases the daily or monthly averaged daily critical intensity ratio $\bar{X}$ is of more interest and is defined as*

$$\bar{X} \equiv U_c \, \overline{\Delta T^+} \, \Delta t_c/\bar{\eta}_0 \bar{I}_c, \qquad (5.46)$$

where $\bar{\eta}_0$ is the daily averaged optical efficiency and $\overline{\Delta T^+}$ is the daily mean temperature difference *during collection*. These can also be expressed

$$\overline{\Delta T} = \frac{1}{\Delta t_c} \int_{t_0}^{t_0+\Delta t_c} (T_c - T_a) \, dt, \qquad (5.47)$$

---

* Note that U_c can be defined to include piping heat loss per collector arrays (B10).

$$\overline{\eta_0} = \frac{\int_{t_0}^{t_0+\Delta t} \eta_0 I_c \, dt}{\int_{t_0}^{t_0+\Delta t_c} I_c \, dt}, \tag{5.48}$$

and

$$\overline{I_c} = \frac{1}{\Delta t_c} \int_{t_0}^{t_0+\Delta t_c} I_c \, dt. \tag{5.49}$$

The collector cut-in time t_0 and cut-off time $t_0 + \Delta t_c$ are described shortly. The time $t = [0,24]$ h and is related to the solar hour angle h_s by $t = (180 - h_s)/15$; Δt_c is the collection period in hours. In Eq. (5.47) T_c can be collector surface, average fluid, inlet fluid, or outlet fluid temperature depending upon the efficiency data basis.

B. The Utilizability

Utilizability (the common but, unfortunately, rather cumbersome word) ϕ has been used to describe the fraction of solar flux absorbed by a collector which is delivered to the working fluid. On a monthly time scale

$$\bar{\phi} \equiv Q_u / F \bar{\eta}_0 \bar{I}_c < 1.0, \tag{5.50}$$

where the overbars denote monthly means and Q_u is the monthly averaged daily total useful energy delivery. $\bar{\phi}$ *is the fraction of the absorbed solar flux which is delivered to the fluid in a collector operating at a fixed temperature* T_c. The $\bar{\phi}$ concept does not apply to a system comprised of collectors, storage, and other components wherein the value of T_c varies continuously. The fixed temperature mode will occur if the collector is a boiler, if very high flow rates are used, if the fluid flow rate is modulated in response to flux variations to maintain a uniform T_c value, or if the collector provides only a minor fraction of the thermal demand. However, if the flow is modulated, note that the value of F (i.e., F', F_R) may not remain constant to engineering accuracy.

When T_c is not constant in time as in the case of a collector coupled to storage, the $\bar{\phi}$ concept cannot be applied directly. However, for most concentrators for CR > 10, the value of U_c is small and the collector is relatively insensitive to a *small* range of operating temperatures. To check this assumption for a particular process, values of $\bar{\phi}$ at the extremes of the expected temperature excursion can be compared.

The value of $\bar{\phi}$ depends upon many system and climatic parameters. However, Collares-Pereira and Rabl (Cl) have shown that only three are of first order—the clearness index $\bar{K}_T$ (See Chapter 2), the critical intensity ratio $\bar{X}$ [Eq. (5.46)], and the ratio r_d/r_T (See Chapter 2). The first is related to insolation statistics, the second to collector parameters and operating conditions, and the last to collector tracking and solar geometry.

Empirical expressions for $\bar{\phi}$ have been developed for several collector types (Cl). For nontracking collectors,

$$\bar{\phi} = \exp\{-[\bar{X} - (0.337 - 1.76\bar{K}_T + 0.55r_d/r_T)\bar{X}^2]\} \qquad (5.51)$$

for $\bar{\phi} > 0.4$, $\bar{K}_T = [0.3, 0.5]$, and $\bar{X} = [0, 1.2]$. Also,

$$\bar{\phi} = 1 - \bar{X} + (0.50 - 0.67\bar{K}_T + 0.25r_d/r_T)\bar{X}^2 \qquad (5.52)$$

for $\bar{\phi} > 0.4$, $\bar{K}_T = [0.5, 0.75]$, and $\bar{X} = [0, 1.2]$

The $\bar{\phi}$ expression for tracking collectors (CR > 10) is

$$\bar{\phi} = 1.0 - (0.049 + 1.44\bar{K}_T)\bar{X} + 0.341\bar{K}_T\bar{X}^2 \qquad (5.53)$$

for $\bar{\phi} > 0.4$, $\bar{K}_T = [0, 0.75]$, and $\bar{X} = [0, 1.2]$. Also

$$\bar{\phi} = 1.0 - \bar{X} \qquad (5.54)$$

for $\bar{\phi} > 0.4$, $\bar{K}_T > 0.75$ (very sunny climate), and $\bar{X} = [0, 1.0]$ for any collector type.

Equations (5.51)–(5.54) were developed using curve-fitting techniques emphasizing large $\bar{\phi}$ values since this is the region of interest for most practical designs. Hence, they should be considered accurate to $\pm 5\%$ only for $\bar{\phi} > 0.4$.

C. Example Calculation

To illustrate the use of the long-term method an example will be worked in stepwise fashion. The several steps used are

(1) Evaluate $\bar{K}_T$ from terrestrial $\bar{H}_h$ data and extraterrestrial $\bar{H}_{o,h}$ data.

(2) Calculate r_d/r_T for the concentration ratio and tracking mode for the collector.

(3) Calculate the critical intensity ratio $\bar{X}$ from Eq. (5.46) using a long-term optical efficiency value $\bar{\eta}_0$ and monthly average collector-plane insolation.

$$\bar{I}_c = (r_T - r_d\bar{D}_h/\bar{H}_h)\bar{H}_h. \qquad (5.55)$$

The collection time Δt_c may need to be determined in some cases for non-tracking, low-concentration collectors by an iterative method as described in the next section.

Example Find the energy delivery of a polar-mounted, parabolic trough collector operated for 8 h per day ($\Delta t_c = 8$) during March in Kabul, Afghanistan ($L = 34.5°N$). The collector has an optical efficiency $\bar{\eta}_o$ of 60%, a heat loss coefficient $U_c = 0.5$ W/m²°C, CR = 20, and heat removal factor $F_R = 0.95$. The collector is to be operated at 150°C. The mean, horizontal solar flux is 450 Ly/day (5.23 kW h/m² day) and the ambient temperature is 10°C.

Solution Following the three-step procedure above, the clearness index is calculated:

$$\bar{H}_{o,h} = 8.15 \quad \text{kW h/m}^2 \quad \text{(Table 2.9)}; \quad \bar{K}_T = 5.23/8.15 = 0.64.$$

The geometric factors r_d and r_T are calculated from expressions in Table 2.11:

$$r_T = (ah_{coll} + b \sin h_{coll})/d \cos L;$$
$$r_d = (h_{coll}/d)(1/\cos L + \cos h_{sr}(\alpha = 0)/(\text{CR})) - \sin h_{coll}/d(\text{CR}),$$

where

$$h_{coll} = 60° = 1.047 \quad \text{rad} \quad (h_{coll} = (\Delta t_c/2) \times 15°$$
$$\text{if the collection period is centered about solar noon}),$$
$$h_{sr}(\alpha = 0) = 90° = 1.571 \quad \text{rad},$$
$$a = 0.409 + 0.5016 \sin 30° = 0.66,$$
$$b = 0.6609 - .4767 \sin 30° = 0.42,$$
$$d = \sin 90° - 1.571 \cos 90° = 1.0,$$

in which case

$$r_T = (0.66 \times 1.047 + 0.42 \times \sin 60°)/1.0 \cos 34.5° = 1.28,$$
$$r_d = (1.047/1.0)[1/\cos 34.5° + \cos 90°/20] - \sin 60°/1.0 \times 20 = 1.31.$$

Finally the critical intensity ratio is

$$\bar{X} = U_c \, \overline{\Delta T}^+ \, \Delta t_c/\bar{\eta}_o \bar{I}_c$$

and the collector plane insolation $\bar{I}_c$ is from Eq. (2.51)

$$\bar{I}_c = (1.28 - 1.31 \times 0.34) \times 5.23 = 5.4 \quad \text{kW h/m}^2$$

so

$$\bar{X} = 0.5 \times (150 - 10) \times 8/0.6 \times 5400 = 0.173.$$

The utilizability $\bar{\phi}$ from Eq. (5.53) is

$$\bar{\phi} = 1.0 - (0.049 + 1.44 \times 0.64)(0.173)$$
$$+ 0.341 \times 0.64 \times (.173)^2 = 0.84.$$

Finally, the useful energy is

$$Q_u = F_R\bar{\eta}_0\bar{I}_c\bar{\phi} = 0.95 \times 0.6 \times 5.4 \times 0.84 = 2.58 \quad \text{kW h/m}^2 \text{ day.}$$

for the month of March on the average.

D. *Collection Period* (Δt_c)

The collection period Δt_c can be dictated either by optical or thermal constraints. For example, with a fixed collector, the sun may pass beyond the acceptance limit or be blocked by another collector and collection would then cease. Alternatively, a high efficiency, solar-tracking concentrator operating at relatively low temperature might be able to collect from sunrise to sunset. A third scenario would be for a relatively low concentration device operating at high temperature to cease to have a positive efficiency during daylight at the time that heat losses are equal to absorbed solar flux. In this case, the cutoff time is dictated by thermal properties of the collector and the operating conditions.

Collares-Pereira and Rabl (Cl) have suggested a simple procedure to find the proper value of Δt_c. Useful collection Q_u is calculated using the optical time limit first, i.e., $\Delta t_c = 2$ min $[h_{sr}(\alpha = 0), h_{sr}(i = 90)]/15$. Second, Q_u is calculated for a time period slightly shorter, say by one-half hour, than the optical limit; if this value of Q_u is larger than that for the first, optically limited case, the collection period is shorter than the optical limit. The time period is then further reduced until the maximum Q_u is reached.

The above method assumes that collection time is symmetric about solar noon. This is almost never the case in practice since the heat collected for an hour or so in the morning is required to warm the fluid and other masses to operating temperature. A symmetric phenomenon does not occur in the afternoon. If the time constant of the thermal mass in the collector loop is known, the collection period may be assumed to begin at t_0, $\Delta t_c/2$ h (from above symmetric calculation) before noon decreased by two or three time constants. Another asymmetry can occur if solar flux is obstructed during low sun angle periods in winter. It is suggested that r_T and r_d from Table 2.11 under asymmetric collection conditions be calculated from

$$r_T = [r_T(h_{s,\text{stop}}) + r_T(h_{s,\text{start}})]/2, \tag{5.56}$$

$$r_d = [r_d(h_{s,stop}) + r_d(h_{s,start})]/2, \tag{5.57}$$

where the collection starting and stopping hour angles account for transients, shading, etc., as described above:

$$h_{s,start} = 180 - 15t_0, \tag{5.58}$$

$$h_{s,stop} = 180 - 15(t_0 + \Delta t_c). \tag{5.59}$$

Example Calculations in the previous example were based on $\Delta t_c = 8$ h. Repeat for 10 h to see the effect of collection time if a symmetric collection period about noon is used.

Solution The values of r_T and r_d for $h_{coll} = 75° = 1.31$ rad are

$$r_T = (0.66 \times 1.31 + 0.42 \times \sin 75°)/1.0 \cos 34.5° = 1.98$$
$$r_d = (1.31/1.0)1/\cos 34.5° - \sin 75°/20 = 1.54.$$

The collector plane insolation is then

$$\bar{I}_c = (1.98 - 1.54 \times 0.34) \times 5.23 = 7.6 \quad \text{kw h/m}^2$$

and

$$\bar{X} = [0.5 \times (150 - 10) \times 10]/0.6 \times 7600 = 0.154.$$

Then $\bar{\phi}$ is 0.86 from Eq. (5.53) and the useful energy delivery is 3.7 kW h/m² day. Hence, it is worthwhile operating the collector for at least 10 h. The calculation can be repeated by the reader for an asymmetric case 4 h before noon and 6 h after to determine the effect of warm up. ∎

E. Long-Term Performance of Collector Systems with Storage

The previous section of this chapter described a method of predicting long-term performance of a solar collector operated at a temporally constant temperature. This situation is a good approximation of the operating conditions experienced by several types of generic thermal systems. Other systems, however, do not operate at constant temperature and the $\bar{\phi}$ method cannot be used. Although there is no simplified performance method now extant for varying temperature systems, Klein and Beckman (K11) have correlated some modeling results on collector–heat exchanger–storage subsystems coupled to a uniform, processlike load operating above some temperature T_{proc}. The method is called the $\bar{\phi},f$ chart by the authors and is described below. Although the method was developed for flat-plate collectors and uses a different $\bar{\phi}$ calculation method than used above, it can be applied equally well to concentrators (K12).

The calculation method requires first the determination of the utilizability $\bar{\phi}$ from Eqs. (5.51)–(5.54) above. This represents the maximum energy deliverable to a load at T_{proc}. When storage is present and collected solar heat is greater than the demand, the temperature of storage and hence the collector inlet temperature will rise. (The $\bar{\phi},f$ method applies only to well-mixed, sensible heat storage with liquid heat transfer fluids and storage media.) Hence, the monthly averaged, daily useful energy collected Q_u will be less than $F\bar{\eta}_0\bar{I}_c\bar{\phi}$ but storage may permit a greater fraction of the demand to be met by solar since the maximum amount of heat $F\bar{\eta}_0\bar{I}_c\bar{\phi}$ collectable may be more than can be used, depending on the demand amount. The $\bar{\phi},f$ method can be employed to find Q_u in a system with storage.

The technical basis for the $\bar{\phi},f$ method lies in the nondimensionalization of governing energy equations for a solar thermal system. Dimensionless groups, so identified, are used to correlate monthly thermal energy delivery-to-load for various systems simulated in various climates by an hourly time-scale computer model. The two dimensionless groups identified for use in the $\bar{\phi},f$ method, in addition to $\bar{\phi}$ ($\bar{\phi}$ is defined in this context relative to the minimum temperature acceptable to the process T_{proc} not relative to the collector temperature as in the previous section) are a solar parameter P_s, a measure of long-term solar gain by the collector receiver, and a collector heat loss parameter P_L, a measure of long-term heat loss at a fixed collector-to-ambient temperature difference of 100°C. This 100°C value does not restrict the generality of the results, however. In equation form,

$$P_s = F\bar{\eta}_0\bar{I}_c A_c N_d/L, \tag{5.60}$$

$$P_L = FU_c A_c N_h 100/L, \tag{5.61}$$

in which F (F', F_R, etc.) is given in Eqs. (4.44)–(4.47), L is the monthly thermal demand, and N_d and N_h are the number of days and hours in a month. The $\bar{\phi},f$ chart predicts the monthly solar load fraction

$$f_s(\bar{\phi}, P_s, P_L) \equiv Q_u/L. \tag{5.62}$$

Figure 5.25 is one $\bar{\phi},f$ chart and is entered with values of $\bar{\phi}P_s$ and P_L to give a monthly value of f_s. The calculation of $\bar{\phi}P_s$ and P_L is done once for each month of an average year and the totals added to give annual performance. An example below shows how the method is used. It is noted that $\bar{\phi}P_s$, the ordinate, is the ratio of maximum possible energy

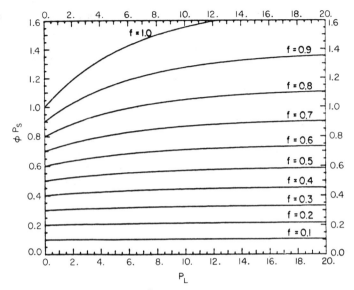

FIGURE 5.25   The $\bar{\phi},f$ chart used to calculate average, monthly solar fraction $f(f_s)$ of solar-thermal systems. [From (K11).]

delivered by a collector operating at fixed T_{proc}, $(F\bar{\eta}_0\bar{I}_c\bar{\phi}A_c)$, to the monthly load L. At values of $f_s > 0.4$, the $\bar{\phi},f$ curves are not independent of X since at progressively higher load fractions the average storage and collector temperatures are higher and collected solar heat per unit area smaller because collector efficiency is lower at higher temperature.

This $\bar{\phi},f$ chart is based upon several limiting assumptions which should be noted in interpreting f_s values:

(a)   The load L is distributed uniformly over the month between the hours of 6:00 and 18:00.

(b)   Storage amount is fixed at 350 kJ/°C m² (about 2 gal of H_2O/ft$_c^2$ or 84 1/m$_c^2$). See second equation in footnote p. 222 for a storage correction.

(c)   No energy is rejected from storage; therefore, the vessel is assumed to be designed for the peak temperature and pressure expected.

(d)   Storage is well mixed and no storage-to-load heat exchanger is used.

(e)   The load device uses solar heat at temperature-independent efficiency to meet the load L. Therefore, the load device cannot be a turbine, for example.

(f)   No parasitic heat losses from storage occur.

Some of these restrictions can be relaxed using work on the $\bar{\phi},f$ method conducted by Klein and his co-workers (K11). Users of the $\bar{\phi}$ methods must exercise caution in the proper choice of the $\bar{\phi}$ time scale. In the method presented here $\bar{\phi}$ and $\bar{I}_c$ are calculated over the collection period Δt_c, not over all daylight hours. The method developed by Klein uses $\bar{\phi}$ and $\bar{I}_c$ for all daylight hours. Although the Collares-Pereira $\bar{\phi}$ value can be used with the $\bar{\phi},f$ chart, the two methods of finding $\bar{\phi}$ itself must not be confused.*

Example Repeat the first example from the previous section for Kabul, Afghanistan, for a monthly averaged load of 260 kW h/day using a collector of 100 m². From the previous example recall that $F_R\bar{\eta}_0 = 0.57$. $F_R U_c = 0.475$ W/m² °C, $\bar{I}_c = 5.4$ kW h/m², and $\bar{\phi} = 0.84$. What is the effect of storage on energy delivery per unit collector area?

Solution First calculate P_s and P_L, then use the $\bar{\phi},f$ chart to find the solar fraction.

$$P_s = F_R\bar{\eta}_0\bar{I}_cA_cN_d/L = 0.57 \times 5.4 \times 100 \times 31/260 \times 31 = 1.18,$$
$$P_L = F_R U_c A_c N_h 100/L$$
$$= 0.475 \times 100 \times (31 \times 24) \times 100/(260 \times 31) \times 1000 = 0.44.$$

The value of f_s from the chart with $P_L = 0.44$ and $\bar{\phi}P_s = 0.99$ is $f_s = 0.97$. Therefore, the energy delivery per unit area is $0.97 \times 260/100 = 2.52$ kW h/m² day, nearly identical to the result using the $\bar{\phi}$ method. This is a result of the low value of U_c for the concentrator and its resulting insensitivity to temperature fluctuations above T_{proc}. The reader may repeat the calculations for a 200-m² collector with $U_c = 2.0$ W/m² °C to show that $\bar{\phi} = 0.43$, $P_L = 1.75$, $P_s = 2.37$, $f_s = 0.85$. The energy delivery per unit area is then 1.11 kW h/m² day compared with 1.32 kW h/m² day predicted by the $\bar{\phi}$ method. Hence the effects of storage reduce the unit energy delivery by 16% for the more lossy collector. ■

---

* The data from which Fig. 5.25 was constructed can be represented by the empirical equation (K11) $f_s = \bar{\phi}P_s - a[e^{3.85/s} - 1][1 - e^{-0.15P_L}]$ where $a = 0.015\left[\dfrac{Mc_p}{350\,A_c}\right]^{-0.76}$ in which $m = $ (kg), $c_p = $ (kJ/kg °C), and $A_c = $ (m²).

HIGH-TEMPERATURE
SOLAR PROCESSES

6

High-temperature solar processes are those which occur above 350–400°C. This definition is developed in Chapter 4 based on thermal performance of one- and two-axis tracking solar concentrators and the temperature level at which good efficiency can be achieved. In this chapter two-axis tracking concentrators are described including parabolic and spherical reflectors, Fresnel lenses, and Fresnel reflectors. These collectors are used for several thermal processes including power production using Brayton or Stirling engines, power production by the central-receiver system, solar furnaces, and thermionic conversion. These high-temperature solar processes are the most difficult technologically since precise tracking of collectors, special materials, and extra safety precautions are essential for reliable, long-term operation. In addition, this area of solar thermal technology is the least well developed. Hence, this chapter will be less design oriented than the others in this book. Physical principles and basic analytical results will be emphasized. In addition, this chapter will contain descriptions of high-temperature collectors, the several applications listed above, and a method for calculating

223

long-term performance of high-temperature collectors. The topics of storage, fluid flow, heat transfer, and ancillary component performance have been treated in Chapters 3 and 4 and will not be repeated in this chapter.

I. COMPOUND-CURVATURE SOLAR CONCENTRATORS

The collectors used for high-temperature solar processes are of the double-curvature type and require a tracking device with two degrees of freedom. Concentration ratios above 50 are generally used. Four collector designs seem to hold promise: spherical mirror, CR = 50–150; paraboloidal mirror, CR = 500–3000; Fresnel lens, CR = 100–1000; Fresnel mirror, CR = 1000–3000. These concentrator types will be discussed briefly in turn in this section.

A. Paraboloidal Concentrators

The surface produced by rotating a parabola about its optical axis is called a paraboloid. The ideal optics of such a reflector are the same, in cross section, as those of the parabolic trough described in Chapter 4. However, owing to the compound curvature, the focus occurs ideally at a point instead of along a line. Figure 6.1 shows a commercial paraboloidal collector.

The optical efficiency η_0 (defined relative to the direct-normal solar flux) of a paraboloid is the product of six terms:

$$\eta_0 = \rho_m \tau_r \alpha_r f_t \delta(\psi_1, \psi_2) F(\psi_3), \qquad (6.1)$$

where ρ_m is the mirror reflectance, τ_r the receiver cover (if any) transmittance, α_r the receiver absorptance, f_t the fraction of the aperture not shaded by supports and absorber, $\delta(\psi_1, \psi_2)$ the intercept factor depending on mirror slope errors ψ_1 and solar beam spread ψ_2, and $F(\psi_3)$ the tracking error where ψ_3 is the angle between the solar direction and the aperture normal.

Mirror reflectance and absorber optical properties have been discussed in Chapters 3 and 4. The optical intercept factor for a paraboloid is given by (D3)

$$\delta(\psi_1, \psi_2) = 1 - \exp[-(\pi D_r^2/4)/\sigma_y^2] \qquad (6.2)$$

FIGURE 6.1   Commercial paraboloidal solar concentrator. The receiver assembly has been removed from the focal zone for this photograph. (Courtesy of Omnium-G Corp., Anaheim, California.)

where D_r is the receiver diameter and σ_y^2 is the beam spread variance at the receiver. For a spherical segment receiver σ_y^2 is given by (D3)

$$\sigma_y^2 = 2A_a(4\sigma_{\psi_1}^2 + \sigma_{\psi_2}^2)(2 + \cos \phi)/3\phi \sin \phi, \tag{6.3}$$

where A_a is the aperture area and ϕ is the paraboloid rim half-angle. The beam variances σ_{ψ_1} and σ_{ψ_2} are described in Chapter 4.

For a flat receiver with paraboloid optics, σ_y^2 is given by (D3)

$$\sigma_y^2 = 2A_a(4\sigma_{\psi_1}^2 + \sigma_{\psi_2}^2)/\sin^2 \phi. \tag{6.4}$$

The concentration ratio of paraboloids can be determined easily from basic geometry. The aperture area is πR^2 where R is the aperture radius and the area for a spherical receiver is $4\pi R_r^2$ where R_r is the spherical receiver radius. If the receiver is only a spherical segment, not a complete sphere, the receiver area to be used below is ΩR_r^2, where Ω is the segment included solid angle, instead of $4\pi R_r^2$. The receiver radius for perfect optics is sized to collect all rays with the acceptance half-angle θ_{max}. Hence,

$$R_r = (R/\sin \phi) \sin \theta_{max}. \tag{6.5}$$

Therefore, the concentration CR is

$$CR = (\pi R^2)/[4\pi \times (R/\sin \phi)^2 \sin^2 \theta_{max}]$$

or

$$CR = \sin^2 \phi/(4 \sin^2 \theta_{max}). \tag{6.6}$$

The small effect of absorber shading of the mirror is ignored. Its inclusion would reduce CR given by Eq. (6.6) by only 0.25. For $\theta_{max} = \frac{1}{4}°$ (the sun's half-angle) CR $\sim$ 13,000 if $\phi = 90°$. For $\theta_{max} = \frac{1}{2}°$, CR $\sim$ 3300 and for $\theta_{max} = 1°$, CR $\sim$ 800. Note that concentrations for this configuration are one-fourth the thermodynamic limit, which is $(\sin \theta_{max})^{-2}$ for compound-curvature collectors [see Chapter 4, Eq. (4.12)].

For a flat absorber the concentration for perfect optics is

$$CR = \sin^2 \phi \, \cos^2(\phi + \theta_{max})/\sin^2 \theta_{max} \qquad (6.7)$$

where the small effect of mirror shading by the absorber is again ignored.

Example Calculate the concentration ratio and diameter of a flat receiver for a 10-m diameter paraboloid which is designed to accept 90% of the incident beam radiation. The rim half-angle is 90° and the mirror surface variance $\sigma^2_{\psi_1}$ is expected to be $(0.25°)^2$. Assume that the standard deviation for the sun's disk is 0.125°.

Solution From Eq. (6.2)

$$\pi D_r^2/4\sigma_y^2 = -\ln(1 - .90) = 2.303$$

and from Eq. (6.4)

$$\begin{aligned}
\sigma_y^2 &= 2A_a(4\sigma^2_{\psi_1} + \sigma^2_{\psi_2})/\sin^2 \phi \\
&= 2 \times (\pi/4 \times 10^2)[4 \times (0.25)^2 + (0.125)^2](\pi/180)^2/\sin^2 90° \\
&= 0.0127 \quad m^2 \quad (\sigma_y = 11.3 \quad cm).
\end{aligned}$$

Solving the first equation for D_r,

$$D_r = [2.303 \times 4 \times \sigma_y^2/\pi]^{1/2} = 0.193 \quad m.$$

The concentration ratio is

$$CR = A_a/A_r = \frac{\pi \times 10^2/4}{\pi(0.193)^2/4} = 2685. \quad \blacksquare$$

Note that the effect of tracking error ψ_3 can be included in the intercept factor $\delta(\psi_1, \psi_2)$ by defining an appropriate $\sigma^2_{\psi_3}$ and adding it to the slope and solar image variances in Eqs. (6.3) and (6.4).

The thermal losses from a paraboloid are quite small and primarily radiative. Since the absorber area is so small, it is generally not worthwhile to use any type of convection suppressing cover. For example, consider the thermal losses from a planar absorber of CR = 1500 paraboloid with a receiver surface temperature of 600°C. For a cavity absorber ($\epsilon_{ir} \sim 1.0$) the radiation heat loss is about 20 W/m² aperture if the ambient temperature is 60°C. For a typical convection coefficient in light winds of 25 W/m² °C, the convection loss is about 10 W/m² aperture. If the insolation is 900 W/m² and the optical efficiency 65%, the total heat loss at 600°C represents less than 6% of the absorbed flux. Stated another way, the collector loss coefficient U_c at 600°C is only 0.064 W/m² °C. The performance of high-concentration paraboloids is therefore much more sensitive to optical properties than to thermal losses.

B. Spherical Concentrators

A second type of compound-curvature collector uses spherical geometry instead of parabolic geometry. Figure 6.2c shows ray traces in a plane of symmetry for normal incidence (two-axis tracking) on a spherical concentrator. Spherical aberration is seen to be present and causes the reflected flux to be along a line instead of at a point as is the case for other compound-curvature mirrors. It is also clear that rays intercepted near the pivot end of the absorber intercept the absorber at very large incidence angles. Therefore, an absorber envelope with very low re-

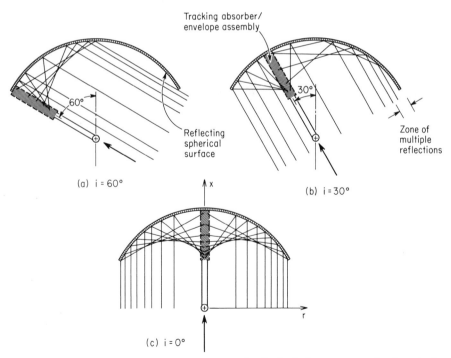

FIGURE 6.2   Ray traces for a fixed spherical concentrator for three incidence angles: (a) normal incidence for two-degree-of-freedom tracking of the mirror; (b) and (c) off-normal incidence for a fixed mirror and a moving absorber.

flectance or no envelope at all is used to avoid severe penalties to the optical efficiency. If no cover is used the optical efficiency η_0 is given by

$$\eta_0 = \rho_m \bar{\alpha}_r f_t \delta(\psi_1, \psi_2) F(\psi_3), \tag{6.8}$$

where the receiver absorptance $\bar{\alpha}_r$ is averaged over all absorber incidence angles.

Figure 6.2 shows that the optical path length from the mirror to the absorber depends on the point of incidence of rays on the aperture plane. As a result the flux distribution is highly nonlinear at the focal line. It can be shown that the absorber flux distribution $q_a(x)$ is given by (K9)

$$q_a(x) = \rho_m \frac{2rI_b}{D_r} \frac{dr}{dx}, \tag{6.9}$$

where x is measured from the geometric center of curvature toward the reflector and r is measured in an orthogonal coordinate system as shown in Fig. 6.2. The value of x, the point of ray intersection on the absorber, is given by (K9)

$$\frac{x}{R_0} = \frac{1}{2\sqrt{1 - (r/R_0)^2}} \left[1 + \frac{D_r}{2R_0} \left(\frac{R_0}{r} - \frac{2r}{R_0} \right) \right], \tag{6.10}$$

where R_0 is the mirror radius of curvature and D_r is the receiver diameter. Substituting Eq. (6.10) into Eq. (6.9) will give the flux distribution $q_a(x)$, which has very large values near the pivot end of the absorber and relatively small values near the reflecting surface end. The highly nonlinear flux distribution makes a single calculation of the intercept function $\delta(\psi_1, \psi_2)$ impossible in closed form. A numerical procedure can be used, however.

The concentration ratio of a spherical concentrator can be calculated from the definition of CR:

$$CR = A_a/A_r = \pi R_0^2 \sin^2 \phi / \pi D_r L_r. \tag{6.11}$$

Since $L_r = R_0/2$ (the receiver length),

$$CR = (2R_0/D_r) \sin^2 \phi. \tag{6.12}$$

If the absorber is a cylinder and is to intercept all singly reflected rays within the acceptance half-angle θ_{max},

$$D_r = 2R_0 \sin \theta_{max} \tag{6.13a}$$

and

$$CR = \sin^2 \phi / \sin \theta_{max} \tag{6.13b}$$

if the acceptance half-angle is $\frac{1}{4}°$ for $\phi = 90°$, CR = 229; for $\theta_{max} = \frac{1}{2}°$, CR = 115 for perfect optics. It is obvious that the spherical concentrator cannot achieve concentration ratios approaching those of paraboloids owing to spherical aberration and hence this concentrator is much more sensitive to operating temperature than the point-focus type.

Since the efficiency of this concentrator is lower than that of other compound-curvature devices, because of smaller CR values, the unit cost of solar heat will be higher. One method of reducing the unit cost is to use the concentrator with a *fixed mirror,* thereby eliminating the expensive mirror tracking and structural components (S9). The collector is kept in "focus" by tracking only the absorber by aiming it directly at the sun's center. However, the effective aperture in this fixed mirror mode is reduced by the cosine of the incidence angle. Hence, the saving in cost by eliminating the reflector tracker is partly offset by reduced energy capture. If a value of rim half-angle $\phi = 60°$ is used as shown in Fig. 6.2 and if the fixed aperture faces the sun directly at noon, then the day-long cosine penalty will reduce the captured flux by about 17%. Of course, if the aperture does not directly face the sun at noon, the cosine loss will be larger.

Figure 6.3 shows a spherical concentrator operating in the fixed mode.

The thermal performance of a spherical collector can be calculated using an approach analogous to that used for other line-focus devices considered in Chapter 4. However, since the flux distribution is very

FIGURE 6.3 A spherical concentrator operating in the fixed-aperture mode. The absorber tube is defocused and the smoke pattern shows the caustic curve or envelope of reflected rays. [Courtesy of B. Authier, Laboratoire d'Astronomie Spatiale (CNRS), Marseille.]

nonlinear along the absorber, a numerical finite difference method must be used to calculate useful energy delivery accurately (K9). The use of a single, average absorber temperature is not satisfactory for careful design.

C. *Other Compound-Curvature Collectors*

Circular Fresnel lens collectors are designed in the same way as linear lenses considered in Chapter 4. The concentration achievable is roughly the square of that for linear lenses of the same facet size and shape. Therefore, thermal losses are reduced by analogy with the paraboloidal collector described earlier in this section. The optical efficiency η_0 of a Fresnel lens is given by

$$\eta_o = \tau_l \alpha_r \delta(\psi_1, \psi_2) F(\psi_3) \tag{6.14}$$

where the terms are defined in the usual way; τ_l is the lens transmittance.

From Chapter 4 it will be recalled that Fresnel lenses are only one-fourth as sensitive to slope errors ψ_1 as are reflecting concentrators. Absorptance α_r and tracking $F(\psi_3)$ effects are of the same order as for reflecting designs. Therefore, since no mirror reflectance effect is present and since $\delta(\psi_1, \psi_2)$ is larger than for reflecting concentrations, the optical efficiency of the lens design is higher by about 15%. However, the chromatic aberration in lenses limits CR to the order of 1000 (F4).

Figure 6.4 shows a point-focus Fresnel lens collector array under construction. The most common lens material is plastic, which is inexpensive, relatively light, and can be precisely cast. Of course, the plastic must be immune to ultraviolet degradation.

Fresnel mirror concentrators consist of independently tracking mirror segments which focus sunlight onto an absorber surface. Fresnel mirrors are used in both solar furnace applications and for central receiver solar-thermal power plants. The Fresnel mirror collector is an integral part of the central receiver power plant and will be described later in this chapter.

II. SOLAR FURNACES

A solar furnace is a large concentrator used to achieve very high temperatures usually used for research or batch processing of refrac-

FIGURE 6.4   Point-focus Fresnel lens array under construction. Many lenses focus on receivers which are then connected to provide the total array output. The entire array tracks about two axes. (Courtesy of Sandia Laboratories.)

tory materials. An absorber with a throughflow of a working fluid is not present but specimens are exposed to very high flux levels in a noncontaminating environment. Most of the work on large solar furnaces has been done since World War II. However, Lavoisier melted platinum in a small furnace in the 18th century. Lavoisier's lens-type furnaces are not amenable to large-scale use owing to the enormous weight of large lenses. Most modern furnaces use reflecting concentrating devices.

A. Single-Heliostat Furnaces

Professor Felix Trombe of the Centre National de la Recherche Scientifique (CNRS) began the design and development of the world's first, large furnace in 1948. The furnace, located at Montlouis in the Pyrennees in the south of France, was completed in 1952 and produced 45 kW$_t$ in bright sun. A single heliostat—a sun-tracking, *planar* reflector consisting of many small, coplanar facets—was used. The sun's rays were reflected from the heliostat onto a parabolic mirror which accomplished the concentration effect. An average flux concentration of about 50 was achieved with an optical efficiency of about 55% (W2).

The Montlouis heliostat contained 540 small contoured facet mirrors with a collective area of 135 m². The parabolic concentrator had a focal length of 6 m and consisted of 3500 small planar mirrors each 16 cm². The total parabolic mirror area was about 90 m². With 900 to 950 W/m² of solar flux, the peak heat flux was measured to be 1.2 × 10⁶ W/m² at the center of the flux pattern at the receiver (focal) plane.

The Montlouis furnace design concept was used by various other groups for construction of other single-heliostat furnaces in Japan, France, and the U.S. The U.S. Army Quartermaster Corps built a furnace in Natick, Massachusetts, in 1958. It was similar to the Montlouis furnace in many ways except that a spherical reflector, not a parabolic one, was used. As a result, the thermal power was less than that of the Montlouis furnace by 25% owing to spherical aberration even though the heliostat surface area was nearly the same at both facilities. The Natick furnace was moved to the White Sands Missile Range in 1973—see Fig. 6.5.

In 1962 a single heliostat furnace was built at Tohuku University at Sendai in Japan. It had the largest heliostat—214 m²—of any single-heliostat furnace and used a parabolic reflector with a 3.2-m focal length. The optical efficiency of this device is about 50% and an average flux concentration of about 40 has been achieved.

The most recent single-heliostat furnace to be built was for the French army in 1972. It is located at Odeillo, Font-Romeu near Montlouis in the Pyrennees. A spherical mirror was used with a 638-m² heliostat. Again, spherical aberration results in reduced performance per

FIGURE 6.5   White Sands solar furnace showing the single heliostat on the right, the concentrator on the left, and flux control shutters in the center. The receiver is located to the right of the shutters in the small, elevated building.

unit heliostat area vis-à-vis the original Montlouis design. Both the French army furnace and the White Sands furnace were developed to simulate the thermal radiation environment produced by a nuclear explosion. The shutters shown in Fig. 6.5 can be used to produce flux variations which model a large range of nuclear devices.

Table 6.1 shows the important characteristics of the four major single-heliostat furnaces.

B. *Multiple Heliostat Furnaces*

Experience gained at the Montlouis site enabled CNRS scientists to design and build the largest solar furnace in the world at the nearby village of Odeillo, Font-Romeu, located at 1800 m above sea level (T6). On the average, about 1200 h of sunlight per year at a level above 700 W/m² are available at this site (T4). The furnace was completed in 1970 after a decade of design, construction, and alignment. A cross section of the furnace is shown in Fig. 6.6 and a photo in Fig. 6.7.

63 7.5-m × 6-m heliostats each made up of 180 facets are used. Total heliostat mirror area is slightly more than 2800 m². As shown in Fig. 6.6, the mirrors are located on eight levels corresponding to the eight floors of the paraboloid concentrator support structure. Therefore, heliostat tracking is such that the sun's rays are reflected horizontally onto a specific portion of the fixed parabolic reflector. Heliostat tracking is accomplished with a 2-df mechanical drive controlled by two pairs of photo diodes in an optical tube 100 cm long and 1.2 cm in diameter. The tracking accuracy is about 1′ of angle.

Paraboloidal optics are used in the fixed reflector, which has a focal length of 18 m. The mirror measures 40 m high and 54 m wide; 9500 mechanically contoured mirror facets make up the reflector and the mirror measures 1920 m² in effective area. About 60% of the flux at the reflector aperture intercepts the absorber in a 0.10-m² zone with a resulting local flux concentration ratio of about 20,000.

Table 6.2 summarizes the Odeillo furnace performance. Various receiver surface diameters are shown from 2–40 cm with corresponding flux amounts and temperatures under conditions of bright sun. It is seen that temperatures up to 3825°C can be achieved on a sunny day at the center of the focal plane. For a 40-cm diameter zone, the mean temperature is about 2950°C. The diameter of the sun corresponds to a 17-cm diameter zone and the mean temperature is 3600°C in this portion of the focal plane. Figure 6.8 shows an isoflux plot measured at the focal plane.

TABLE 6.1

Large Solar Furnace Properties and Performance, 35–50-kW Thermal Capacity[a]

	CNRS, Montlouis, France	Tohoku University, Sendai, Japan	French Army, Odeillo, Font-Romeu, France	U.S. Army, nuclear weapon effects Lab.[b] White Sands Missile Range, New Mexico
Date first operated	1952	1962	1972	1974
HELIOSTAT				
Size				
Meters	10.5 × 13	14 × 15.5	13.2 × 17.5	11 × 12.2
Feet	34 × 43	46 × 51	43 × 57	36 × 40
Number of mirror facets	540	238	638	356
Mirror size				
Centimeters	50 × 50	90 × 100	50 × 50	62 × 62
Inches	19.7 × 19.7	35.4 × 39.4	19.7 × 19.7	24.4 × 24.4
Total mirror area				
Square meters	135	214	159.5	137
Square feet	1453	2306	1717	1473

CONCENTRATOR	Parabolic	Parabolic	Spherical	Spherical
Configuration	Parabolic	Parabolic	Spherical	Spherical
Size				
Meters	9 × 11	10 (diam.)	10 × 10	8.5 × 8.5
Feet	29.5 × 36	33 (diam.)	33 × 33	28 × 28
Focal length				
Meters	6	3.2	10.75	10.9
Feet	19.7	10.5	35.3	35.7
Number of mirrors	3500	181	384	180
Mirror size				90
Centimeters	16 × 16	80 × 75	50 × 50	62 × 62 64 × 66
Inches	6.3 × 6.3	31 × 30	19.7 × 19.7	24.4 × 24.4 25.2 × 26
Total mirror area				
Square meters	89.6	78.5	96	72.6
Square feet	964	845	1033	781
THERMAL PERFORMANCE[c]				
Total thermal power	45 (est.)	35 (est.)	42.5	32 (est.)
Thermal efficiency (%)	55 (est.)	50 (est.)	48	50 (est.)
Max. heat flux (W/cm²)	1200		580	400

[a] From "Energy Technology Handbook" by J. D. Walton. Copyright © 1977. Used with permission of McGraw-Hill Book Company.

[b] Previously operated by the U.S. Army, Quatermaster Corps, Natick, Massachusetts, 1958.

[c] Based on insolation = 900 to 950 W/m².

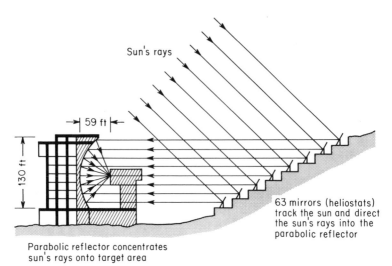

FIGURE 6.6   Cross section of the Odeillo 1000-kW$_t$ furnace.

FIGURE 6.7   Photograph of the Odeillo furnace showing the south-facing receiver tower and the eight-story parabolic reflector.

TABLE 6.2

Summary of Thermal Performance of the Odeillo Furnace[a,b]

Diameter of radiation:								
cm		2	6	12	16.8[c]	20	30	40
(in.)		(0.79)	(2.36)	(4.72)	(6.61)	(7.87)	(11.81)	(15.75)
% of total energy in area of								
radiation		0.5	4.50	15.5	27	35	58	75
Energy in area of radiation, kW:		5	45	155	270	350	580	750
Minimum	Watts cm^{-2}	1600	1472	1200	912	800	400	192
Heat	Cal cm^{-2} s^{-1}	383	352	287	218	191	96	46
flux	(Btu ft^{-2} s^{-1})	(1410)	(1297)	(1057)	(804)	(705)	(352)	(169)
Average	Watts cm^{-2}	1600	1595	1370	1215	1115	820	595
Heat	Cal cm^{-2} s^{-1}	383	381	328	294	266	196	142
flux[d]	(Btu ft^{-2} s^{-1})	(1410)	(1405)	(1207)	(1071)	(982)	(723)	(524)
Temperature of radiation:								
Minimum	°C	3825	3740	3540	3285	3170	2625	2140
	°F	(6915)	(6765)	(6405)	(5945)	(5740)	(4755)	(3885)
Average	°C	3825	3805	3665	3585	3465	3185	2950
	°F	(6915)	(6880)	(6630)	(6485)	(6270)	(5765)	(5340)

[a] From "Energy Technology Handbook" by J. D. Walton. Copyright © 1977. Used with permission of McGraw-Hill Book Company.

[b] For incident energy of 950 W m^{-2} ± 5%.

[c] Diameter of solar image.

[d] Heat flux calculated from water calorimeter and radiometry measurements.

Another landmark multiple heliostat furnace was built by G. Francia in Genoa in 1968 (F3). The 135-kW$_t$ furnace shown in Fig. 6.9 consists of 271 1.0-m diameter heliostats which reflect sunlight onto a downward-facing cavity receiver. The flat heliostats are mechanically linked and are moved by a single, centralized clock-drive mechanism. Tests in bright sunshine show that a maximum thermal efficiency of 68% can be achieved corresponding to 135-KW$_t$ delivery. Table 6.3 compares the major design parameters and performance data for the Genoa and Odeillo furnaces.

Recently, two solar furnaces of the types described above have been built in the U.S. A Francia-type furnace rated at 400 KW$_t$ has been operated at Georgia Institute of Technology in Atlanta for several years. It is a test facility for solar thermal receivers to be used in several types of systems described in this chapter and in Chapter 5.

A second furnace the Central Receiver Solar Thermal Test Facility (CRSTTF) rated at 5 MW$_t$ is located near Albuquerque, New Mexico. The furnace is an experimental facility used for evaluating heliostat and receiver designs to be used in the U.S. central receiver,

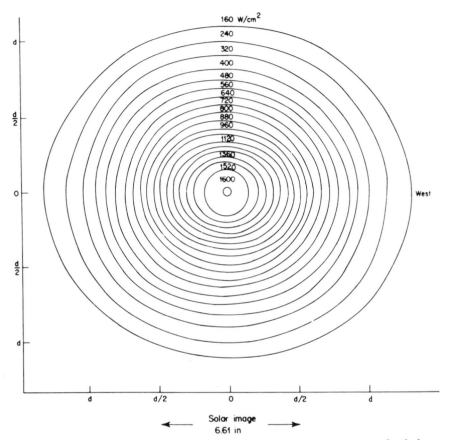

FIGURE 6.8   Isoflux lines measured on the Odeillo furnace focal plane (tilted at 25° from the vertical) where *d* is the solar image diameter. [From (W2).]

thermal power program. Since the purpose of the 5-MW$_t$ facility is to test components and materials, not to produce power, it is considered to be a furnace. However, the central receiver power plants described later in this chapter are similar in design to the 5-MW$_t$ test facility. Figure 6.10 shows the Albuquerque facility.

In a preliminary test at the 1.7-MW$_t$ level 71 1.4-m² heliostats, each with 25 facets, produced temperatures above 1650°C on a 1.8-m diameter target plane. A 60-cm × 90-cm hole was melted in a 6-mm steel target in less than 2 min. The focal point was 35 m above the ground, about halfway up the receiver tower.

The CRSTTF tower has four test bays at various positions above the surface. The topmost one is used for the largest experiments re-

FIGURE 6.9   Francia multiple heliostat 135-kW$_t$ solar plant located at St. Ilario near Genoa. Sunlight is reflected from the tracking circular mirrors into the downward-facing cavity receiver located above the array on the tower.

quiring the full complement of 222 heliostats. Heat collected by receivers under test is rejected from a dry cooling tower which can use air, a liquid coolant, or steam as the test receiver working fluid.

The planar heliostats used at the CRSTTF consist of 5 × 5 arrays of 1.22-m² mirrors. The total reflective surface of the heliostats is 37.2 m². Silvered float glass is used as the reflective medium. As described later in this chapter it is sometimes required to focus heliostats to achieve an additional increment of concentration. At the CRSTTF this is accomplished by a separate structure at each mirror. A threaded stud at the center of each mirror facet is used with a diagonal structure to provide the slight curvature needed for focusing. Heliostat alignment is accomplished by use of a laser and the alignment process is automated. The total flux at the receiver from all mirrors on peak conditions is about 2.5 MW/m². Additional focusing and secondary concentration can increase local fluxes above this level. Figure 6.11 shows a CRSTTF heliostat.

The spatial distribution of flux at the receiver aperture is an important factor in the design of receivers. The CRSTTF has developed a system to measure aperture flux for any number of active heliostats

TABLE 6.3

Multiple-Heliostat Solar Furnace Properties and Performance[a]

	CNRS 1000-kW solar furnace, Odeillo-Font-Romeu, France	University of Genoa, Genoa, Italy
HELIOSTAT		
Number of heliostats	63	271
Heliostat Size:		
Meters	6 × 7.5	1 (diameter)
Feet	19.7 × 24.6	3.3 (diameter)
Number of mirrors in each heliostat	180	1
Mirror size:		
Centimeters	50 × 50	100 (diameter)
Inches	19.7 × 19.7	39.4 (diameter)
Mirror area in each heliostat:		
Square meters	45	0.785
Square feet	484	2.69
Total number of heliostat mirrors	11,340	271
Total heliostat area:		
Square meters	2,835	212.7
Square feet	30,515	2,291
CONCENTRATOR		
Size:		
Meters	40 × 54	
Feet	131 × 177	
Number of mirrors:	9,500	No concentrator
Mirror size:		
Centimeters	45 × 45	
Inches	17.7 × 17.7	
Total mirror area:		
Square meters	1,923	
Square feet	20,707	
THERMAL PERFORMANCE[b]		
Total thermal power (kW)	1000	130
Average thermal power/8 hours (kW)		115
Maximum thermal efficiency (%)	58	68
Maximum heat flux (W/cm²)	1600	30

[a] The data shown for the Univ. of Genoa installation are for Unit No. 1, first operated in 1967. A second comparable unit was first operated in 1972. The mirror element of Unit No. 2 is 86 cm in diameter. The total thermal power is 100 kW, average thermal power per 8 h is 88 kW, and maximum thermal efficiency is 70%. From "Energy Technology Handbook" by J. D. Walton. Copyright © 1977. Used with permission of McGraw-Hill Book Company.

[b] Based on insolation = 900 to 950 W/m².

FIGURE 6.10   Photograph of the 5-MW$_t$ (6.5-MW$_t$ peak) solar test facility located at Sandia Laboratories showing the north heliostat field and the receiver tower.

from one to the full field. The flux system is a rectangular frame which can be mounted at several locations. When a flux scan is needed, a chilled water-cooled bar traverses the aperture and a real time display of flux at 10-cm steps is generated from signals from the heat flux spectral photon gage on the bar. A full meteorological station is also at the CRSTTF site.

The first receiver test at the CRSTTF was carried out in late 1978. The device tested was a receiver manufactured by Boeing for use in the Electric Power Research Institute solar power program.

C.   Tracking Requirements

Heliostats used on most furnaces other than the Francia type direct sunlight horizontally to the paraboloid reflector. For such furnace systems the altitude and azimuth angles *of the line normal to the heliostat* and measured as described in Chapter 2—α_h and a_h—are given by (S5)

FIGURE 6.11   A CRSTTF heliostat assembly consisting of 25 1.2-m² facets. Each heliostat can be made slightly concave to more sharply focus the reflected beam onto the receiver. (Courtesy of Sandia Laboratory.)

$$\sin \alpha_h = \sin \alpha/[2(1 + \cos \alpha \cos a_s]^{1/2} \qquad (6.15)$$

and

$$\sin a_h = \tan \alpha_h \cot \alpha \sin a_s \qquad (6.16)$$

where α and a_s are the solar altitude and azimuth angles from Chapter 2.

D.   Solar Furnace Applications

Solar furnaces were originally conceived as sources for very high temperatures for materials research. Specifically they are used for (W2) high-temperature chemistry, high-temperature property measurement, thermal shock research, solar power plant receiver testing. Fused ceramic materials such as zirconia (ZrO_2) aluminum (Al_2O_3), silica (SiO_2) have been produced in large quantities at Odeillo at the rate of 150–200 kg/h (T5). Although solar furnaces were not economical sources for those materials in the past, they may become economical as fossil fuel prices rise. Figure 6.12 shows a cross section of a crucible used for refractory research.

The Odeillo furnace has recently (1977) been operated as a 75-kW$_e$ solar-thermal power plant. Although the efficiency is quite low, (7–8%) considerable experience has been collected on the interface problems between solar plants and a large utility grid of the Electricité de France. A schematic diagram of the plant is shown in Fig. 6.13. The heat transfer fluid serves also as the storage medium in the accumulator.

Temperature measurements inside a solar furnace cannot be made directly. Therefore, an indirect method is used. A time–

FIGURE 6.12   Section of crucible used in refractory materials research at the Odeillo furnace.

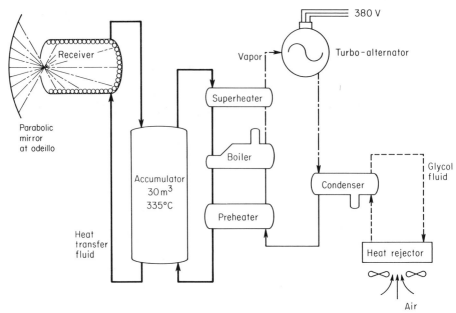

FIGURE 6.13   Schematic diagram of the CNRS power plant at Odeillo, Font-Romeu.

temperature profile can be measured with a pyrometer beginning immediately after the solar radiation is cut off. A rapid response pyrometer is used for the purpose, but the results are not entirely satisfactory. Another method proposed by Sakurai (S8) uses infrared pyrometry beyond the long-wavelength limit of the solar spectrum. By comparing radiation at 8–15 μm (the emittance of most refractories is about 1.0 in this range) to a reference source in an electric oven, the sample temperature can be determined.

III. CENTRAL RECEIVER SOLAR-THERMAL POWER SYSTEMS

An alternative to the transmission of solar-heated fluid through pipe networks, as in distributed power systems, is the transmission of sunlight itself from an array of heliostats to a central receiver. At the receiver site, heat and power production take place. Central receiver solar thermal power (CRSTP) plants are an outgrowth of the multiple-heliostat solar furnaces described in the previous section. Several CRSTP plants have been designed by various firms in the U.S. Although none at the 100-MW$_e$ scale will be built until the mid-1980s at the earliest, some general results of these design studies are given in this section but not all details of each design will be described. A thorough review of the formative years of the U.S. CRSTP program is contained in (V2).

Figure 6.14 shows an artist's schematic representation of a 50-MW$_e$ plant. In this particular design a square field of heliostats is used. However, various asymmetric or circular fields have also been specified in other designs. The major design variables include 0.5 km² of heliostats (15,400 units) covering 1.3 km² of land, an optical efficiency of 62%, and an overall plant efficiency of roughly 27% (44% cycle efficiency) at 815°C using a Brayton cycle.

Two generic types of power production systems can be identified—the closed Rankine cycle plant and the open or closed Brayton cycle plant (see Chapter 3). Rankine cycles are common in the electric industry and are described in Chapter 5. Open Brayton cycles are similar to gas turbines used in the utility industry for peaking power or in

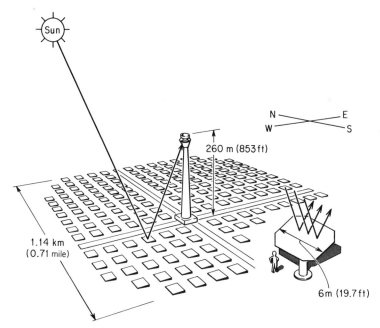

FIGURE 6.14   Schematic diagram of a 50-MW$_e$ central receiver power plant. A single heliostat is shown in the inset to indicate its human scale. [From Electric Power Research Institute (EPRI).]

small generating plants. The closed Brayton cycle is not common in the U.S. but has been infrequently used in Europe in the 10–50-MW$_e$ range.

The major components of a CRSTP include the heliostat field, receiver subsystem, thermal storage, engine-generator plant, and other ancillary components and controls. In this section the heliostat field (Fresnel mirror) receiver and generator are described. High-temperature thermal storage has been described in Chapters 4 and 5.

A. The Heliostat Field

The purpose of a heliostat, as in the case of a solar furnace, is to direct beam radiation which is incident on the heliostat onto the surface of a receiver in an accurate and reliable fashion. Since the Fresnel mirror, which the heliostats represent, is made up of several thousand facets, flat elements can be used. However, additional concentration can be achieved if a slight concavity is added to each heliostat. For smaller plants of the order of 10 MW$_e$ focused heliostats are probably required.

Many heliostat designs are possible. Four concepts which have been built in prototype form in the U.S. are shown in Figure 6.15. The Honeywell design module consists of four mechanically linked reflectors which track with two degrees of freedom. The Boeing concept uses an aluminized Mylar reflector protected by a Tedlar dome whereas the McDonnell–Douglas heliostat uses a glass, front-surface, silvered octagonal mirror with acrylic coating for protection. Finally, the Martin–Marietta approach uses nine fixed focus mirrors attached to the

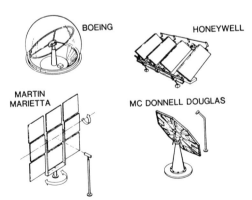

FIGURE 6.15 Heliostat designs proposed for a 10-MW$_e$ prototype CRSTP plant by four U.S. contractors.

heliostat frame. Each concept has been tested and is functional and provides an adequate level of optical efficiency.

In the scope of this book it is not possible to treat mirror field design in detail. However, some macroscale features will be described. The law of specular reflection requires that the normal to the heliostat surface bisect the angle between the sun's rays and a line from the heliostat surface to the target. This tracking mode is somewhat different from that for a heliostat directing a ray horizontal and due south to a specific zone of a solar furnace [Eqs. (6.15) and (6.16)]. Since rotations about two axes are not commutative, a mirror orientation is not unique but depends upon the rotation sequence.

In vector notation the angle θ between a heliostat to receiver vector $\mathbf{H}$ and the heliostat normal $\mathbf{n}$ is given by the scalar product

$$\cos \theta = \mathbf{n} \cdot \mathbf{H}/|\mathbf{n}|\,|\mathbf{H}|. \tag{6.17}$$

This equation is equated to Eq. (2.17), which gives the incidence angle between a solar ray $\mathbf{S}$ and the heliostat normal $\mathbf{n}$. The second equation relating the two heliostat rotation angles (surface tilt and azimuth angles) is the requirement that the incident ray, $\mathbf{S}$, $\mathbf{n}$, and $\mathbf{H}$ all be coplanar, i.e.,

$$(\mathbf{S} \times \mathbf{n}) \cdot \mathbf{H} = 0. \tag{6.18}$$

Unit vectors $\mathbf{S}$, $\mathbf{n}$, and $\mathbf{H}$ are shown in Fig. 6.16.

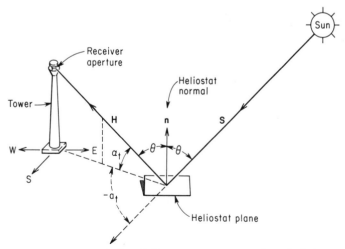

FIGURE 6.16 Sun–heliostat–tower geometry for calculating tracking requirements for CRSTP heliostats. Tower unit vector **H** altitude and azimuth angles are shown.

Riaz (R13) has solved Eqs. (6.17), (2.17), and (6.(8) to provide the components of **n**, **H**, and **S**:

$$\mathbf{n} = \begin{bmatrix} \cos \alpha_n & \sin a_n \\ -\cos \alpha_n & \cos a_n \\ \sin \alpha_n \end{bmatrix}, \tag{6.19}$$

$$\mathbf{H} = \begin{bmatrix} \cos \alpha_t & \sin a_t \\ -\cos \alpha_t & \cos a_t \\ \sin \alpha_t \end{bmatrix}, \tag{6.20}$$

$$\mathbf{S} = \begin{bmatrix} \cos \alpha & \sin a_s \\ -\cos \alpha & \cos a_s \\ \sin \alpha \end{bmatrix} \tag{6.21}$$

where α and a_s are the solar altitude and azimuth angles, α_t and a_t are the fixed tower-vector altitude and azimuth angles shown in Fig. 6.16, and α_n and a_n are the heliostat-normal altitude and azimuth angles measured in the same way as α and a_s. The Cartesian coordinate system in which all vectors are defined is oriented such that (x,y,z) correspond to east, north, and vertical.

A heliostat positioned to reflect flux onto the central receiver can generally cast a shadow on other heliostats "behind" it or can interrupt—"block"—light reflected from another heliostat. It is not always cost effective to completely eliminate shadowing and blocking in winter when the sun is low in the sky. If this were done, the optical performance in summer would be penalized excessively and an unnecessarily sparse field would result. An economic analysis is necessary to resolve this question. If Φ is the ratio of mirror-to-ground area, Vant Hull and Hildebrandt (V1) have shown that

$$\Phi = 1.06 - 0.23 \tan \phi \tag{6.22}$$

where ϕ is the Fresnel mirror (i.e., heliostat field) rim half-angle. A typical value of Φ is 40–50%.

Depending upon the instantaneous position of the sun, a given heliostat can direct either more or less energy to the receiver than strikes an equivalent horizontal surface. P is defined as the ratio of redirected flux for perfect optics and reflectance ($\rho_m = 1.0$), to horizontal flux. It has been shown that P is given by (H7)

$$P = 0.78 + 1.5(1 - \alpha/90)^2 \tag{6.23}$$

where α is the solar altitude in degrees. P in Eq. (6.23) is for an entire array optimized for low sun angles in winter at 2 p.m. Obviously, other expressions apply for other configurations. For optimum summer performance, the high solar altitude dictates a dense field symmetrically located about the receiver tower. For winter optimization, the tower is in the southern half of the field (for sites north of the equator). Mirrors placed far south of the towers are relatively ineffectual since low sun angles and corresponding large incidence angles cause drastic foreshortening of the mirrors. In addition, long shadows result in a rather sparse field. Likewise, afternoon or morning demand peaks cause a western or eastern shift of the tower from the center of the field. Of course, the local ground cover ratio Φ_{local} (heliostat area-to-ground area) will vary throughout the field depending on the daily and seasonal peaking time. For winter optimized fields it will range from 0.2 to 0.5; for summer fields from 0.4 to 0.6 (V2). A method for field design is presented in (V2).

The optical efficiency η_0 of a CRSTP heliostat field is then given by

$$\eta_0 = \Phi P \rho_m \tau_r \alpha_r f_t \delta \tag{6.24}$$

where the intercept factor δ includes solar beam spread, mirror surface errors, and tracking inaccuracies; f_t represents the fraction of receiver area not shaded by supports. Depending upon the tower and heliostat design f_t may range from 0.94 to nearly 1.0. Typical values of optical efficiency are about $\eta_0 \sim 0.67\Phi P$ (here, in accordance with the convention of CRSTP analysts, η_0 is based on *horizontal* beam radiation and total ground area).

B. Receiver Concepts

Very high flux rates on the order of 1–2 MW/m² are present at CRSTP receivers. These flux levels are not beyond the state of the art for fossil fuel boilers but require careful design. One technique for effective heat transfer to a fluid is to use a cavity which has a small aperture for good thermal efficiency but a relatively large internal area and hence lower average flux density than at the cavity aperture.

Figure 6.17 shows schematically the cavity proposed for an EPRI CRSTP plant (Fig. 6.14). It is designed to accommodate the flux incident from a circular heliostat field. Each ray shown in Fig. 6.17 represents the same amount of redirected solar flux and the angle represents the direction of each flux unit. It is seen that most of the flux strikes the wall of the cavity and then intercepts the heat exchanger surface (located at the cavity top) after one or more internal reflections. Hence, a fairly uniform flux density exists at the exchanger surface and no more than 10% of the incident flux can strike the heat exchanger directly.

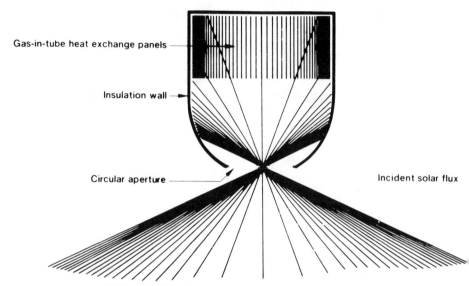

Gas-in-tube heat exchange panels

Insulation wall

Circular aperture

Incident solar flux

FIGURE 6.17   EPRI receiver flux patterns. The heat exchanger is located near the top of the cavity and cannot be directly illuminated by more than 10% of the incident flux. Line spacing depicts spatial distribution of solar energy.

Figure 6.18 shows sketches of several other receiver designs proposed by U.S. firms involved in CRSTP development. The McDonnell-Douglas concept uses an external absorber, not a cavity. However, with a concentration of several thousand, heat losses per unit aperture are small, as shown in the first section of this chapter. The external receiver concept has been selected by the U.S. Department of Energy for installation in the first U.S. CRSTP plant to be located in the Mojave Desert of California and to go on line in the mid-1980s.

C. Rankine Cycle Generators

Any CRSTP receiver and mirror field described above can be used to operate either a Rankine or Brayton cycle generator station. Most utilities currently use Rankine steam cycles and several solar CRSTP systems have been designed around these engines. Figure 6.19 shows the operating characteristics of a solar Rankine cycle designed to deliver 150 MW_e. It is seen that the operating temperature is 500°C, a constraint imposed by turbine materials (carbon-moly steel limits) and available storage media which can be used up to only 400°C for the long term (see Chapter 4). Plant efficiency is necessarily reduced when operated from storage, therefore a dual admission turbine is used.

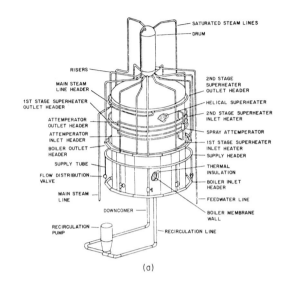

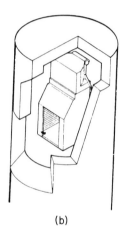

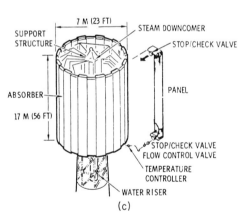

FIGURE 6.18   Several CRSTP receiver concepts developed by U.S. contractors: (a) Honeywell cavity; (b) Martin-Marietta cavity; (c) McDonnell-Douglas external surface receiver.

Contrary to current utility practice, an automatic admission turbine could be used for solar application. Since the turbine will operate both from storage and from the receiver, markedly different steam conditions may be encountered. In addition, a backup fossil-fuel boiler may interface with receiver or storage steam to further increase the range of throttle (i.e., inlet) conditions. If a single pressure turbine were used, it

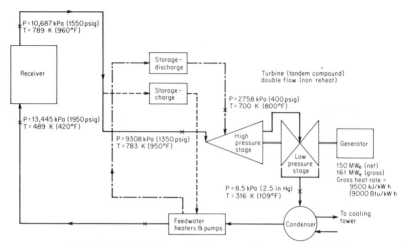

FIGURE 6.19   Schematic diagram of proposed 150-MW$_e$ commercial CRSTP plant using a Rankine cycle. The plant can operate from the receiver directly or from storage at reduced efficiency. A two-stage turbine is used with a subatmospheric condenser.

would need to be overrated to operate effectively from storage. Other throttle pressures result in somewhat decreased turbine efficiency but the cost is reduced since the turbine diameter can be made smaller. As pressure is increased, throttle temperature is limited by turbine materials and thermal storage degradation rates.

Figure 6.20 shows the throttle pressure and temperature characteristics of several commercial turbines which can be used in CRSTP plants. The turbines specified in most current system concepts operate at about 500°C (~950°F), which is near the upper temperature limit of commercial machinery.

D.   Brayton Engine Generators

Brayton cycles offer the opportunity for increasing the efficiency of solar-powered heat engines to produce intermediate load electric power. As shown in Fig. 6.21, the increased operating temperature of these engines can result in a 50% increase in efficiency versus the highest temperature Rankine engines. A design point of about 1500°F (~800°C) is achievable recognizing the limits imposed by receiver materials and blading metals. The high thermodynamic availability of sunlight is, therefore, used more effectively in the Brayton cycle.

Two types of Brayton cycles are feasible—closed and open. A closed cycle (E1) is shown in Fig. 6.22. In concept it is identical to the

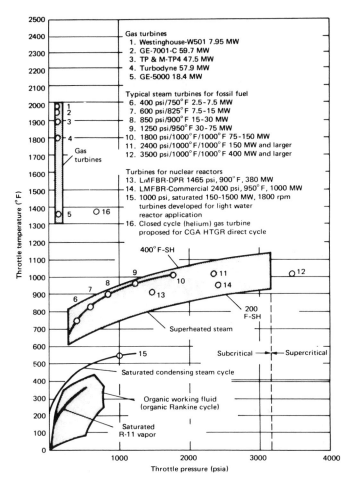

FIGURE 6.20 Throttle temperature and pressure characteristics of commercial gas and steam turbines.

familiar gas turbine except that solar firing is used in place of fuel firing and the turbine exhaust gas is reused. The open-cycle concept eliminates the precooler and solid-phase thermal storage. Instead of relatively low energy density solar-thermal storage, fuel is used in a burner in parallel with the receiver to provide the backup function for periods of solar outage.

Table 6.4 summarizes the major positive and negative features of open and closed Brayton cycles and steam Rankine cycles. Although the closed cycle has good first law efficiency, there is no experience in the U.S. and very little in Europe with such machines. Open Brayton cycles are much more familiar to utilities but efficiency is lower

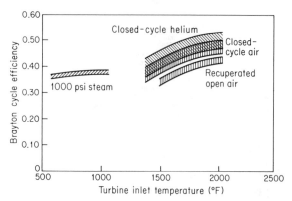

FIGURE 6.21   Relative cycle efficiencies of steam Rankine and various Brayton cycles vs. inlet temperature. Ambient temperature 27°C. [From (E1).]

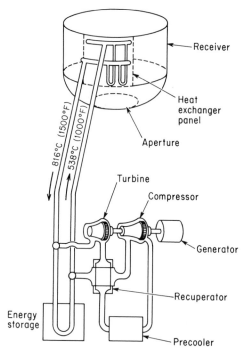

FIGURE 6.22   Schematic diagram of a closed Brayton cycle as proposed by the EPRI.

and the flexibility needed in solar-fired systems is restricted. Rankine cycles have been proven reliable by long experience in the utility industry but are more complex than are Brayton cycles.

Brayton cycles have an additional positive feature when considered for use in the sunny but arid regions of the world. Since water is scarce and expensive in these areas, a dry, air-based cooling tower is highly desirable. If a dry tower is mandatory, the performance penalty for a Rankine cycle is greater than for a closed Brayton cycle. Of course, the open Brayton cycle requires no cooling tower.

One of the most difficult problems in the design of Brayton cycle CRSTP plants is the receiver design since they operate at higher temperatures than for a Rankine cycle. Fluid conduits must have high absorptance, low emittance, low cost, high strength, high thermal shock resistance, and acceptable thermal conductivity at temperatures up to 1000°C. The basic heat transfer problem lies in heat transfer coefficients from receiver tubes to the working gas which are quite low, of the order of $200-400$ W/m^2 °C. Silicon carbide and cordierite appear to be satisfactory materials for the receiver tubes in Brayton cycle CRSTP plants.

Conclusions of studies comparing Rankine and Brayton cycles have shown that (E1)

(1) A commercial-sized Brayton plant can operate at a daily, overall efficiency of 21–22%.

(2) Efficiency is greater than for a Rankine cycle.

(3) Parallel fuel- and solar-firing of the working gas in an open cycle is feasible. This hybrid concept is considered to be the only

TABLE 6.4

Relative Advantages and Disadvantages of Brayton and Rankine Cycles

System	Advantages	Disadvantages
Brayton Closed cycle	■ High efficiency ■ Smallest turbomachinery ■ Operational flexibility ■ Good bottoming cycle potential	■ No operating experience in U.S. ■ Large precooler ■ High Brayton cycle costs
Brayton Open cycle	■ No precooler required ■ Similar to combustion turbine design ■ System simplicity	■ Limited operational flexibility ■ Large turbomachinery ■ Large recuperator
Rankine Steam	■ Proven systems available ■ Utility familiarity ■ High heat flux receiver	■ Large precooler ■ Lower efficiency ■ Complex system

practical approach to thermal storage for Brayton cycles because of the impracticality of very high-temperature storage.

(4)   Location of the turbine at the top of the receiver tower appears to be feasible and necessary to avoid excessive gas pressure drop.

E.   Thermal Performance of CRSTP Plants

The performance of the Rankine and two Brayton cycle concepts is summarized below in Table 6.5. Note that comparisons from one system type to another cannot be made directly from this table since different plant sizes and solar data were used to generate the data. Also, assumptions made in the calculations differ among the several systems. Since the optical performance of all systems, if identically designed and operated, would be the same to lowest order, cycle differences occur owing only to operating temperature differences. High temperatures result in better heat engine efficiency.

F.   Costs

Since no CRSTP plant has been constructed and the economy of the U.S. is experiencing a period of sustained inflation, it is impossible to make a reliable projection of the ultimate cost of such a

TABLE 6.5

Fraction of Solar Flux Leaving Various Subsystems of Several Types of CRSTP Plants[a]

Loss mechanism	Closed helium Brayton cycle (%)	Hybrid open air Brayton cycle (%)	Rankine (%)
Tracking, shading, blocking, tower shadow, cosine	72.9	86.3	—
Reflectance	64.1	66.0	67.3
Receiver	45.7	59.4	—
Turbine–generator	18.1 (20.5% if hybrid with fuel backup)	21.7	19.8

[a] From (E1) and Honeywell Report F3419-DR-302A-1. Table entries are percent of incident beam radiation, pumping power not included.

system. However, it is possible to estimate *relative* costs to find the first- and second-order cost components. Table 6.6 (E1) summarizes the costs of a 100-MW$_e$ CRSTP facility for both Rankine and closed-Brayton cycle plant types.

The major cost is for heliostats (35–45% of the total). Note that heliostats are priced at $60/m², which represents the projected cost of a heliostat based upon the average cost of manufactured goods in the U.S., which has historically been about $1.00/lb–$1.50/lb ($2.20/kg–$3.30/kg). However, no heliostats have been made to date for this price. Storage also represents a major cost which can be reduced by using a hybrid (fuel backup) design although a boiler plant must be purchased.

A detailed net energy analysis of a 100-MW$_e$ CRSTP plant has been conducted by Meyers and Vant-Hull (M4). Heliostats, the tower and piping, 6-h storage, and the receiver were included with energy inputs calculated back to the point of mining of raw materials. It was found that the collector subsystem (heliostats, tower, and receiver) required 81% of the input or capital energy with storage accounting for 19%. The time required to "pay back" the capital energy in plant output was found to be 540 days. Therefore, if the plant were to operate for 30 yr, it would return its capital energy more than twenty times. In addition, if the plant were recycled after its 30-yr life, about 40% of its capital energy could be recovered.

G. Other Concepts

Although the majority of emphasis in CRSTP plants has been with point-focus collectors coupled to Rankine or Brayton cycles, two other ideas have been analyzed preliminarily. Liquid metal technology from the LMFBR program could be adopted to the production of power from CRSTP collector systems. As shown in Fig. 6.23 sodium can be used as the working fluid in the collector loop. The excellent heat transfer properties of liquid metals at near-ambient pressure could increase cycle efficiency to some extent but the hazard and cost of using this very reactive metal may offset the benefits.

A second concept which has been suggested uses a *linear* central receiver with a parabolic Fresnel mirror instead of a paraboloidal mirror. A cavity receiver could be used but very high concentration (CR > 50–60) could not be achieved. A major attribute of the system is the ease of cleaning the heliostats. Since they are linear in this design, a wash mechanism or vehicle could be designed to move along the mirror

TABLE 6.6

Costs ($1975) of Components of Rankine- and Brayton-Based 100-MW$_e$ CRSTP Plants[a]

Plant type	Stand alone[b]		Hybrid[b]	
	Steam Rankine	Helium	Steam Rankine	Helium
Collector area (km²)	1	0.84	0.5	0.42
Storage time (h)	6	6	0.5	0.5
Account				
Land	2	2	1	1
Structures and facilities	44	44	51	51
Heliostats[c]	600	505	300	258
Central receiver/tower[d]/heat exchanger	95	197	68	98
Storage/tanks[e]	180	164	15	15
Boiler plant	—	—	73	73
Turbine plant equipment	80	119	80	105
Electric plant equipment	21	20	21	20
Misc plant equipment	4	4	4	4
Allowance for cooling towers	20	15	20	15
Total direct cost	1,046	1,070	633	640
Contingency allowance and Spare parts allowance (5%)	52	53	32	32
Indirect costs (10%)	105	107	63	64
Total capital investment (1975)	1,203	1,230	728	736
Interest during construction (15%)	180	185	109	110
Total cost at year of comm'l operation (1975 $)	1,383	1,415	837	846

[a] From (E1), costs are $/kW.

[b] "Stand alone" refers to no backup whereas "hybrid" refers to a fossil fuel backup with minimal storage.

[c] Collector cost—$60/m²

[d] Tower height—260m (2, and 1 tower(s), respectively)

[e] Thermal storage cost—$30/kWh$_e$

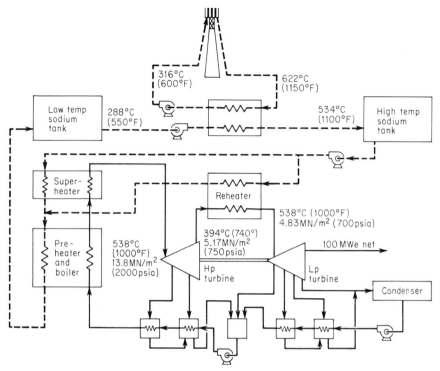

FIGURE 6.23 Liquid Sodium CRSTP concept. A two-stage Rankine turbine with reheat is used. The liquid metal is used to produce steam in the preheater, boiler, and superheater as well as in the reheater. Liquid metal is also used in the solar tower receiver loop. Legend: ---, sodium; —, water/steam.

elements and perform a washing operation of all segments in serial order. Cleaning of heliostats in the desert environment has become more of a problem than early investigators projected.

IV. HIGH-TEMPERATURE DISTRIBUTED SOLAR POWER PRODUCTION

Distributed systems for power production at temperatures above 500°C appear promising if solar-heated fluid transport from the collector field is replaced with electric power transport. This can be accomplished if small heat engines ($\sim 10-12$ kW$_e$) are located at each concentrator with power production also occurring at each dish. These small engines could be of the Brayton or Stirling cycle type operating at $\sim 800°C$ with local heat rejection to ambient air.

Distributed systems have several advantages over central receiver systems described in the previous section:

(1)   low visual impact;

(2)   compatible with irregular collector field shapes;

(3)   modular; stepwise capacity buildup possible;

(4)   relatively quick startup (a few minutes);

(5)   waste heat usable to power a central Rankine plant at 300°C (of course, the fluid transport problem to a central site is thereby reintroduced);

(6)   partial power production during construction;

(7)   waste heat can be rejected to air, no water is required for this purpose;

(8)   low initial energy investment.

Caputo (C5) has analyzed the performance and economics of high-temperature distributed solar-thermal power (DSTP) systems. Some of his major results are summarized in this section.

A.   The Heat Engine

Several engines including Brayton, Stirling, liquid metal, Rankine, and thermionic devices can be used to produce power at high temperature. Since a wide range of Brayton engines is commercially available, this design will be considered in this section although development is underway for the Stirling engine. Thermionic systems are described in the next section of this chapter.

To minimize heat loss from high-temperature ducting, a small Brayton engine is located at each paraboloidal dish collector. If the dish is 11 m in diameter (95-m² area) about 75 kW$_t$ will be intercepted per day. For typical optical properties of components, an optical efficiency of 70% can be achieved if a nonevacuated cavity design is used. Therefore, about 50 kW$_t$ are absorbed per day per collector. A Brayton cycle operating at $\sim 800°C$ has a conversion efficiency of about 33% at pressure ratios of 2.0–2.4. Therefore, 17 kW$_e$ could be produced at each unit on a sunny day. Figure 6.24 shows the dish Brayton system schematically.

It appears that location of the engine itself at the focal point

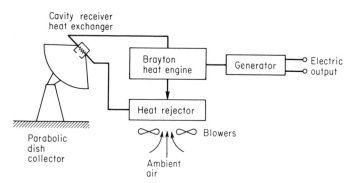

FIGURE 6.24   Schematic diagram of a DSTP unit with electrical transport from each unit.

of the dish collector may be practical if weight can be controlled. If not, several alternatives exist. The engine can be mounted on the ground near the collector with flexible ducts carrying the inert cycle gas from the absorber cavity to the engine. However, care is required to minimize the pressure drop between the collector and the receiver—only tens of kilopascals drop are possible without incurring severe cycle losses with pressure ratios of the order of 2.0.

A second energy transport method uses a secondary reflector at the dish focus to reflect energy to the heat engine suspended at the counterweight position of the collector. This Cassegrain approach is commonly used in telescopes. However, at least a 15% optical penalty will be incurred. A third method is to use a heat pipe with its boiler at the dish focus but, as described in Chapter 3, the varying angle of the heat pipe attached to a tracking collector may cause wicking problems in the liquid phase of the pipe. A liquid metal heat pipe is indicated for this application.

Closed-cycle Brayton engines operable at high temperature are commercially available. Sizes range from 5 kW$_e$ to 250 kW$_e$. In the 10–50 kW$_e$ range of interest for DSTP systems, radial compressors and recuperators are used with an overall engine efficiency of about 35%. The cost is about $400/kW$_e$ [$1975 (C5)]. Performance can be increased (Chapter 3) by increasing the operating temperature. Above 800°C Inconel 713 is no longer usable in the turbine and refractory blading would be required. It is expected that the upper efficiency limit for small machines is of the order of 45% at ~1200°C, a temperature within the reach of accurate paraboloids with good thermal efficiency. If a porous ceramic receiver cavity were used, an open, air standard Brayton engine could be operated at high temperature without the cost of a heat rejector. Heat rejection from a focus-mounted engine can be difficult because of weight constraints.

Figure 6.25 shows the expected thermal efficiency of a Brayton engine as a function of inlet temperature. These data are for low pressure devices ($p < 750$ kP$_a$) with a 90% regenerator effectiveness. No compressor stage intercooling or reheat is used in these small engines, however.

B. Energy Transport

The transport of heat energy in pipe networks is considered in detail in Chapter 5. This subsection considers the transport of collector-produced electricity instead of heated fluid to a central site. This approach eliminates the high parasitic thermal losses from high-

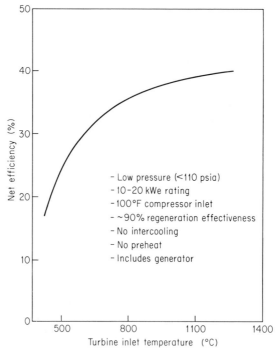

Net efficiency (%)

- Low pressure (<110 psia)
- 10–20 kWe rating
- 100°F compressor inlet
- ~90% regeneration effectiveness
- No intercooling
- No preheat
- Includes generator

Turbine inlet temperature (°C)

FIGURE 6.25   Small (10–20-kW$_e$) Brayton cycle engine efficiency to produce electric power. [From (C5).]

temperature pipe networks. Some electrical power is lost, however, due to I^2R heating.

Several types of generators can be coupled to the local Brayton engines: (a) synchronous ac; (b) induction ac; (c) dc. Dc shunt generation is well suited to operation of a parallel field since start-up is simple. The voltage V is given by

$$V \propto V_c \omega, \qquad (6.25)$$

where ω is the armature angular velocity and V_c the excitation voltage. Therefore, voltage control is simply accomplished by excitation control to compensate for variable speed ω. Dc generators are about 50% more costly than ac generators, however (C5). Also, inversion to ac is required at extra cost and parasitic loss.

A more common machine is the synchronous ac generator. Establishing and maintaining synchronization is the main difficulty. Induction ac generators are compatible with solar heat engine pairing and are particularly easy to control. This generator can be used as a motor to spin the Brayton rotor at synchronous speed. As solar power is applied, the engine rotor will operate slightly above synchronous speed and generate power. Efficiency of the induction generator is slightly less than for the synchronous generator but reliability is good.

Transport of power from dish generators to the central site can be accomplished by buried cable which uses a routine construction method ($10/lineal foot). Aluminum is cheaper than copper but has a higher resistivity so more material is required. This is not altogether undesirable since the larger aluminum wires lose heat to the ground more rapidly than smaller copper wires and the electrical resistance rise with wire temperature is reduced. For the 20-kW$_e$ (20-kV A) generator size considered here, 440-V transport is indicated. Of course, a central transformer would be needed to boost voltage to the grid level of ~ 230 kV. Other standard components of the electrical energy transport system include circuit breakers, capacitors for power factor control, motor/generator starters, and miscellaneous parts. The overall efficiency of the electrical transport system including two step-up transformer stages is estimated to be about 93% (C5).

C. Total System Performance

As in all solar-thermal power applications the operating temperature of a dish electric plant is a first-order design variable. It is constrained by materials available for cavity heat exchangers and various turbine components. For the Brayton engine efficiency shown in Fig. 6.25 and the performance of a CR = 1000 paraboloid, the optimum overall efficiency versus temperature curve exhibits a broad optimum at about 800°C. As noted above, this temperature corresponds to the limit of current commercial turbines in the 10–20-kW$_e$ range. The overall collector engine efficiency at 800°C is about 25% and the efficiency of these engines is quite insensitive to off-design conditions—for example, at one-third the nominal heat rate, cycle efficiency drops only 5%.

The optimum collector module size is likewise determined by the total of engine, dish, and auxiliary component costs. For the prices used in Caputo's study, the cost-optimal dish size was shown to be about 11 m diameter for a probable range of dish and engine costs. As shown above this size corresponds to a 10–20-kW$_e$ generator size.

For these rough size guidelines, the expected overall system efficiency η at the design point will be

$$
\eta = \underset{\substack{\text{optical} \\ \text{efficiency}}}{0.7} \times \underset{\substack{\text{engine} \\ \text{efficiency}}}{0.36} \times \underset{\substack{\text{transport} \\ \text{efficiency}}}{0.93} \times \underset{\substack{\text{spillage and} \\ \text{off-design} \\ \text{operation}}}{0.96}
$$

$$
= 22.5\%.
$$

The spillage term refers to peak insolation periods for which energy collection rates are beyond the capacity of the engine. Since thermal storage is not considered for this local, high-temperature DSTP concept, this excess peak, solar heat collection cannot be used. However, storage at a central site using mechanical, chemical, hydro, or other concepts discussed elsewhere in this book could be used.

The ground cover ratio for two-axis tracking dishes is about the same as for heliostat fields. As shown in the previous section this ratio will typically be of the order of 40–45%. Some shading and blocking will occur at this density and the overall efficiency calculated above will be reduced by 1%, absolute. For a ground cover ratio of about 20%, no shading or blocking will occur; for 46% coverage, 7.2% annual average shading occurs.

D. Costs

The single largest major cost factor—50% of the total—in a DSTP system is the collector with the small Brayton engines costing about one-half the collector cost. The cavity receiver, electric power network, land, and other costs account for the remaining 25% of the total. Caputo suggests that the total cost of the high-temperature DSTP, dish electric system to be about $1400/kW_e in $1975. This cost is twice the cost of fossil-fuel steam plants recently constructed.

Figure 6.26 indicates the relative costs of the several power generator methods described in this book. It is seen that the dish electric system (using a Stirling engine which is expected to be more efficient than the Brayton engine described above) shows the lowest busbar power cost of the four contenders. However, this system has the greatest uncertainty and requires more development than some other systems. Reference (J3) describes research efforts under way directed at developing advanced Stirling engines and other components. All systems are very sensitive to collector cost. It is also notable that the greater economies of scale for CRSTP plants cause a crossover of the cost curves for the linear and dish steam systems at about 10-MW_e plant size. The exact crossover point is sensitive to CRSTP scaling assumptions and requires further study to identify the point at which *current technology* distributed systems become less economic than CRSTP systems.

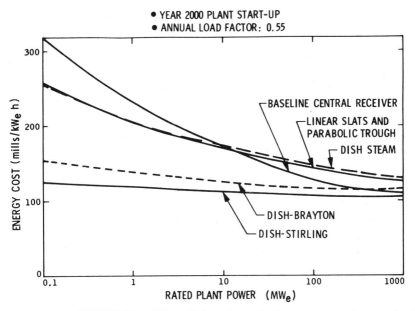

FIGURE 6.26   Effect of plant size on busbar power cost for central and distributed solar power systems for a 2000 A.D. start-up. [From (F2).]

E. Other High-Temperature DSTP Concepts

Although the emphasis in this section has been on one specific concept which appears to be technically feasible, there are other ideas to consider. For example, a high-temperature collector could be used in an endothermic reaction to dissociate a chemical such as SO_3 into its components SO_2 and O_2. These can be recombined at a central power plant in the presence of a catalyst to release heat. This heat of reaction then fires the boiler plant. A major advantage of this idea is that there are no parasitic heat leaks and no requirement for special high-pressure piping used in all other distributed collector, central generator devices. In addition, storage is relatively straightforward since the reactants can each be stored separately until heat is required although volume requirements for gaseous reactants may be large. (See Chapter 5 for details.)

Other heat engines can be considered. Since the Stirling cycle theoretically approaches the Carnot efficiency, it will exhibit better performance than a Brayton engine at the same temperature. Research under way in the automotive industry indicates that efficiencies of 42% could be achieved at 800°C using a Stirling engine. Overall conversion of heat to power could increase by about one-fifth (F2) compared to a Brayton cycle.

V. SOLAR THERMIONICS

Thermionic conversion of heat directly into electric power is accomplished by using a hot *emitter* electrode and a cooler *collector* electrode sealed in an enclosure.

Electrons are emitted or boiled off from the emitter, flow across the emitter–cathode gap, and are condensed at the collector. The electrons then return to the emitter via the external circuitry. The thermionic (TI) device may be viewed as a heat engine operating between the emitter and collector with an electron gas as the working fluid. This section will present a survey of past thermionic work and a review of the governing equations for TI converters.

Several reasons for interest in the TI converter may be identified:

(a) it has no moving parts;
(b) heat rejection at low temperature is not critical;

(c)   it has good potential for efficiency up to 40%;

(d)   temperatures achievable by solar concentrators (1000–1500°C) are adequate to operate the converter.

Until recently most thermionic research was directed at its use as a space power source using a nuclear reactor and as a topping cycle to recover the thermodynamic availability lost in the superheater of fossil power plants. Solar TI research has been quite limited but the odds of success are good since established high-temperature technology can be used. There also exists a history of steady improvement in solving the electrode and control problems of TI devices.

A.   The History of Thermionics

In 1883 Thomas Edison noted that he had observed the thermionic effect in his development work on the incandescent lamp. He also observed the effect that higher emitter (lamp filament) temperature caused higher current flow. This work was set aside, however, because of his involvement in electric lighting research. Elster and Geitel conducted research in TI devices between 1882 and 1889. They made the fundamental observation that the emitter–collector current was a strong function of temperature as well as temperature difference. J. J. Thompson identified the charge carriers as electrons shortly before the turn of the century.

Langmuir (L10) developed the theoretical base for thermionics in 1933. In the preceding years he and his coworkers examined the emission characteristics of cesium films of the type used today. For the next few decades there was little work on the concept, however, in the U.S. In 1956 Hatsopolous (H8) first analyzed the TI converter as a heat engine showing that its efficiency is constrained by the second law to the Carnot limit. He also examined the role of small electrode spacings in efficient converters. The results of the Hatsopolous analysis are summarized in the next section.

B.   Theoretical Analysis

If the TI converter (Fig. 6.27) is considered to be a heat engine (H8,A4), it is relatively simple to calculate the current density J (A/cm^2) and efficiency η of the system. The current density at the collector or emitter is given by (for zero reflectance)

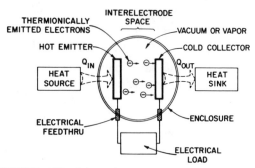

FIGURE 6.27 Schematic diagram of a thermionic converter. [From (H9).]

$$J = A_1 T^2 \exp(-\phi/kT), \qquad (6.26)$$

where T is the electrode temperature, k is the Boltzmann constant (1.38 × 10^{-23} J/K or 1.38 × 10^{-16} erg/K), and A_1 is called the emission constant, whose theoretical value is 120 A/cm² K². The work function ϕ in Eq. (6.26) is related to the heat of vaporization of electrons:

$$\phi \simeq kh_0/R, \qquad (6.27)$$

where h_0 is the molecular heat of vaporization at 0 K and R is the universal gas constant. A typical range of values for ϕ is 1.5–2.0 V (H9).

Figure 6.28 shows the potential diagram of a TI converter. From Eq. (6.26) the emitter current density J_e is

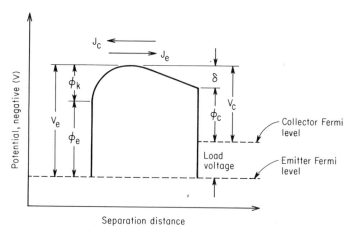

FIGURE 6.28 Potential map of the emitter and collector of a TI converter showing the emitter and collector work functions ϕ_e and ϕ_c, the kinetic barrier ϕ_k, and the potential barrier δ.

$$J_e = A_1 T_e^2 \exp(-eV_e/kT_e), \qquad (6.28)$$

where e is the electron charge (1.6×10^{-19} C). The emitter potential V_e is the sum of the work function ϕ_e and the kinetic energy ϕ_k required to traverse the electrode gap. Analogously, for the collector electrode

$$J_c = A_1 T_c^2 \exp(-eV_c/kT_c), \qquad (6.29)$$

where the collector potential V_c is the collector work function ϕ_c plus the potential barrier δ which tends to retard the flow of electrons to the emitter. The net current J is the difference $J_e - J_c$ and the potential difference V across the load is the difference

$$V = (\phi_e - \phi_c) + (\phi_k - \delta). \qquad (6.30)$$

The power level P is then

$$P = (JA)V, \qquad (6.31)$$

where A is the electrode facial area (cm²). The first group in Eq. (6.30) is a physical constant of the device whereas the second depends on the current level J.

   Equations (6.28)–(6.31) can be used to find the efficiency of a TI device. For simplicity, assume that no space charge barriers exist so that $\phi_k = \delta = 0$. Following Angrist (A4), it can be shown that the peak efficiency η is given by

$$\eta = \frac{\beta_e(1 - \theta)}{(\beta_c + 2)} \left[\frac{(1 - \theta^2)}{1 - \theta^2(\beta_e + 2\theta)/(\beta_c + 2)} \right], \qquad (6.32)$$

where only the energy loss from the emitter due to electron flux has been considered. The quantity β_e is eV_e/kT_e and $\theta \equiv T_c/T_e$. An average value of β_c over a broad range of conditions is about 18. It is seen that the efficiency η is in form very much like the Carnot efficiency (the quantity in brackets is roughly unity). For $\theta = 0$, $\eta = 90\%$; for a Carnot engine if $\theta = 0$, $\eta = 100\%$.

   The quantity V_c in Fig. 6.28 (with $\phi_k = 0$) is sometimes called the barrier index and is an alternative measure of the performance of a TI unit. Under the peak efficiency conditions expressed by Eq. (6.32), the barrier index is given by (A4)

$$V_c = T_c \beta_e/11,606. \qquad (6.33)$$

The lower V_c the higher the efficiency. Current values of the barrier index are about 2.0 (H9), down from 3.0 eV in 1960. Further drops are expected. For a given V_c value Eq. (6.33) gives the TI device temperature above which back emission from the collector becomes significant.

C. Practical Problems

The preceding simplified analysis is adequate to identify first-order effects in TI converters. However, there are practical problems to contend with before these analytic predictions of performance will be achieved. The emitter space charge ($\phi_k - \delta$ in Fig. 6.28) problem is a result of the finite time required for electrons to cross from the emitter to the collector. Unless there is sufficient kinetic energy ϕ_k for emitted electrons to cross the gap, they are repelled to the emitter. Two approaches are used to reduce this problem. The first uses very small gaps (<0.001 in) to suppress the space charge. The difficulty with this method lies in the consistent maintenance of the small gap for long periods.

A second method of space charge control involves the use of a cesium vapor in the gap. Ionized cesium introduces positive ions in the gap and the space charge barrier is reduced. It is possible to just neutralize the space charge if proper conditions are used. Cesium vapor seems to be the one practical method of charge control.

The second problem in TI systems is in the identification of suitable materials. For example, the emitter must be a good electron emitter and a poor thermal emitter to control heat loss, and be stable at high temperature. The first criterion is particularly important if the space charge has been neutralized. Collectors must have low work functions ϕ_c but electron emission properties are of little consequence. Materials research is the key to practical and economic TI systems. Solar thermionics research has not been given high priority in the U.S. and its use in the near future on a substantial scale is doubtful.

VI. LONG-TERM PERFORMANCE OF HIGH-TEMPERATURE SOLAR PROCESS

The long-term performance calculation method described in Chapter 5 can also be used with high-concentration devices. To review, the method consists of three steps:

(1)   Calculate the clearness index $\bar{K}_T$ (Eq. 2.49).
(2)   Calculate the geometric factors r_T and r_d (Table 2.11).
(3)   Calculate the critical intensity ratio $\bar{X}$ from

$$\bar{X} = U_c \,\overline{\Delta T}\, \Delta t_c / \bar{\eta}_0 \bar{I}_c. \tag{6.34}$$

The long-term average energy delivery $\bar{Q}_u$ on a monthly basis is then

$$\bar{Q}_u = F\bar{\eta}_0 \bar{\phi}\bar{I}_c, \tag{6.35}$$

where F is the heat extraction coefficient depending upon the definition of the efficiency curve abscissa.

The utilizability $\bar{\phi}$ is given by (C1)

$$\bar{\phi} = 1.0 - (0.049 + 1.44\bar{K}_T)\bar{X} + 0.341\bar{K}_T\bar{X}^2 \tag{6.36}$$

for $\bar{\phi} > 0.4$, $\bar{K}_T = [0,0.75]$, and $\bar{X} = [0,1.2]$.

In the case of the central receiver collector, the ground cover fraction is to be included in the optical efficiency $\bar{\eta}_0$. Since this fraction is seasonally variable, appropriate monthly values of $\bar{\eta}_0$ may not be the same [see Eq. (6.24)].

ECONOMIC ANALYSIS OF SOLAR-THERMAL SYSTEMS

7

In nature, everything is free.
But if you use it,
You must pay for it.

V. A. Baum

I. ECONOMIC ANALYSIS OF SOLAR-THERMAL SYSTEMS

If there is to be a long-term use of the solar resource, the major incentive for its use in "free" economies must be to reduce the cost of energizing a specific process. Most energy sources are cost-free in their primitive states—oil in a geologic formation, coal in a mine, geothermal fluids locked in rock formations, and terrestrial solar flux intercepted by a surface. The cost of energy in a useful form can be assigned by using one of two philosophical approaches. The first requires that the cost of a unit of energy be the same as the cost of producing an identical replacement unit of energy, e.g., a barrel of oil. The second cost approach assigns to energy the costs of acquisition, distribution, and conversion to the required use form. The resource itself is considered to be cost-free.

In this book the second approach is used to assign a cost to solar heat in accordance with common microeconomic theory. The costs which are assigned to the solar resource by a system purchaser to identify its unit cost include: solar system capital, design, acquisition, and installation, land and other useful space consumed, interest charges, maintenance, insurance, depreciation, operating costs—personnel, auxiliary power, taxes—federal, state, and local—and salvage—may be positive or negative.

Many of the costs listed occur at various points in time through the life of the solar system. Therefore, these costs must be reduced to an equivalent set of costs based on uniform currency values at one specific point in time to make valid comparisons possible with competing energy sources. The discipline of engineering economics provides a rational method of cost comparison using the principles of discounted cash flow. These principles are described in the first section of this chapter.

The unit cost, e.g., $/GJ, of solar energy requires not only the evaluation of discounted cash flows over a project's life, but also the prediction of what useful energy delivery to the load can be expected. Methods of calculating solar delivery have been given in Chapters 5 and 6. It was shown that most solar systems are subject to the law of diminishing returns. A convenient way of quantifying solar system performance, useful in the optimization of solar systems, is by means of a technical production function which relates the magnitude of all system inputs—collector and storage sizes, fluid flow rates, etc.—to useful energy delivery. The concept of production functions is described in a section of this chapter. Its use as one means of optimizing system component sizes is also described.

This chapter deals with microeconomic phenomena, i.e., on the scale of the individual firm. Macroeconomic impacts of solar energy use are only beginning to be quantified. These include raw and finished material requirements, need for labor and capital, impacts on other energy sources and their costs, and the like. Reference (K7) summarizes preliminary estimates of macroeconomic impacts of solar energy for the U.S. through the turn of the century.

II. SELECTED TOPICS IN ENGINEERING ECONOMICS

Engineering economics provides a method whereby the costs of various alternative methods for providing the same result can be determined in an unbiased and rational way. For this purpose the concepts of the time value of money and discounted cash flow have been almost universally accepted, even in controlled economies by a different name. This section summarizes the ideas of engineering economics germane to solar system economic analysis. For further study the reader may consult (G4,O2,R8,S7).

A. Interest and the Time Value of Money

Interest is the cost of using capital and is a concept whose roots lie in antiquity. At the time of the Greeks and Romans interest rates were more or less standardized with rates of about 10% to good risks. However, it is interesting to note that the Israelites did not permit interest charges and canon law also forbade it into the Middle Ages.

The cost of interest includes the cost of loan administration, risk of nonrepayment, and the loss of earnings which could have been realized if the lender had invested the money for productive purposes. Of course, interest also represents an opportunity to the borrower to achieve a goal at once which otherwise would be delayed. That is, from the borrower's viewpoint interest is the time value of money.

The interest rate is a measure of the productivity of capital in real terms. The market interest rate includes both this productivity and inflation effects which have been significant in the mid- and late-1970s. Interest rates usually reflect the minimum rate of capital productivity.

Interest rates vary depending on the mix of debt (loans, bonds) to equity (common or preferred stock) financing. In equation form the before-tax rate i is

$$i = r_d i_d + r_p i_p + r_c i_c, \tag{7.1}$$

where r_d, r_p, and r_c are the ratios of debt, preferred stock, and common stock financing to the total; and i_d, i_p, and i_c are the interest rates for debt, preferred stock, and common stock. The cost of debt i_d is usually less than i_p or i_c since the risk, and hence return, for stockholders is larger than for lenders. A typical value in *real* dollars for i_d might be 4–5%, whereas i_p and i_c may be 8–10% (B8).

B. Discounted Cash Flow

Cash flows have both a monetary value and a time value. In order to reduce cash flow at various times to equivalent flows at a fixed time, the value of future cash flows must be reduced to the present by the time value of money, usually denoted by *i*. For example, if the after-tax time value of money is 10% per year, $100 today has the same value to the investor as $110 one yr from today. This section develops several useful formulas for discounting of future cash flows.

The discount rate is related to the minimum rate of return which a firm expects to receive on its investments. It is usually higher than the cost of capital or interest to compensate for risks and overhead. If the discount rate were the same as the interest rate, a firm would retire its own debts rather than make another investment. The discount rate is usually the minimum acceptable rate of return on an investment made by a firm. Hence, the rate of return on a solar system must always be greater than the discount rate. Typical required rates of return in the early 1970s were about 13% for replacement or modernization projects (G6). In the late 1970s the "hurdle" rate of return is 15–16%, reflecting the effect of inflation.

Nominal and Effective Interest Rates Interest rates are generally quoted on an annual basis. This *nominal* rate *r* may, however, be compounded more than once a year. If so, the *effective* annual rate *i* will be greater than *r*. The effective rate is defined as the annual interest charge divided by the principal or amount borrowed (or loaned). If there are *m* compounding periods per year for a nominal rate *r*, the effective rate can be calculated from

$$1 + i = [1 + (r/m)]^m. \qquad (7.2)$$

For example, a savings account compounding daily at a nominal rate of 7% has an effective annual rate of $7\frac{1}{4}\%$.

If the number of periods $m \to \infty$, interest is said to be compounded continuously. In the limit as $m \to \infty$, Eq. (7.2) reduces to

$$i = e^r - 1 \qquad (7.3)$$

since

$$e = \lim_{k \to \infty} [1 + (1/k)]^k. \qquad (7.4)$$

Therefore, the effective rate for continuous compounding is $(e^r - 1)$. For example, if interest were compounded continuously at 10% per year, the effective rate would be 10.52%. Continuous compounding is used if a series of cash flows can be approximated by a quasi-continuous pattern.

The Compound Amount Factor (F/P,i,N) There are several basic numerical factors which recur in engineering economics. It is convenient to calculate their value once at the outset to avoid unneeded repetition. The first factor is called the *compound amount factor*. It represents the future effect of compound interest on a present single payment P. In accordance with the recommendations of the Engineering Economy Division of the American Society for Engineering Education (ASEE), it is denoted by $(F/P,i,N)$ where F denotes the future value, P the present value, i the interest rate, and N the number of *equal* periods between payments or credits. In this book, end of year compounding is used; therefore,

$$F/P \equiv (F/P,i,N) = (1 + i)^N. \tag{7.5}$$

The Present Worth Factor $(P/F,i,N)$ The present worth factor represents the effect of the time value of money on a single payment F made in the future. It is simply the inverse of $(F/P,i,N)$; therefore,

$$P/F \equiv (P/F,i,N) = (1 + i)^{-N}. \tag{7.6}$$

Alternatively,

$$(P/F,i,N) = 1/(F/P,i,N). \tag{7.7}$$

Table 7.1 contains a summary of present worth factors.

TABLE 7.1

Present Worth Factors $(P/F, i, N)^a$

n \ i	0%	2%	4%	6%	8%	10%	12%	15%	20%	25%
1	1.0000	0.9804	0.9615	0.9434	0.9259	0.9091	0.8929	0.8696	0.8333	0.8000
2	1.0000	0.9612	0.9246	0.8900	0.8173	0.8264	0.7972	0.7561	0.6944	0.6400
3	1.0000	0.9423	0.8890	0.8396	0.7938	0.7513	0.7118	0.6575	0.5787	0.5120
4	1.0000	0.9238	0.8548	0.7921	0.7350	0.6830	0.6355	0.5718	0.4823	0.4096
5	1.0000	0.9057	0.8219	0.7473	0.6806	0.6209	0.5674	0.4972	0.4019	0.3277
6	1.0000	0.8880	0.7903	0.7050	0.6202	0.5645	0.5066	0.4323	0.3349	0.2621
7	1.0000	0.8706	0.7599	0.6651	0.5835	0.5132	0.4523	0.3759	0.2791	0.2097
8	1.0000	0.8535	0.7307	0.6274	0.5403	0.4665	0.4039	0.3269	0.2326	0.1678
9	1.0000	0.8368	0.7026	0.5919	0.5002	0.4241	0.3606	0.2843	0.1938	0.1342
10	1.0000	0.8203	0.6756	0.5584	0.4632	0.3855	0.3220	0.2472	0.1615	0.1074
11	1.0000	0.8043	0.6496	0.5268	0.4289	0.3505	0.2875	0.2149	0.1346	0.0859
12	1.0000	0.7885	0.6246	0.4970	0.3971	0.3186	0.2567	0.1869	0.1122	0.0687
13	1.0000	0.7730	0.6006	0.4688	0.3677	0.2897	0.2292	0.1625	0.0935	0.0550
14	1.0000	0.7579	0.5775	0.4423	0.3405	0.2633	0.2046	0.1413	0.0779	0.0440
15	1.0000	0.7430	0.5553	0.4173	0.3152	0.2394	0.1827	0.1229	0.0649	0.0352
16	1.0000	0.7284	0.5339	0.3936	0.2919	0.2176	0.1631	0.1069	0.0541	0.0281
17	1.0000	0.7142	0.5134	0.3714	0.2703	0.1978	0.1456	0.0929	0.0451	0.0225
18	1.0000	0.7002	0.4936	0.3503	0.2502	0.1799	0.1300	0.0808	0.0376	0.0180
19	1.0000	0.6864	0.4746	0.3305	0.2317	0.1635	0.1161	0.0703	0.0313	0.0144
20	1.0000	0.6730	0.4564	0.3118	0.2145	0.1486	0.1037	0.0611	0.0261	0.0115
25	1.0000	0.6095	0.2751	0.2330	0.1460	0.0923	0.0588	0.0304	0.0105	0.0038
30	1.0000	0.5521	0.3083	0.1741	0.0994	0.0573	0.0334	0.0151	0.0042	0.0012
40	1.0000	0.4529	0.2083	0.0972	0.0460	0.0221	0.0107	0.0037	0.0007	0.0001
50	1.0000	0.3715	0.1407	0.0543	0.0213	0.0085	0.0035	0.0009	0.0001	–
100	1.0000	0.1380	0.0198	0.0029	0.0005	0.0601	–	–	–	–

[a] For interest rates i of from 0 to 25 percent and for periods of analysis n of from 1 to 100 yr.

***The Sinking Fund Factor** (A/F,i,N)* The size of a series of uniform payments of amount A made to a fund to provide a future sum F can be evaluated using the sinking fund factor. The value of this factor can be derived by repeated applications of $(F/P,i,N)$ to the uniform series of payments A:

$$F = A(1 + i)^{N-1} + A(1 + i)^{N-2} + \cdots + A(1 + i)^{N-(N-1)} + A. \quad (7.8)$$

Using the formula for the sum of a geometric series with first term unity and ratio $1 + i$ it is easy to show that

$$A/F \equiv (A/F,i,N) = i/[(1 + i)^N - 1] \quad (7.9)$$

***The Series Compound Amount Factor** (F/A,i,N)* The total future worth F of a series of uniform payments A can be calculated from the series compound amount factor. This factor is simply the reciprocal of $(A/F,i,N)$; therefore,

$$F/A \equiv (F/A,i,N) = [(1 + i)^N - 1]/i \quad (7.10)$$

or

$$(F/A,i,N) = 1/(A/F,i,N) \quad (7.11)$$

***The Capital Recovery Factor** (A/P,i,N)* The capital recovery factor relates a series of annual payments A to the total present value P. It is similar to the sinking fund factor except that the present value P is required rather than the future value F. The value of $(A/P,i,N)$ can be calculated by applying the factor $(P/F,i,N)$ to the series of payments A:

$$P = A(1 + i)^{-1} + A(1 + i)^{-2} + \cdots + A(1 + i)^{-N} \quad (7.12)$$

using the formula for a geometric series,

$$A/P \equiv (A/P,i,N) = i/[1 - (1 + i)^{-N}]. \quad (7.13)$$

The capital recovery factor could be used to calculate the annual mortgage payment on a loan of amount P at rate i, for example. Table 7.2 contains a summary of capital recovery factor values.

Example Use the capital recovery factor concept to find the optimum thickness of insulation on a storage tank if the initial insulation investment is given by $a + bt$ where t is the insulation thickness and a and b ($/ft/ft^2$) are constants. The value of heat lost through the insulation per year is given by $Q_L C_f$, where Q_L is the heat loss per year (kJ/yr) and C_f is the fuel cost ($/kJ).

Solution The total cost of insulation plus the cost of heat loss through it should be a minimum to optimize the investment in insulation. If the insulation is very thick, the cost of the heat loss will be very

TABLE 7.2

Capital Recovery Factors $(A/P, i, N)^a$

n	0%	2%	4%	6%	8%	10%	12%	15%	20%	25%
1	1.00000	1.02000	1.04000	1.06000	1.08000	1.10000	1.12000	1.15000	1.20000	1.25000
2	0.50000	0.51505	0.53020	0.54544	0.56077	0.57619	0.59170	0.61512	0.65455	0.69444
3	0.33333	0.34675	0.36035	0.37411	0.38803	0.40211	0.41635	0.43798	0.47473	0.51230
4	0.25000	0.26262	0.27549	0.28859	0.30192	0.31547	0.32923	0.35027	0.38629	0.42344
5	0.20000	0.21216	0.22463	0.23740	0.25046	0.26380	0.27741	0.29832	0.33438	0.37184
6	0.16667	0.17853	0.19076	0.20336	0.21632	0.22961	0.24323	0.26424	0.30071	0.33882
7	0.14286	0.15451	0.16661	0.17914	0.19207	0.20541	0.21912	0.24036	0.27742	0.31634
8	0.12500	0.13651	0.14853	0.16101	0.17401	0.18744	0.20130	0.22285	0.26061	0.30040
9	0.11111	0.12252	0.13449	0.14702	0.16008	0.17364	0.18768	0.20957	0.24808	0.28876
10	0.10000	0.11133	0.12329	0.13587	0.14903	0.16275	0.17698	0.19925	0.23852	0.28007
11	0.09091	0.10218	0.11415	0.12679	0.14008	0.15396	0.16842	0.19107	0.23110	0.27349
12	0.08333	0.09156	0.10655	0.11928	0.13270	0.14676	0.16144	0.18148	0.22526	0.26845
13	0.07692	0.08812	0.10014	0.11296	0.12652	0.14078	0.15568	0.17911	0.22062	0.26454
14	0.07143	0.08260	0.09467	0.10758	0.12130	0.13575	0.15087	0.17469	0.21689	0.26150
15	0.06667	0.07783	0.08994	0.10296	0.11683	0.13147	0.14682	0.17102	0.21388	0.25912
16	0.06250	0.07365	0.08582	0.09895	0.11298	0.12782	0.14339	0.16795	0.21144	0.25724
17	0.05882	0.06997	0.08220	0.09544	0.10963	0.12466	0.14046	0.16537	0.20944	0.25576
18	0.05556	0.06670	0.07899	0.09236	0.10670	0.12193	0.13794	0.16319	0.20781	0.25459
19	0.05263	0.06378	0.07614	0.08962	0.10413	0.11955	0.13576	0.16134	0.20646	0.25366
20	0.05000	0.06116	0.07358	0.08718	0.10185	0.11746	0.13388	0.15976	0.20536	0.25292
25	0.04000	0.05122	0.06401	0.07823	0.09368	0.11017	0.12750	0.15470	0.20212	0.25095
30	0.03333	0.04465	0.05783	0.07265	0.08883	0.10608	0.12414	0.15230	0.20085	0.25031
40	0.02500	0.03656	0.05052	0.06646	0.08386	0.10226	0.12130	0.15056	0.20014	0.25003
50	0.02000	0.03182	0.04655	0.06344	0.08174	0.10086	0.12042	0.15014	0.20002	0.25000
100	0.01000	0.02320	0.04081	0.06018	0.08004	0.10001	0.12000	0.15000	0.20000	0.25000
∞		0.02000	0.04000	0.06000	0.08000	0.10000	0.12000	0.15000	0.20000	0.25000

^{a} For interest rates i of from 0 to 25 percent and for periods of analysis n of from 1 to 100 yr.

small but the insulation will be too expensive. Also, if the insulation is thin, its cost is low but the cost of heat lost through the insulation is too high to be optimal.

The total cost C_T of insulation plus heat loss per square foot is, on an annual basis

$$C_T = (A/P,i,N)(a + bt) + Q_L C_f; \tag{7.14}$$

the heat loss per square foot is

$$Q_L = \Delta T (R_0 + t/k)^{-1} \Delta t, \tag{7.15}$$

where ΔT is the temperature drop from the tank fluid to the environment, R_0 is the thermal resistance if no insulation is present (See Chapter 3), k is the insulation thermal conductivity, and Δt is the number of hours of heat loss per year. Then

$$C_T = (A/P,i,N)(a + bt) + C_f \Delta T (R_0 + t/k)^{-1} \Delta t. \tag{7.16}$$

The derivative $\partial C_T/\partial t$ can be found and equated to zero to find t_{opt}:

$$\frac{\partial C_T}{\partial t} = 0 = (A/P,i,N)b - \frac{C_f \Delta T \Delta t}{k} (R_0 + t_{opt}/k)^{-2}. \tag{7.17}$$

Solving for t_{opt},

$$t_{opt} = \{C_f \Delta T \Delta t \, k/[(A/P,i,N)b]\}^{1/2} - R_0 k. \tag{7.18}$$ ∎

The Series Present Worth Factor $(P/A,i,N)$ The total present worth of a series of payments A can be calculated using the series present worth factor. It is simply the inverse of the capital recovery factor, or

$$P/A \equiv (P/A,i,N) = [1 - (1 + i)^{-N}]/i. \tag{7.19}$$

Alternatively,

$$(P/A,i,N) = 1/(A/P,i,N) \tag{7.20}$$

The Gradient Factor $(A/G,i,N)$ Instead of a uniform series of payments A, it is frequently a uniformly increasing or decreasing series of payments changing by a fixed amount G per time period which is of interest. The gradient factor converts a G series into a uniform A series for convenience. The gradient factor can be shown to be given by

$$A/G \equiv (A/G,i,N) = \{(1/i) - [(N/i)(A/F,i,N)]\}. \tag{7.21}$$

The seven discounted cash flow factors developed above are summarized in Table 7.3 along with the analogous expressions for

TABLE 7.3

Summary of Interest Factors for Cash Flows with End of Period Compounding[a]

Interest factor	Quantity known	Quantity to be found	Discrete compounding of a discrete flow expression	Continuous compounding of a discrete flow expression
$(F/P, i, N)$	P	F	$(1 + i)^N$	e^{rN}
$(P/F, i, N)$	F	P	$(1 + i)^{-N}$	e^{-rN}
$(A/F, i, N)$	F	A	$\dfrac{i}{(1 + i)^N - 1}$	$\dfrac{e^r - 1}{e^{rN} - 1}$
$(F/A, i, N)$	A	F	$\dfrac{(1 + i)^N - 1}{i}$	$\dfrac{e^{rN} - 1}{e^r - 1}$
$(A/P, i, N)$	P	A	$\dfrac{i}{1 - (1 + i)^{-N}}$	$\dfrac{e^r - 1}{1 - e^{-rN}}$
$(P/A, i, N)$	A	P	$\dfrac{1 - (1 + i)^{-N}}{i}$	$\dfrac{1 - e^{-rN}}{e^r - 1}$
$(A/G, i, N)$	G	A	$[1/i - N/i(A/F, i, N)]$	—[b]

[a] P is a discrete present amount, F a future amount, A a uniform end of period payment, and G the uniform increase in an amount; i is the effective interest rate and r is the nominal rate; N is the number of payment periods.
[b] Substitute $i = e^r - 1$ into the previous column expression for the uniform gradient factor.

continuous compounding. The various factors are related by several expressions which may be convenient if tables of a desired factor are not available (R8):

$$(F/P,i,N) \times (P/A,i,N) = (F/A,i,N); \qquad (7.22)$$

$$(F/A,i,N) \times (A/P,i,N) = (F/P,i,N); \qquad (7.23)$$

$$(A/F,i,N) + i = (A/P,i,N). \qquad (7.24)$$

III. THE SOLAR SYSTEM COST EQUATION

Cash flow factors derived in the previous section can be used to reduce the various costs associated with the purchase and operation of a solar system into a series of uniform yearly payments—called the *annual worth*, the *annualized cost*, or the *levelized cost*—over the expected economic or mechanical life of the system. Alternatively, the various cost components could be accumulated into a sum representing the *total present worth* of all future cash flows. Annual worth and present worth analyses will always give the same result when used to select a cost-optimal solar system since the two are related term-by-term by the capital recovery factor. In this book the annual worth approach is used since it is customary for a firm to review a year's gains and losses as a measure of economic viability. In addition, most depreciation, tax, accounting, and other summaries are generally made on a time scale of a year.

The reason for considering the annual cost of a solar system is obviously to make a comparison with one or more alternative choices of energy source which would accomplish the same goal. It is essential that such comparisons all be based upon achieving precisely the same goal or end result.

A. Solar System Cost Classes

Two generic cost types may be identified in solar economic analyses. The first are initial, capital costs and include: mechanical components-collector; storage vessels, materials, supports, and insulation; piping, pumps, fans, valves; heat exchangers; backup energy source equipment; working fluids; controls; labor, construction, and construction interest; design fees, testing, and shakedown; profit and overhead. These costs may either be related to system size—variable costs—or independent of system size—fixed costs. Many variable costs can be related to collector size; therefore, collector size is frequently used as a measure of the size of a solar system.

The second class of costs recurs through the system's life and includes: interest on loans—for both debt and equity financing; maintenance; operating costs; replacements and repairs; insurance; taxes. These costs can usually be represented as discrete cash flows during the life of the solar system. Specific values of initial and distributed costs must be determined for each type of system considered from consultants familiar with solar designs.

B. Solar Cost Equations

The preceding costs can be combined into a single equation representing the yearly cost (sometimes called the *levelized* cost) C_y of owning and operating a solar system above the cost of the conventional backup system if present. One such *after-tax* cost equation for a profit-making firm is given below. Note that C_y is the cost to the firm of a solar system, not the revenue required to pay for it—the *before-tax* cost. All dollar values and interest rates are based on constant dollars, i.e., net of inflation:

$$
\begin{aligned}
C_y = &\ [C_{s,\text{initial}} - \text{ITC}](A/P,i,N) \\
&\qquad\qquad\qquad \text{(initial cost less investment tax credit)} \\
&\ - C_{s,\text{salv}}(A/F,i,N)(1 - 0.3) \\
&\qquad\qquad\qquad \text{(salvage value at 30\% tax rate; see p. 287.)} \\
&\ + \left[\sum_{k=1}^{N} R_k(P/F,i,k)\right](A/P,i,N)(1 - T_{\text{inc}}) \quad \text{(replacements)} \\
&\ + C_e(1 - T_{\text{inc}}) \quad \text{(operating energy cost)} \\
&\ + T_p(1 - T_{\text{inc}})C_{s,\text{assess}} \quad \text{(property tax)} \\
&\ + M(1 - T_{\text{inc}}) \quad \text{(maintenance)} \\
&\ + I(1 - T_{\text{inc}}) \quad \text{(insurance)} \\
&\ - T_{\text{inc}}i_m[\Sigma\, P_k(P/F,i,k)](A/P,i,N) \quad \text{(interest tax deduction)} \\
&\ - T_{\text{inc}}\left[\sum_{k=1}^{n} D_k(P/F,i,k)\right](A/P,i,N) \quad \text{(depreciation).} \qquad (7.25)
\end{aligned}
$$

The symbols used in Eq. (7.25) are discussed shortly and are defined as follows

$C_{s,initial}$ initial net solar system costs total above that of a conventional-fueled backup, if present,

$C_{s,assess}$ assessed value of solar system for property taxes,

$C_{s,salv}$ salvage value after N years,

C_e annual cost of operating energy,

i real discount rate net of inflation,

D_k depreciation amount in year k,

i_m market loan interest rate (8–13%) multiplied by the percent of debt financing r_d from Eq. (7.1)

I insurance cost per year ($\sim 0.25\%$),

ITC   investment tax credit (if any) (0–20%),

k year during period N, $k = [1,N]$,

M maintenance cost per year (1–2%),

N number of years of economic analysis (5–30 yr),

P_k principal at the end of year k,

$(A/P,i,N)$ capital recovery factor based on the cost of money i prorated in accordance with Eq. (7.1) to include debt and equity financing

R_k replacement cost in year k,

T_{inc} income tax rate (48–52% for profit making firm), federal rate + state rate − state rate × federal rate,

T_p property tax rate (0–3%).

Several of the terms listed above will be described in more detail below.

    The solar cost equation for a nonprofit firm (or a private residence) is different from that given above. For a nonprofit firm, ITC = 0, $D_k = 0$, and no tax deductions for replacements R_k, power costs C_e, property taxes T_p, maintenance M, and insurance I are permitted. In addition, backup fuel costs are not deductible.

    System Lifetime   In (Eq. 7.25) the period of analysis N can have several meanings. The service life may be less than the physical life which measures the period for which the facility is sound but is no longer used owing to obsolescence or environmental constraints, for example.

    The accounting life of a facility is determined by the tax code and is used for calculating depreciation as described below. The accounting lifetime may differ for state and federal tax purposes. The economic life is a third common period of analysis and is the optimal replacement interval. The economic life ends when an asset is recycled either because its service is no longer needed or because there is a more profitable replacement. It is the economic life which is generally used in making life-cycle costing studies except for the depreciation calculation.

Depreciation   Depreciation of a system results from a decrease in worth arising from obsolescence, wear, or capital consumption. Several generic types of depreciation occur including physical depreciation from everyday wear and tear or accidental damage. Functional depreciation occurs if new requirements are imposed which are beyond the capacity of the equipment; or contrariwise, the demand for the service may cease to exist. Technological depreciation can occur if entirely new and more efficient approaches to accomplishing a task make the present method of operation uneconomical. In addition, physical depletion of an exhaustible resource is considered to be depreciation. Depletion costs are high for the scarcer resources.

An additional type of depreciation, which could be called monetary depreciation, decreases the value of an asset. Since depreciation rates apply to the original asset and not its replacement, the significant impacts of inflation are not included. That is, an identical replacement for a given asset will cost more, in a period of inflation, than the asset cost when first purchased. This type of depreciation cannot be deducted from gross profits for tax purposes. All other depreciation types are theoretically accounted for by several depreciation methods described below.

Straight Line Depreciation   The simple method of linearly depreciating a solar system over a period of N years leads to the following expression for D_k, the annual depreciation amount in Eq. (7.25):

$$D_k = (C_{s,\text{initial}} - C_{s,\text{salv}})/N. \qquad (7.26)$$

The book value (initial cost less accumulated depreciation) BV_k is given by

$$BV_k = C_{s,\text{initial}} - kD_k. \qquad (7.27)$$

Example   A solar system has a first cost of $100,000 and an estimated salvage value of $10,000. If the estimated life is 10 yr, the depreciation charge per year will be $\frac{1}{10}(100,000 - 10,000) = \$9000$; that is, the depreciation rate is 10%. The table shows the yearly depreciation charge and book value.

Year	Depreciation charge ($)	Book value ($)
0	—	100,000
1	9000	91,000
2	9000	82,000
3	9000	73,000
4	9000	69,000
5	9000	55,000
6	9000	46,000
7	9000	37,000
8	9000	28,000
9	9000	19,000
10	9000	10,000

■

Sum of Digits Depreciation Federal tax laws permit depreciation schedules which accelerate depreciation during the early years of a solar system's life. This practice enables the discounted life-cycle value of the total depreciation amount to be higher than with the simple, straight line method. Of course, the total amount depreciated in current dollars is the same for any method.

The sum of digits method can be used if an asset is depreciated for a period greater than 7 yr. The annual depreciable amount D_k is the ratio of the digit corresponding to the remaining years of life to the sum of digits for the entire life. In equation form,

$$D_k = [2(N - k + 1)/N(N + 1)](C_{s,\text{initial}} - C_{s,\text{salv}}). \qquad (7.28)$$

The book value is

$$BV_k = [(N - k + 1)(N - k)/N(N + 1)](C_{s,\text{initial}} - C_{s,\text{salv}}) + C_{s,\text{salv}}. \qquad (7.29)$$

Example Consider an asset with 5-yr life having first cost $10,000 and salvage value $2,000. The sum of digits is $1 + 2 + 3 + 4 + 5 = 15$, and the depreciation during the first year will be $\frac{5}{15}$ ($10,000 − $2,000) = $2,667. Annual depreciation charges and book values are given in the table.

Year	Depreciation charge ($)	Book value ($)
0	—	10,000
1	$\frac{5}{15}$ (8000) = 2,667	7,333
2	$\frac{4}{15}$ (8000) = 2,133	5,200
3	$\frac{3}{15}$ (8000) = 1,600	3,600
4	$\frac{2}{15}$ (8000) = 1,067	2,533
5	$\frac{1}{15}$ (8000) = 533	2,000

■

Declining Balance Depreciation Another accelerated depreciation method, called the declining balance method, is used in the U.S. A constant *rate* of depreciation r_d is applied to the book value each year. The book value is given by

$$BV_k = C_{s,\text{initial}}(1 - r_d)^k. \tag{7.30}$$

The depreciation rate $r_d = 1 - (C_{s,\text{salv}}/C_{s,\text{initial}})^{1/N}$ so the depreciation amount is

$$D_k = C_{s,\text{initial}}(C_{s,\text{salv}}/C_{s,\text{initial}})^{k/N}[1 - (C_{s,\text{salv}}/C_{s,\text{initial}})^{1/N}].$$

The declining balance method requires the knowledge of the salvage value, which is rarely known accurately. In addition, the salvage value cannot be negative or zero. A more widely used version of the declining balance method is based on a fixed rate independent of the salvage value. The rate is established as a multiple of the straight-line rate $1/N$. The maximum rate permitted in the U.S. is $2/N$—the double declining balance rate.

Example Calculate the depreciation amount and book value of a $10,000 asset depreciated over 5 yr at the double declining rate.

Year	Depreciation charge ($)	Book value ($)
0	—	10,000
1	$\frac{2}{5}$ (10,000) = 4,000	6,000
2	$\frac{2}{5}$ (6,000) = 2,400	3,600
3	$\frac{2}{5}$ (3,600) = 1,440	2,160
4	$\frac{2}{5}$ (2,160) = 864	1,296
5	$\frac{2}{5}$ (1,296) = 518	778

∎

Depreciation charges are deducted from the gross income of a firm prior to taxes and are, therefore, the same as operating expenses for tax purposes. Figure 7.1 shows the relative size of the three depreciation modes. Note that land is not considered to be a depreciable asset.

Taxes Taxes are imposed on income and property as a tool of fiscal policy and to support (a part of) the operation of governments. Property taxes are charged by local governments based upon the assessed value of taxable assets. Relative to income taxes, they are quite small in some states. Income taxes are levied on corporate income and are based on gross income reduced by various types of deductions for operations, depreciation, etc. Direct tax credits on the tax amount are sometimes used to stimulate the use of solar systems. Therefore, corporate taxes can be expressed as

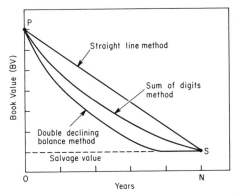

FIGURE 7.1   Book values of an asset depreciated by the straight line, sum of digits, and double declining balance methods over a period of N years. S denotes the salvage value and P the present value of the asset when purchased.

corporate tax = effective tax rate x [gross income
  − expenses − debt interest − depreciation
  − bad debts] − tax credits.   (7.31)

The corporate tax rate in the U.S. is 22% plus a 26% surtax on all income over \$25,000. For purposes of this book an effective rate of 48% is used since it applies to engineering accuracy for taxable incomes over \$500,000. States generally charge an income tax as well. If the federal tax is deductible from state taxes, the effective tax rate T_{inc} is

T_{inc} = effective federal rate + state rate
  − state rate × effective federal rate.   (7.32)

A typical rate T_{inc} of 50% for solar analysis is often used. Note that long-term capital gains, such as net income from salvage sale of an asset, are taxed at a maximum rate of 30% (R8).

Investment tax credits are used to stimulate investment. Solar systems for industrial use claim an initial 20% credit on federal taxes if the lifetime is greater than seven years. Solar systems for space heating or cooling for industrial facilities are subject to a 10% tax credit enacted as part of the 1978 National Energy Act. The extra 10% credit expires in 1985. Also, some states have enacted state tax credits for solar systems. However, since state tax payments are relatively much smaller than federal taxes, the tax credit must be able to be distributed over several years if an appreciable saving is to be realized.

Inflation   The annual costs used in Eq. (7.25) do not include the effect of differential inflation in future costs of energy, maintenance, etc. As such they are numerically smaller than future costs which inflate and are involved in year-by-year cash flows. Inflation effects can be included using modified compound interest factors.

For example, positive *differential inflation* may apply to fuel costs in a period of general inflation. That is, fuel prices may increase more rapidly than other goods and services. If the differential rate is e, the discounted life-cycle cash flow in constant dollars net of *general inflation* for fuel costs is given by

$$F_0 \left[\left(\frac{1+e}{1+i}\right) + \left(\frac{1+e}{1+i}\right)^2 + \cdots \right],$$

where F_0 is the fuel cost in the first year. An equivalent annualized cost $\bar{C}_f$ for fuel can be identified so that the same life cycle total will be accumulated by payments of $\bar{C}_f$ per year:

$$\bar{C}_f \left[\left(\frac{1}{1+i'}\right) + \left(\frac{1}{1+i'}\right)^2 + \cdots \right].$$

Equating these two series and solving for $\bar{C}_f$,

$$\bar{C}_f = F_0(A/P,i,N)/(A/P,i',N), \tag{7.33}$$

where the inflation-modified discount rate is

$$i' = (i - e)/(1 + e). \tag{7.34}$$

(Note that i' may be positive, zero, or negative.) The annualized cost $\bar{C}_f$ is sometimes called the *levelized* cost. Equation (7.33) can be used to find the levelized cost of maintenance, replacements, insurance, and other future costs if they inflate at rates different from the general inflation rate.

Example If the cost of oil is $F_0 = \$16.50/$bbl today, what is the levelized cost over 15 yr if the discount rate is $i = 10\%$ and the price of oil increases differentially at $e = 5\%$ per year?

Solution Find i' and the capital recovery factors from Eq. (7.13):

$$i' = (0.1 - 0.05)/1.05 = 0.0476;$$
$$(A/P,0.0476,15) = 0.0947;$$
$$(A/P,0.10,15) = 0.1315.$$

Then from Eq. (7.33) the levelized cost is

$$\bar{C}_f = \$16.50 \times 0.1315/0.0947 = \$22.90/\text{bbl}. \qquad \blacksquare$$

The solar cost equation (7.25) can be rewritten using the modified rate i' to account for inflation:

$$C_y = [C_{s,\text{initial}} - \text{ITC}](A/P,i,N)$$
$$- C_{s,\text{salv}}|_{k=0}(1 - 0.3)(A/F,i_s',N)$$
$$+ \left[\sum_k R_k \bigg|_{k=0} (P/F,i_R',k)\right](1 - T_{\text{inc}}) \bigg|_{k=0} (A/P,i,N)$$
$$+ C_e(1 - T_{\text{inc}})|_{k=0}(A/P,i,N)/(A/P,i_e',N)$$
$$+ T_p C_{s,\text{ass}}(1 - T_{\text{inc}})|_{k=0}(A/P,i,N)/(A/P,i_p',N)$$
$$+ M(1 - T_{\text{inc}})|_{k=0}(A/P,i,N)/(A/P,i_m',N)$$
$$+ I(1 - T_{\text{inc}})|_{k=0}(A/P,i,N)/(A/P,i_i',N)$$
$$- T_{\text{inc}}i_m[\Sigma P_k(P/F,i,k)](A/P,i,N)$$
$$- T_{\text{inc}}[\Sigma D_k(P/F,i,k)](A/P,i,N). \tag{7.35}$$

It is assumed that property taxes T_p, maintenance M, replacements R_k, power costs C_e, insurance I, and salvage value inflate differentially and the effective discount rates i_p', i_m', i_R', i_e', i_i', i_s' evaluated from Eq. (7.34) apply. Replacements R_k made in year k are evaluated at $t = 0$ in Eq. (7.35). It is also assumed that the effective tax rate is not affected by inflation. (In the case of an individual's personal returns, the rate may increase.)

Each term in Eq. (7.35) represents an annual payment at time $t = 0$ (i.e., at the *start* of year $k = 1$) which, when compounded and discounted, will give the same total life-cycle discounted cash flow as the actual year-by-year cash flows made by the system owner. Solar capital, tax deduction for interest, and depreciation payments are discounted year-by-year into the future and are not affected by inflation.

Depending on the relative magnitudes of inflation and discount rates, the annual, discounted flows may increase or decrease into the future; that is, i' may be negative or positive. Salvage value and replacement costs also are assumed to inflate in the future and are discounted to the present using the effective rate i'. Although the salvage value may increase owing to inflation effects, the depreciation rate D_k is based on the value at $t = 0$ and is not subject to differential inflation.

Equation (7.25) can also be expressed in *current* dollars, i.e., in dollars subject to general inflation instead of constant dollars whose amount is net of inflation. Real or constant dollars can be viewed as dollars of constant purchasing power whereas current or future dollars are dollars of reduced purchasing power.

Example Calculate the terms in Eq. (7.35) and annual cash flows in constant and discounted dollars for a solar system costing $10,000 net of investment tax credits. The salvage value is $1000 at $t = 0$ (for simplicity, ignore the long-term capital gains tax on salvage sale). Insurance, maintenance, and property taxes total $400/yr at $t = 0$. A repair costing

$100 (based on $t = 0$) is made in year $k = 2$ and straight-line depreciation is used. The tax rate is 50%, and tax-deductible interest charges are $400, $300, and $200 for years 1, 2, and 3. The discount rate is 15%, the differential inflation rate is 10%, and the analysis is to be carried out for a 3-yr period.

Solution The effective interest rate net of inflation for inflating costs is $i' = (0.15 - 0.10)/1.10 = 0.04545$. Capital recovery factors are $(A/P,i,3) = 0.437977$ and $(A/P,i',3) = 0.3641$. The annual cost terms in Eq. (7.35) are then

solar capital $[10,000 \times (A/P,i,3)]$	4379.77
salvage value $[1,000 \times (A/F,i',3)]$	− 318.63
replacements $[100 \times (P/F,i',2)(A/P,i,3) \times 0.5]$	20.04
operating costs $[400 \times 0.5] \times (A/P,i,3)/(A/P,i',3)$	240.54
interest tax deductions	− 154.65
depreciation $[3000 \times 0.5]$	− 1500.00
Annual Cash Flow	$2667.07

Additional insight into the cash flow terms can be gained by examining year-by-year components. The next table summarizes annual flows in constant, nondiscounted dollars. The amounts shown represent actual after-tax dollar outlays made in years 1–3.

Cash Flow Summary—Constant Dollars

					Interest	
Year	Solar capital[a]	Salvage[b]	Repairs[b]	Operating costs[b]	Tax savings[a]	Depreciation[a]
1	4379.77	—	—	220.00	−200	−1500
2	4379.77	—	60.50	242.00	−150	−1500
3	4379.77	−1331.00	—	266.20	−100	−1500
	13139.31	−1331.00	60.50	728.20	−450	−4500

[a] Not affected by differential inflation.
[b] Inflate at 10% per year, differentially.

Note that the 10% and 15% rates used in this example are for illustrative purposes only and are both higher than rates usually experienced in the U.S. economy in *constant* dollars. The salvage value, repair cost, and operating costs have been inflated at the rate of 10% per year above the values given in the problem statement (for time $t = 0$).

Discounted cash flows at $i = 15\%$ are shown in the table below.

Discounted Cash Flows Summary—Constant, Discounted Dollars

Year	Solar capital	Salvage	Repairs	Operating costs	Tax savings	Depreciation
1	3808.50	—	—	191.30	−173.91	−1304.35
2	3311.73	—	45.75	182.99	−113.42	−1134.22
3	2879.77	−875.15	—	175.03	−85.75	−986.27
	10,000.00	−875.15	45.75	549.32	−353.08	−3424.84

The column totals represent the total *present worth* of cash flows occurring over the 3-yr period. The various terms in Eq. (7.35) and the first table in this example are these present worth amounts P multiplied by the capital recovery factor $(A/P,i,N)$ except for the salvage value, which is multiplied by $(A/P,i',N)$. ■

Before-Tax Analysis Equations (7.25) and (7.35) represent the after-tax cost of solar energy to a profit-making firm. However most firms use before-tax calculations to determine the *before-tax revenue* which must be produced to pay for and operate a plant. The before-tax cost factor is usually called the fixed charge ratio (FCR) and is defined as the ratio of the levelized, annual, before-tax cost of solar energy to the solar system initial capital cost.

The annual or levelized revenue R in constant dollars is given by [using the notation of Eq. (7.25)]

$$R = (1 - T_{\text{inc}})[C_e + T_p C_{s,\text{assess}} + M + I + (1 - \bar{f}_s)C_f Q_y]$$
$$+ T_{\text{inc}}(R - \text{DEP}) + (C_{s,\text{initial}} - \text{ITC})(A/P,i'',N) \qquad (7.36)$$

where the cost of backup fuel has been included; $\bar{f}_s$ is the annual solar load fraction, i'' is the effective after-tax interest rate, C_f is the unit cost of fuel, and Q_y is the annual plant energy demand; DEP represents the levelized depreciation amount and a zero salvage value is used. If the prices in brackets in the first term are expected to inflate differentially, they will be multiplied by the ratios of capital recovery factors in accordance with Eq. (7.33). The difference between Eq. (7.25) and (7.36) is that the former represents the cost of a solar system to a firm and the latter is the revenue which must be produced *before taxes* to pay for a solar system *and the backup fuel.*

Solving for the revenue R,

$$R = C_e + T_p C_{s,\text{assess}} + M + I + (1 - \bar{f}_s)C_f Q_y$$
$$+ [1/(1 - T_{\text{inc}})][(C_{s,\text{initial}} - \text{ITC})(A/P,i'',N) - T_{\text{inc}} \times \text{DEP}]. \qquad (7.37)$$

The average unit cost of solar energy C_{se} in the absence of auxiliary energy can be calculated from Eq. (7.37) by dropping the fifth term:

$$C_{se} = R(\bar{f}_s = 1)/(PL_f 8760) \equiv FCR \times C_{s,\text{initial}}, \tag{7.38}$$

where P is the solar plant name plate rating, and L_f is the *load factor* defined as the number of full-load hours of solar system operation divided by the number of hours in a year. Typical units of C_{se} are ¢/kW h, for example.

IV. PRODUCTION FUNCTIONS

The previous sections of this chapter have described the methods for calculating the discounted, annual cost of owning and operating a solar system, after taxes. However, the benefit from owning the system must also be known in order to assess the costs-to-benefits ratio. The benefit from owning a solar system is the displacement of fuel and the associated cost savings. It is usually assumed that a solar system does not replace a conventional, backup system but only displaces part of its fuel. That is, the solar system is a fuel saver.

Most solar energy systems are constrained by the law of diminishing returns, which prohibits the economical existence of full-load solar system which can provide 100% of the demand at all times. The law of diminishing returns states that as equal amounts of a variable are added to other inputs which remain fixed, the system output does not increase proportionately beyond some point.

Figure 7.2 shows an example of the law of diminishing returns. The annual energy output of a solar system Q_s is seen to continuously increase with collector area A_c, but at a decreasing rate, all other inputs remaining unchanged. That is, the marginal output decreases. The saturation effect exhibited in Fig. 7.2 is associated with most design variables for a solar system including flow rates, storage sizes, heat exchanger areas, insulation thickness, etc. The shape of the Q_s curve indicates the general functional nature of production functions.

A *production function* is the technical relationship specifying the maximum amount of output capable of being produced by each and every set of inputs (S7). For example a 200-m² solar collector with 5 tons of thermal storage, along with other components, produces 400 GJ/yr. The production function relates all the technical inputs to the technical level of output for a range of all variables around the nominal design

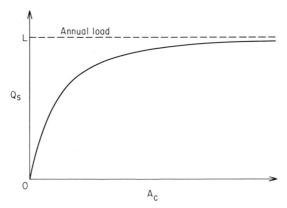

FIGURE 7.2   Example of the law of diminishing returns. As extra collector area A_c is added to a system, the annual energy delivery per unit area Q_s/A_c decreases, while Q_s itself increases, all other system inputs remaining fixed.

point. It is also possible to assign costs to each component and the output to devise an economic production function.

There are several sorts of production functions including analytical and tabular. Production functions are usually provided from mathematical modeling efforts which can predict output for a broad range of inputs. Although full-scale physical systems can be used to check points on a production function, they are not an economic means of formally defining the function space. The performance prediction methods described in Chapters 5 and 6 are well suited to calculation of production functions.

Inputs to production functions are of two types. Endogenous parameters include collector area and efficiency, collector orientation, storage size, control function, flow rates, heat exchanger performance, and many other variables. Exogenous inputs impacting production functions include climatic variables, solar flux level, and end use system energy demands. Several types of production functions will be described.

A.   Analytic Production Functions

Functions which can be expressed in equation form are particularly useful since they can be used with cost equations relating the size and costs of various inputs to calculate the optimal system configuration. However, analytical production functions cannot be written down *a priori* since the governing equations of solar systems are nonlinear and must respond to the vagaries of an unpredictable climatic environment. Therefore the functional forms are usually empirical fits of computer simulations or, rarely, of field test data.

One simple functional form is a sum of polynomials in the principle inputs, for example,

$$Q_s = P_1(A_c) + P_2(m_s) + P_3(\dot{m}_f) + P_4(\epsilon_{hx}) + P_5(R_{ins}) + \cdots, \quad (7.39)$$

where A_c is the area of a collector of a given type, m_s the mass of storage of a given type, $\dot{m}_f$ the fluid flow rate, ϵ_{hx} the heat exchanger effectiveness, and R_{ins} the thermal resistance of insulation. The polynomial form has the advantage that it is differentiable, which is a useful feature, as shown below. However, an equation of the type of Eq. (7.39) is rather cumbersome if many terms are included.

The shape of the curve in Fig. (7.2) also indicates that exponential decay, logarithmic, or power law functions might be used. Functional forms could be, respectively,

$$\begin{aligned} Q_s = B \, \exp[-f_1(A_c)] &\times \exp[-f_2(m_s)] \\ &\times \exp[-f_3(\dot{m}_f)] \cdots + \text{const} \end{aligned} \quad (7.40)$$

or

$$Q_s = \ln \prod_{i=1}^{N} a_i X_i^{n_i}, \quad (7.41)$$

where the X_i denote system component sizes.

Other forms include

$$Q_s = \prod_{1=1}^{n} (a_i \ln X_i) \quad (7.42)$$

and

$$Q_s = a_0 \prod_{1=1}^{N} X_i^{\nu_i}. \quad (7.43)$$

Power law and logarithmic functions are easy to manipulate mathematically and the coefficients contain information not present in the polynomial forms. For example, a power law function can be differentiated to show that

$$\frac{dQ_s}{Q_s} = \sum_{1=1}^{n} \nu_i \frac{dX_i}{X_i}. \quad (7.44)$$

The exponents ν_i are called the returns to scale and represent the percentage change in output Q_s for a percentage change in input. The sum of the

ν_i represents the total returns to scale. If the sum is greater than one, returns to scale are said to be increasing, if less, decreasing.

  Example A production function which closely predicts the performance of an industrial water heater for a specific location has been shown by computer simulations to be of the form

$$Q_s = L \left[0.35 - \frac{F_R U_c}{100(F_R \eta_0)} \right] \ln \left[1 + \frac{20(F_R \eta_0) A_c}{L} \right]. \qquad (7.45)$$

Find the percentage change in delivery Q_s if the load L, area A_c, and collector loss coefficient $F_R U_c$ increase individually by 1% from their nominal values:

$$L = 100 \quad \text{GJ/yr;}$$
$$A_c = 50 \quad \text{m}^2;$$
$$F_R U_c = 2.0 \quad \text{W/m}^2 \, {}^\circ\text{C;}$$
$$F_R \eta_0 = 0.7.$$

The delivery of the nominal system is from Eq. (7.45) 66.8 GJ/yr. The expected percentage effects in output can be estimated by differentiating Eq. (7.45) as follows. For the collector loss U_c effect,

$$\frac{dQ_s}{Q_s} \sim - \frac{d(F_R U_c)}{F_R U_c} \{ [35 F_R \eta_0 / (F_R U_c)] - 1 \}^{-1}$$

$$\frac{dQ_s}{Q_s} \sim -1\% [35 \times 0.7/2 - 1] = -0.089\%.$$

Therefore, a 1% increase in collector loss coefficient will decrease performance by less than $\frac{1}{10}\%$. For the collector area,

$$\frac{dQ_s}{Q_s} \sim \frac{dA_c}{A_c} \, 1/(L/20 F_R \eta_0 A_c + 1) \ln(1 + 20 F_R \eta_0 A_c / L)$$

$$\sim 1\% \frac{1}{\frac{8}{7} \ln(8)}$$

$$\frac{dQ_s}{Q_s} \sim 0.42\%$$

Therefore, a 1% increase in area causes a 0.42% increase in energy delivery.

  Finally, for a change in load,

$$\frac{dQ_s}{Q_s} \sim \frac{dL}{L} \left\{ 1 - \frac{1}{(1 + L/20 F_R \eta_0 A_c)(\ln(1 + 20 F_r \eta_0 A_c / L))} \right\}$$

$$= 1\% \left\{ 1 - \frac{1}{\frac{8}{7} \ln(8)} \right\}$$

$$= 0.58\%.$$

Hence, the load shows the largest impact of the three variables considered. The load is an exogenous variable and increases the energy delivery since the system load factor is increased. For this example, if the collector area were doubled, a 1% increase in load would cause a 0.65% increase in energy delivery. The energy delivery for a system with double the nominal area would be 87.0 GJ/yr. ∎

B. Tabular Production Functions

System performance is usually calculated at discrete points. If the points are sparse in the performance space, an empirical curve fit for the production function may not be appropriate. In that case, a look-up table may represent the production function. Either tables or analytical production functions may be used in the following section on optimization.

V. OPTIMAL SYSTEM SELECTION

Very many types and sizes of solar systems can be designed to displace the majority of fuel required to energize a given process. In general, one member of this large class will exhibit a discounted annual total energy cost (solar energy plus fuel energy) which is less than the discounted annual cost of all other options. The purpose of solar system economic optimization is to identify the minimum annual cost (MAC) system. An equivalent optimization criterion is to minimize the total present worth of all solar and fuel cash flows. As shown earlier, the discounted annual and total present worth approaches are a duality related by the capital recovery factor.

A third method of optimization is to select the solar plus fuel mix which assures the maximum internal rate of return on investment (IRR). The IRR method usually correlates with the discounted cash flow methods—annual or total—but under certain circumstances two or more IRR values can exist which will solve the cash flow equation. This feature of the IRR method plus the necessity for an inefficient iterative solution to find the IRR tends to make the MAC method preferable. The MAC method is the principal method recommended in this chapter but a brief description of the IRR method is included.

A. The Minimum, Discounted Annual Cost (MAC) Method

The MAC method seeks to find the solar system configuration with the minimum cost of accomplishing a given task by use of a mix of solar- and fuel-based energy. The minimum value of the annual worth of solar plus fuel energy is the optimization criterion with constraints imposed by physical or economic limits. For a specific solar system size the total annual energy cost C_T after taxes is given by

$$C_T = C_y(Q_s) + C_f(Q_f)^*. \tag{7.46}$$

C_y, the annual solar cost, is given by Eq. (7.35) and C_f represents the annual worth of auxiliary fuel costs to provide the fraction of heat Q_f not supplied by solar energy. Note that Q_f will vary year by year depending upon *future* plant schedules and local weather. However, these effects cannot be known in advance during the system design phase when economic analyses are needed. Therefore, optimization calculations are generally based on features of the average *past* history which are deemed to be most appropriate.

The minimization of Eq. (7.46) is subject to the constraint that the annual total energy demand Q_T be met by the sum of solar and nonsolar energy. In equation form,

$$Q_T = Q_s + Q_f. \tag{7.47}$$

Note that the magnitudes of Q_s and Q_f effect the prices C_y and C_f. Obviously, a larger solar system to provide a larger Q_s will have a larger cost C_y. The costs of all extra solar system components not a part of the nonsolar system are totaled and discounted to find C_y.

The method of Lagrange multipliers may be used to find the minimum cost $C_{T,\min}$ from Eqs. (7.46) and (7.47). Equation (7.46) can be unconstrained as follows:

$$C_T = C_y(Q_s) + C_f(Q_f) + \lambda(Q_T - Q_s - Q_f). \tag{7.48}$$

The stationary point of Eq. (7.48) can be calculated if C_y and C_f are differentiable:

$$dC_T = 0 = (dC_y/dQ_s)\, dQ_s + (dC_f/dQ_f)\, dQ_f + \lambda(-dQ_f - dQ_s). \tag{7.49}$$

Since the differentials dQ_f and dQ_s are unconstrained, we have

$$\left.\frac{dC_y}{dQ_s}\right|_{C_{T,\min}} = \left.\frac{dC_f}{dQ_f}\right|_{C_{T,\min}} = \lambda. \tag{7.50}$$

---

* Note that C_f is the after-tax cost of fuel given by $Q_f \bar{C}_f(1 - T_{\text{inc}})$ where $\bar{C}_f$ is the *levelized* unit cost of fuel ($/GJ, for example).

The term dC_y/dQ_s is the marginal cost of solar energy and dC_f/dQ_f is the marginal cost of nonsolar fuel. Equation (7.50) states that the minimum total cost $C_{T,min}$ occurs when the marginal costs of fuel and solar energies are equal. Figure 7.3 shows a typical set of total and marginal cost curves. It is seen that the marginal costs are equal when the slopes of the annualized cost curves are equal.

The minimum total cost may not occur in the open interval of solar system delivery from zero to $Q_{s,max}$. The zero delivery point corresponds to no solar system and this may be the cost-optimal configuration if the marginal cost of solar heat is everywhere greater than the marginal cost of fuel. Note that the solar marginal cost increases with solar heat amount Q_s owing to the effect of the law of diminishing returns.

It is also possible for the minimum cost point to lie at the maximum solar system size even though marginal costs are not equal there. This will occur if the marginal cost of fuel is very high, but the budget or space required for solar is limited by economic and/or physical constraints. Figure 7.4 shows the two types of boundary minima for total

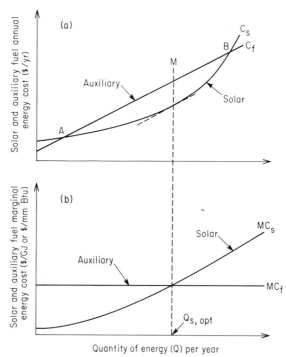

FIGURE 7.3   (a) Annual cost curves for solar and nonsolar energy. (b) Marginal cost curves intersecting at the minimum cost point.

cost curves along with the more typical curve which exhibits a minimum cost in the open interval $(0, Q_{s,\max})$.

Figure 7.3 shows another property of solar versus fuel costs. It is seen that solar energy is less costly than fuel at any configuration lying between points A and B. Of course, the difference is a maximum at point M but a firm could show a net cost savings anywhere between points A and B. Or the solar system could be designed to save the maximum amount of fuel while incurring no net cost increase. Thus point B would represent the optimal design under these conditions. In this book, however, point M is usually taken to represent the optimal configuration.

It should be noted that the cost expression $C_y(Q_s)$ is related to the solar system production function described in the previous section. Since the energy delivery in a given location Q_s is related to component sizes X_i, the annual cost C_y is seen to depend on the magnitudes of the X_i—for example, collector area, storage size, flow rate, etc. The annual solar cost is given by

$$C_y = C_{X_1} + C_{X_2} + C_{X_3} + \cdots \tag{7.51}$$

where C_{X_i} are annualized, discounted costs of components X_i. The La-

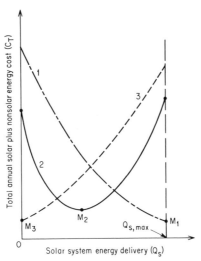

FIGURE 7.4   Total cost curves C_T plotted vs. Q_s, the solar system delivery amount. Curve 1 occurs if the marginal cost of solar heat is everywhere less than that for fuel and curve 3 represents the case where fuel marginal costs are everywhere less than solar marginal cost. Curve 2 shows a minimum at M_2, where solar and fuel marginal costs are equal.

grange multiplier operation leading to Eq. (7.50) can be repeated using
Eq. (7.51) to show that

$$\frac{\partial C_f}{\partial Q_f} = \frac{\partial C_{X_1}}{\partial X_1} \Big/ \frac{\partial Q_s}{\partial X_1} = \frac{\partial C_{X_2}}{\partial X_2} \Big/ \frac{\partial Q_s}{\partial X_2} = \cdots = \lambda. \tag{7.52}$$

Note that the numerator in each fractional term is the marginal cost of that
component while the denominator is the marginal productivity of each
component, which is calculated from the production function. The use of
Eq. (7.52) to optimize a system is shown by the following example.

Example   The production function of a specific process heat
system in San Antonio, Texas, has been found to be

$$Q_s = 100\{0.7 + \ln[(A_c/100)^{1/3}(m_s/100)^{1/15}]\}, \tag{7.53}$$

where A_c is the collector area (m²) and m_s the storage mass (kg). The annualized after-tax cost of the collector is $10A_c$ and for storage is $5m_s^{1/2}$. The
total heat demand is 100 GJ/yr and the discounted marginal cost of
backup fuel after taxes is $6/GJ (constant). Find the optimum storage size
and collector area.

From the data provided, the cost function is

$$C_y = 10A_c + 5m_s^{1/2}. \tag{7.54}$$

The numerators of fractional terms (marginal costs) in Eq. (7.52) are then

$$\frac{\partial C_{A_c}}{\partial A_c} = 10, \qquad \frac{\partial C_{m_s}}{\partial m_s} = \frac{2.5}{m_s^{1/2}}$$

and the marginal productivities from Eq. (7.53) are

$$\frac{\partial Q_s}{\partial A_c} = \frac{100}{3A_c}, \qquad \frac{\partial Q_s}{\partial m_s} = \frac{100}{15m_s}.$$

Substituting in Eq. (7.52) we have

$$6 = \frac{10}{100/3A_c} = \frac{2.5/m_s^{1/2}}{100/15m_s};$$

solving,

$$A_c = 20 \quad m^2, \qquad m_s = 256 \quad kg.$$

The annual energy delivery from Eq. (7.53) is 22.6 GJ/yr and the cost C_y
is $280/yr. The total energy cost C_T is $744.40/yr.

Exercise   Verify for several other (A_c, m_s) pairs that the
$744.40 annual cost is the minimum for the given cost and production
functions. ∎

The preceding analysis and examples assume that a functional form for a production function exists. For relatively simple systems such functions have been derived for specific configurations and may be used directly as shown above. It may not be possible to solve for component sizes X_i, however, owing to the complexity of the functional forms used. Iterative solution and numerical differentiation methods for calculating the marginal productivity may need to be used on a computer for complex problems. The marginal cost terms $\partial C_{X_i}/\partial X_i$ are usually quite easy to calculate in closed form, however.

If an analytic production function does not exist, numerical optimization methods must be used. Linear programming, integer programming, dynamic programming, and large-scale programming are standard numerical methods used for optimization (K7). If a numerical technique is used, it is frequently embodied into a systems model so that the search pattern may be directed by the optimization model on a step-by-step basis. The reader is referred to systems analysis texts for descriptions of the several optimization methods.

B. The Internal Rate of Return (IRR) Method

The internal rate of return is defined as that discount rate i which when used in Eq. (7.25) for discounting, will cause the magnitude of the annualized after-tax cost C_y to exactly equal the value of the annual savings in fuel. In equation form on an annual basis,

$$C_y(i_{\mathrm{IRR}}) = \bar{c}_{\mathrm{f}}\bar{f}_{\mathrm{s}}Q_{\mathrm{T}}, \tag{7.55}$$

where $\bar{c}_{\mathrm{f}}$ is the average unit cost of fuel after taxes, $\bar{f}_{\mathrm{s}}$ the annual solar load fraction, and Q_{T} the annual heat demand. Total life cycle costs can also be used instead of annual costs. Since i enters the solar cost equation in various forms such as $(P/F,i,N)$ and $(A/P,i,N)$, it is rarely possible to find a closed form solution for the value of i_{IRR}. An iterative solution is nearly always required; the Newton–Raphson procedure can be used, for example.

The method of finding the optimum system by using the IRR method therefore requires much more calculation than the annual worth method prescribed earlier. Note that an internal rate of return of 10% corresponds to a dollar rate of return of 16.6% if the general inflation rate is 6%, for example. An example illustrates the method.

Example A solar total energy system is expected to save $125,000 per year over the next 10 yr (ignore inflation for simplicity). The system costs \$750,000. What is the rate of return on the investment if the project is public and therefore tax exempt and if annual operating costs are expected to be \$3,000? Replacements and the salvage value are taken to be zero. Depreciation is not considered since there are no taxes involved. The life-cycle discounted cash flows are given by

$$125,000(P/A,i,N) \overset{?}{=} 750,000 + 3000(P/A,i,N).$$

Solving for $(P/A,i,N)$,

$$(P/A,i,N) = 750,000/(125,000 - 3,000) = 6.14.$$

Referring to the expression for $(P/A,i,N)$, we have

$$(P/A,i,N) = \{i/[1 - (1 + i)^{-N}]\}^{-1}.$$

Try

$$
\begin{aligned}
i &= 12\%, & (P/A,12,10) &= 5.605, \\
i &= 11\%, & (P/A,11,10) &= 5.889, \\
i &= 10\%, & (P/A,10,10) &= 6.145.
\end{aligned}
$$

Hence, the internal rate of return on the \$750,000 investment was 10%. Note that this simple example permitted a relatively direct solution. However, if salvage value, taxes, interest, tax deductions, or depreciation had been involved, present worth factors $(P/F,i,N)$ of various time scales would have been used and a much more complex trial and error solution would have been needed to find the IRR. ■

   The IRR method must be used with care to avoid erroneous conclusions. For example, two alternatives may change rankings when examined by the discounted cash flow and IRR methods.

   *Example** A small solar water heater can be purchased for \$1,000 to save gas in one plant of a firm (G) or to save electricity in another (E). Which solar system should be used if the cash flows are as follows:

Cash Flow-Fuel Savings per Year

System	0	1	2	3	4
E	-\$1000	\$100	\$350	\$600	\$850
G	-\$1000	\$1000	\$200	\$200	\$200

The cash flows for the gas system G reflect the fact that the gas price is regulated but a gas tap was unavailable the first year so expensive LP gas

---

* Adapted from (R8).

was used. If the minimum rate of return is 10%, the total, present worth of the two savings flows is

$$PW_E = -1000 + [100 + 250(A/G,10,4)](P/A,10,4) = \$411.56,$$
$$PW_G = -1000 + [1000 + 200(P/A,10,3)](P/F,10,1) = \$361.27.$$

(Note: The annualized cash flows can be calculated from $PW_G(A/P,10,4)$, for example). Therefore, the electric system is ranked highest.

When an IRR ranking is made we have

$$PW_E = 0 = -1000 + [100 + 250(A/G,i,4)](P/A,i,4).$$

Trial and error shows that $IRR_E = 23.4\%$. By an analogous procedure, $IRR_G = 34.5\%$. Therefore, contrary to the MAC ranking, the IRR method ranks the gas option ahead of the electric option!

The reason for this result is the assumption inherent in the IRR method that funds recovered during the system life can be invested at the IRR value. Another difficulty of the IRR method occurs if the running total of annual cash flows change sign more than once—for example, a negative cash flow for solar system purchase followed by several years of energy savings followed by a large cost for collector replacement. If the cash flows change sign more than once, multiple rates of return (multiple roots to the equation) may occur. In that event the IRR method is rendered useless. ■

C. *Payback Period*

One common but deficient economic index is the payback period, which is defined as the time required to recoup an initial investment in net savings. That is,

$$POP = \frac{C_{s,\text{initial}} - ITC}{(1 - T_{\text{inc}})[\bar{f}_s C_F - OMRIP] + T_{\text{inc}} C_{s,\text{initial}}/N} \tag{7.56}$$

where C_F the annual cost of fuel to operate the process *without* solar energy, and OMRIP includes annual operating, maintenance, replacement, insurance costs, and property taxes. Linear depreciation is assumed. Usually discounted cash flows are not used and no salvage value is considered. Although the method is popular, it is inadequate for careful economic studies since it ignores the time value of money and cash flows occurring after the payback period, penalizes long-lifetime projects, and cannot rank competing systems. It should not be used for solar system economic optimization.

REFERENCES*

A1 ASHRAE, "Handbook of Fundamentals." New York, 1977.

A2 ASHRAE, "Handbook of Fundamentals." New York, 1972.

A3 S. I. Abdel-Khalik, H-W. Li and K. R. Randall, *J. Heat Transfer* **100**, 199 (1978).

A4 S. W. Angrist, "Direct Energy Conversion," 3rd ed. Allyn-Bacon, Boston, Massachusetts, 1976.

A5 F. Aerstin and G. Street, "Applied Chemical Process Design." Plenum Press, New York, 1978.

A6 M. Abramowitz and I. A. Stegun, "Handbook of Mathematical Functions." Dover, New York, 1965.

B1 E. C. Boes *et al.*, Availability of Direct, Total, and Diffuse Radiation for Fixed and Tracking Collectors in the USA. NTIS Rep. No. SAND 72-0885 (1977).

B2 L. S. Marks, "Mechanical Engineer's Handbook," (T. Baumeister, ed.), 8th ed. McGraw-Hill, New York, 1977.

B3 O. E. Balje, Dr. O. E. Balje, Inc., Sherman Oaks, California.

B4 R. E. Barber, The Use of Solar Energy for the Cooling of Buildings (F. deWinter, ed.), ERDA Rep. SAN/1122-76/2, p. 91 (1976).

B5 Battelle Columbus Laboratories, Survey of the Applications of Solar Thermal Energy Systems to Industrial Process Heat, ERDA Rep. No. TID/27348/1, 3 volumes (1977).

B6 F. Benford and J. E. Bock, *Trans. Illum. Eng. Soc.* **34**, 200 (1939).

B7 L. R. Bush, STES Tech. Prog. Rep. No. 1, Aerospace Corp. ATR-77 (7692-01)-1 (1977).

B8 L. R. Bush, STES Tech. Prog. Rep. No. 2, Aerospace Corp. ATR-77 (7692-01)-2 (1977).

B9 L. R. Bush, STES Final Tech. Summary Rep., Aerospace Corp. ATR-78 (7692-01)-1, 2 vols. (1978).

B10 W. Beckman, *Solar Energy* **21**, 531 (1978).

C1 M. Collares-Pereira and A. Rabl, Simple Procedure for Predicting Long Term Average Performance of Nonconcentrating and of Concentrating Solar Collectors, Argonne Laboratory Rep. No. ANL-78-67, (1978); to be published in *Solar Energy*.

C2 S. W. Chi, "Heat Pipe Theory and Practice; A Sourcebook." McGraw-Hill, New York, 1976.

C3 J. D. Chase, "Chemical Engineering Report Series." McGraw-Hill, New York, 1970.

C4 S. W. Churchill, *Chem. Eng.* **99**, 91 (1977).

C5 R. S. Caputo, An Initial Study of Solar Power Plants Using a Distributed Network of Point Focusing Collectors, JPL Rep. No. 900-724 (1975); see also, by the same author, An Initial Comparative Assessment of Orbital and Terrestrial Central Power Systems, JPL Rep. No. 77-44, JPL, Pasadena, California (1977).

C6 M. Collares-Pereira, A. Rabl, and R. Winston, Lens Mirror Combinations With Maximal Concentration. Enrico Fermi Institute Rep. No. EFI 77-20 (1977).

C7 M. Collares-Pereira *et al.*, *Proc. Ann. Meeting*, p. 36-1. Am. Sec. ISES, Orlando, Florida (1977).

* In this list NTIS refers to the National Technical Information Service, 5285 Port Royal Rd, Springfield, VA 22161. NTIS reports may be ordered by the report number in paper copy or microfiche.

C8 T. A. Chubb, *Solar Energy* **17**, 129 (1975).

C9 M. Collares-Pereira and A. Rabl, *Solar Energy*, **22**, 155 (1979).

D1 B. de Jong, "Net Radiation Received by a Horizontal Surface on Earth." Delft Univ. Press, Delft, 1973.

D2 S. Dushman, "Scientific Foundations of Vacuum Technology," 2nd ed., Wiley, New York, 1962.

D3 W. S. Duff and G. F. Lameiro, ASME Paper 74-WA/Sol-4, New York (1974).

D4 W. S. Duff, ASME Paper 76-WA/HT-15, New York (1976).

D5 F. deWinter, *Solar Energy* **17**, 335 (1975); (see also Ref. K1).

D6 V. E. Dudley and R. M. Workhoven, Summary Report: Concentrating Solar Collector Test Results, Sandia Laboratories Rep. No. SAND78-0815, (1978).

D7 Department of Energy, *Thermal Storage Applications Workshop, II*, Rep. No. 5102-78, Golden, Colorado (1978).

E1 Electric Power Research Institute, *Proc. Semiann. EPRI Solar Program Rev. Meeting Workshop* Rep. No. ER-371-SR (1977).

E2 B. G. Eldridge *et al. Proc. Solar Ind. Process Heat Conf., Denver, Colorado* p. 141 (1978).

E3 ERDA, Solar Total Energy System: Large Scale Exp., Shenandoah, Georgia, Final Technical Progress Rep., NTIS Rep. No. ALO/3987-1/1 and -1/2 (1977).

F1 J. Fourier, "The Analytical Theory of Heat." Dover, New York, 1955.

F2 T. Fujita *et al.*, Projection of Distributed-Collector Solar-Thermal Electric Power Plant Economics to Years 1990–2000, JPL Rep. No. DOE/JPL-1060-77/1 (1977).

F3 G. Francia, *Solar Energy* **12**, 51–64 (1968).

F4 T. Fujita *et al.*, Techno-Economic Projections for Advanced Dispersed Solar-Thermal Electric Power Plants to Years 1990–2000, JPL Rep. No. DOE/JPL-1060-78/4 (1978).

G1 J. W. Gibbs, "The Collected Works of J. W. Gibbs," Vol. 1, p. 77. Yale Univ. Press, New Haven, Connecticut, 1948.

G2 K. M. Guthrie, "Chemical Engineering Reprint Series." McGraw-Hill, New York, 1969.

G3 W. Goldstern, "Steam Storage Accumulators." Pergamon, Oxford, 1970.

G4 E. L. Grant, W. G. Ireson, and R. S. Leavenworth, "Principles of Engineering Economy," 6th ed. Ronald Press, New York, 1976.

G5 D. F. Grether *et al., Proc. Conf., AS-ISES Winnipeg* p. 363 (1976); see also other Lawrence-Berkeley Laboratory Rep. by the same authors.

G6 A. Greenspan, *Fortune* p. 140 (February 13, 1978).

G7 General Electric, Solar Total Energy, Phase II, NTIS Rep. No. TID-27995 (1978).

H1 D. G. Holloway, "The Physical Properties of Glass." Springer-Verlag, Berlin and New York, 1973.

H2 J. P. Holman, "Heat Transfer," 4th ed. McGraw-Hill, New York, 1976.

H3 H. Hottel and A. Sarofim, "Radiative Transfer." McGraw-Hill, New York, 1967.

H4 C. D. Hodgman *et al.*, "Handbook of Chemistry and Physics," 10th ed. CRC, Cleveland, Ohio, 1959.

H5 D. Handley and P. J. Heggs, *Trans. Inst. Chem. Eng.* **46**, 251 (1968).

H6 M. Harrison and C. Pelanne, *Chem. Eng.* **84**, 62 (December 19, 1977).

H7 A. F. Hildebrandt and L. L. Vant-Hull, *Mech. Eng.* 23 (September 1974).

H8 G. N. Hatsopolous, The Thermoelectron Engine, Ph.D. dissertion, MIT, Cambridge, Massachusetts (1956).

H9 G. N. Hatsopolous and F. N. Huffman, *IECEC Proc.* (1975).

I1 Intertechnology Corporation, Analysis of the Economic Potential of Solar Thermal

Energy to Provide Industrial Process Heat, ERDA Rep. No. COO/2829-1, 3 volumes (1977); see also Ref. (R12).

J1   P. D. Jose, *Solar Energy* **1**, 19 (1957).

J2   M. Jakob, "Heat Transfer." Wiley, New York, 1956.

J3   Jet Propulsion Laboratory, Advanced Systems Development, Second Progress Rep. No. DOE/JPL-1060-78/6 (1978).

K1   S. A. Klein, A Design Procedure for Solar Heating Systems, Ph.D. dissertation, Univ. of Wisconsin, Madison, Wisconsin (1976).

K2   F. Kreith and J. F. Kreider, "Principles of Solar Engineering." McGraw-Hill, New York, 1978.

K3   S. A. Klein, *Solar Energy* **13**, 325 (1977).

K4   Kestin, J. (ed.), "The Second Law of Thermodynamics." Halsted Press, 1976.

K5   F. Kreith, "Principles of Heat Transfer." Intext Ed. Publ., Scranton, Pennsylvania, 1973.

K6   S. S. Kutateladze and V. M. Borishanskii, "A Concise Encyclopedia of Heat Transfer." Pergamon, Oxford, 1966.

K7   Kreider, J. F. (Editor-in-Chief), "Handbook of Solar Energy." McGraw-Hill, New York, 1980.

K8   W. M. Kays and A. L. London, "Compact Heat Exchangers." McGraw-Hill, New York, 1964.

K9   J. F. Kreider, *J. Heat Transfer* **97**, 451 (1975).

K10 G. R. Kent, *Chem. Eng.* **85**, 119 (September 25, 1978).

K11 S. A. Klein and W. A. Beckman, *Proc. Conf. Syst. Simulat. Modeling Solar Heating Cooling, San Diego, California* p. 171 (1978); see also by the same authors, *Proc. the Ann. Meeting, AS-ISES, Orlando, Florida* p. 8-1 (1977) and *Solar Energy* **22**, 269 (1979).

K12 S. A. Klein, Private communication (1978); also, A. Rabl, *ibid.*

K13 K. Ya. Kondratyev and M. P. Fedorova, *Proc. UNESCO/WMO Symp.* WMO Rep. No. 477, p. 36 (1976).

K14 B. V. Karlekar and R. M. Desmond, "Engineering Heat Transfer." West Publ., St. Paul, Minnesota, 1977.

K15 D. Q. Kern, "Process Heat Transfer." McGraw-Hill, New York, 1950.

L1   G. O. G. Löf *et al.*, World Distribution of Solar Energy, Univ. of Wisconsin Rep. No. 21, Madison, Wisconsin (1966).

L2   B. Y. H. Liu and R. C. Jordan, "Application of Solar Energy for Heating and Cooling of Buildings." ASHRAE, New York, 1977.

L3   B. Y. H. Liu and R. C. Jordan, *Solar Energy* **7**, 53 (1963).

L4   B. Y. H. Liu and R. C. Jordan, *Trans. ASHRAE* **67**, 526 (1961).

L5   B. Y. H. Liu and R. C. Jordan, *ASHRAE J.* **3**, 53 (1961).

L6   B. Y. H. Liu and R. C. Jordan, *Solar Energy* **4**, 1, (1960).

L7   P. T. Landsberg, *Photochem. Photobiol.* **26**, 313 (1977).

L8   P. T. Landsberg and J. R. Mallinson, *Inter. Colloq. Solar Elec.* p. 27. Toulouse C.N.E.S., 1976.

L9   J. R. Lloyd and W. Moran, *J. Heat Transfer* **96**, 443 (1974).

L10 I. Langmuir, "Collected Works," Vols. *3* and *5*. Pergamon, Oxford, 1961.

M1   H. Y. B. Mar *et al.*, Optical Coatings for Flat Plate Solar Collectors, NTIS PB-252383 (1975); See also by the same authors, Optical Coatings for Flat Plate Collectors, Final Report, NTIS COO/2625-75/1 (1975).

M2   D. M. Mattox, *Journal of Vac. Sci. Technol.*, **13**, 127, 1976.

M3   G. R. Morgan *et al.*, *Proc. IES Seminar Testing Solar Energy Mater. Syst., Gaithersburg, Maryland* p. 69 (1978).

M4   A. C. Meyers and L. L. Vant-Hull, *Proc. AS-ISES Ann. Meeting, Denver, Colorado* **2.1**, 786 (1978).

M5   S. L. Milora and J. W. Tester, "Geothermal Energy as a Source of Electric Power." MIT Press, Cambridge, Massachusetts, 1976.

O1   M. N. Ozisik, "Boundary Value Problems of Heat Conduction." International Textbook Co., Scranton, Pennsylvania, 1968.

O2   P. Ostwald, "Cost Estimating for Engineering and Management." Prentice-Hall, Englewood Cliffs, New Jersey, 1974.

P1   R. B. Pettit, *Solar Energy* **19**, 733 (1977).

P2   J. E. Parrott, *Solar Energy* **21**, 227, 1978.

P3   R. H. Perry *et al.*, "Chemical Engineer's Handbook." McGraw-Hill, New York, 1963.

P4   R. E. Peterson and J. W. Ramsey, *J. Vac. Sci. Technol.* **12**, 174 (1975).

P5   C. L. Pitman and L. L. Vant-Hull, *Proc. Ann. Meeting AS-ISES* **2.2**, 701 (1978).

P6   P. J. Potter, "Power Plant Theory and Design." Ronald Press, New York, 1959.

P7   W. Place, "Passive Solar State of the Art," p. 769. Mid-Atlantic Solar Energy Assoc., Philadelphia, Pennsylvania, 1978.

R1   C. M. Randall and S. L. Leonard, Report and Recommendations of the Solar Energy Data Workshop, NTIS NSF-RA-N-74-062 (1973).

R2   W. Rohsenow and J. Hartnett, "Handbook of Heat Transfer." McGraw-Hill, New York, 1973.

R3   R. Richter, Solar Collector Thermal Power System, Vol. 1, Xerox Corp., 1974 (available from NTIS, AD/A 000940).

R4   A. Rabl, *Solar Energy* **18**, 93 (1976).

R5   A. Rabl, *Solar Energy* **18**, 497 (1976).

R6   J. W. Ramsey *et al.*, *J. Heat Transfer* **99**, 163 (1977).

R7   A. Rabl, *Solar Energy* **19**, 215 (1977).

R8   J. L. Riggs, "Engineering Economics." McGraw-Hill, New York, 1977.

R9   K. A. Reed, "Concentrating Collectors" (J. R. Williams, ed.), pp. 5–59. Georgia Institute of Technology, 1977.

R10 A. Rabl, *Proc. Sol. Thermal Conc. Coll. Tech. Symp., Denver, Colorado* (B. Gupta, ed.), pp. 1–42, (June 1978).

R11 J. L. Russell *et al.*, Principles of the Fixed Mirror Solar Concentrator, General Atomics Rep. No. GA-A12902 (February 1977) (to be published in *Solar Energy*).

R12 G. M. Reistad, Analysis of Potential Non-Electric Applications of Geothermal Energy and Their Place in the National Economy, Livermore Lab Rep. No. UCRL-51747 (1975).

R13 M. Riaz, ASME Paper No. 75-WA/Sol-1, New York (1975).

R14 B. Rydgren, *Tellus* **20**, 55 (1968).

R15 A. Rabl *et al.*, *Solar Energy* **22**, 373 (1979).

S1   G. G. Stokes, *Proc. R. Soc. London* **11**, 546 (1860–1862).

S2   P. Siegel and J. R. Howell, "Thermal Radiation Heat Transfer." McGraw-Hill, New York, 1972.

S3   E. M. Sparrow and R. D. Cess, "Radiation Heat Transfer." McGraw-Hill, New York, 1978.

S4   B. O. Seraphin, "Optical Properties of Solids—New Developments." North-Holland Publ., Amsterdam, 1975.

S5   W. P. Schimmel *et al.*, ASME Paper 77 WA/Sol-8, New York (1977).

S6   J. Schröder, *J. Eng. Ind.* **97**, 893 (1975).

S7   P. Samuelson, "Economics," 10th ed. McGraw-Hill, New York, 1976.

S8   J. Sakurai, "Solar Energy Engineering" (A. Sayigh, ed.), p. 233, Academic Press, New York, 1977.

S9   W. G. Steward, *Proc. Solar Heating Cooling Buildings Workshop, NTIS* p. 24 (1973).

S10 E. M. Sparrow and K. K. Tien, *J. Heat Transfer* **99**, 507 (1977).

S11 E. M. Sparrow *et al.*, *J. Heat Transfer* **101**, (1979).

S12 E. M. Sparrow, R. R. Schmidt, and J. W. Ramsey, *J. Heat Transfer* **100**, 11 (1978).

S13 W. W. Schertz, R. Winston, and R. Matlock, Status Report for Non-Imaging Collectors. Argonne National Laboratory Rep. No. ANL-SOL-77-01 (1977).

S14 R. Siegel, *Solar Energy* **15**, 273 (1973).

S15 A. A. M. Sayigh, "Solar Energy Engineering." Academic Press, New York, 1977.

T1   M. P. Thekaekara, "Solar Energy Engineering" (A. Sayigh, ed.), p. 37. Academic Press, New York, 1977.

T2   M. P. Thekaekara, *Proc. ISES Conf.* **1**, 313 (1976).

T3   H. Tabor, *Solar Energy* **20**, 293 (1978).

T4   F. Trombe, Personal conversation.

T5   F. Trombe, *Proc. U.N. Conf. New Sources Energy, Rome* (1961).

T6   F. Trombe and L. P. Vinh, *Solar Energy* **15**, 57–61 (1973).

T7   M. P. Thekaekara, *Solar Energy* **14**, 109 (1973).

U1   University of Minnestoa, Research Applied to Solar Thermal Power Systems, a series of reports (1972–1975).

V1   L. L. Vant-Hull and A. F. Hildebrandt, *Proc. Solar Thermal Conversion Workshop, January, 1973* NTIS Rep. No. PB 239277 (1974).

V2   L. L. Vant-Hull, *Proc. SPIE Solar Energy Utilization Conf. II* **85**, 104 (1976); see also by the same author in the same volume p. 111.

W1   A. M. Weinberg, *Am. Sci.* 153 (1978).

W2   Jesse D. Walton, "Energy Technology Handbook" (R. Considine, ed.). McGraw-Hill, New York, 1977.

W3   R. W. Wood, *Philos. Magazine* **17**, 319 (1909).

W4   W. W. Willoughby, *Chem. Eng.* 146 (September 11, 1978).

W5   R. Winston, Seminar at Solar Energy Res. Inst. (June 13, 1978).

W6   D. Watt, On the Nature and Distribution of Solar Radiation. U.S. Dept. of Energy Rep. No. HCP/T2552-01 (1978).

W7   W. Welford and R. Winston, "Optics of Non-imaging Concentrators." Academic Press, New York, 1978.

W8   Westinghouse Electric Co., Solar Total Energy: Large Scale Exp., Phase II, Fort Hood, Texas, Final Technical Progress Rep., NTIS Rep. No. TID-28040 (1977).

W9   J. M. Weingart, Systems Aspects of Large Scale Solar Conversion. Int. Inst. for Appl. Syst. Analysis, Laxenburg, Austria, Rep. No. RM-77-23 (1977).

Z1   Zentner, R. C., *Solar Energy* **19**, 15 (1977).

A | *TABLES*

Solar Irradiance for Various Air Masses[a]

Wave-length		$\mu = 0.66, \beta = 0.085$				$\mu = 0.66, \beta = 0.170$			
	0	1	4	7	10	1	4	7	10
0.940	847.0	313.4	95.0	39.6	18.5	286.9	66.7	21.3	7.6
0.950	837.0	296.5	86.3	35.0	16.0	271.6	60.7	18.9	6.7
0.955	828.5	321.1	102.3	44.1	21.2	294.2	72.1	23.9	8.8
0.965	811.5	344.4	120.4	55.1	27.8	315.7	85.0	30.0	11.6
0.975	794.0	576.9	346.0	224.6	150.1	529.1	244.9	122.7	63.2
0.985	776.0	544.6	316.1	201.2	132.4	499.8	224.2	110.3	56.1
1.018	719.2	617.5	391.0	247.5	156.7	567.8	279.4	137.5	67.6
1.082	620.0	512.9	290.4	164.4	93.1	473.1	210.3	93.5	41.5
1.094	602.0	464.1	303.1	210.8	149.9	428.4	220.0	120.3	67.3
1.098	596.0	503.7	304.1	183.6	110.9	465.0	220.9	105.0	49.9
1.101	591.8	504.8	362.7	267.3	198.8	466.1	263.6	153.0	89.5
1.128	560.5	135.1	27.7	9.1	3.6	124.9	20.2	5.3	1.6
1.131	557.0	152.2	35.3	12.6	5.3	140.7	25.8	7.3	2.4
1.137	550.1	143.1	31.7	11.0	4.5	132.4	23.2	6.4	2.1
1.144	542.0	191.2	57.4	24.2	11.6	176.9	42.1	14.1	5.3
1.147	538.5	174.5	48.2	19.3	8.8	161.5	35.3	11.2	4.1
1.178	507.0	399.3	195.1	95.4	46.6	370.0	143.8	55.9	21.7
1.189	496.0	402.2	214.5	114.4	61.0	372.8	158.4	67.3	28.6
1.193	492.0	424.0	310.8	233.3	176.6	393.1	229.7	137.4	82.9
1.222	464.3	391.8	235.3	141.3	84.9	363.6	174.7	83.9	40.3
1.236	451.2	390.8	254.1	165.2	107.4	363.0	189.1	98.5	51.3
1.264	426.5	329.2	209.7	140.0	94.3	306.1	156.7	84.1	45.5
1.276	416.7	342.6	238.6	172.6	126.3	318.7	178.6	104.0	61.2
1.288	406.8	347.3	216.1	134.4	83.7	323.2	162.1	81.3	40.7
1.314	386.1	298.3	137.6	63.5	29.3	277.9	103.6	38.6	14.4
1.335	369.7	190.6	85.0	46.7	27.7	177.6	64.2	28.6	13.7
1.384	343.7	5.7	0.1	0.0	0.0	5.3	0.1	0.0	0.0
1.432	321.0	44.6	5.4	1.3	0.4	41.7	4.1	0.8	0.2
1.457	308.6	85.4	20.6	7.7	3.3	80.0	15.8	4.8	1.7
1.472	301.4	77.4	17.4	6.2	2.6	72.4	13.3	3.9	1.3
1.542	270.4	239.3	165.9	115.0	79.7	224.5	128.5	73.6	42.1
1.572	257.3	222.6	168.1	130.4	102.1	209.0	130.7	83.9	54.4
1.599	245.4	216.0	166.7	131.5	104.5	203.0	129.9	85.0	56.0
1.608	241.5	208.5	157.4	122.1	95.7	195.9	122.7	79.1	51.4
1.626	233.6	206.7	160.7	127.5	101.9	194.3	125.6	82.8	55.0
1.644	225.6	197.9	152.4	120.1	95.5	186.1	119.3	78.3	51.8
1.650	223.0	195.7	150.9	119.1	94.7	184.1	118.2	77.6	51.4
1.676	212.1	181.9	114.8	72.4	45.7	171.2	90.1	47.5	25.0
1.732	187.9	161.5	102.5	65.1	41.3	152.2	80.9	43.0	22.9
1.782	166.6	136.7	75.6	41.8	23.1	129.0	59.9	27.9	12.9

1.862	138.2	4.0	0.1	0.0	0.0	3.8	0.1	0.0	0.0
1.955	112.9	42.7	14.5	6.8	3.6	40.5	11.6	4.6	2.1
2.008	102.0	69.4	35.8	17.7	6.4	65.7	28.9	12.2	3.8
2.014	101.2	74.7	45.5	28.8	17.8	70.8	36.7	19.8	10.4
2.057	95.6	69.5	41.3	25.3	14.8	66.0	33.4	17.5	8.7
2.124	87.4	70.0	35.9	18.4	9.5	66.4	29.2	12.8	5.6
2.156	83.8	66.0	32.3	15.8	7.7	62.7	26.3	11.0	4.6
2.201	78.9	66.1	49.1	38.0	29.7	62.8	40.1	26.7	17.9
2.266	72.4	61.6	46.8	36.8	29.3	58.6	38.4	26.0	17.9
2.320	67.6	57.2	43.8	33.8	26.8	54.4	35.5	24.0	16.5
2.338	66.3	54.7	39.9	30.4	23.4	52.1	32.8	21.6	14.4
2.356	65.1	52.0	36.3	26.5	19.6	49.5	29.9	18.9	12.1
2.388	62.8	36.0	18.7	11.7	7.8	34.3	15.5	8.3	4.8
2.415	61.0	32.5	15.8	9.4	6.0	31.0	13.0	6.7	3.7
2.453	58.3	29.6	13.7	7.9	5.0	28.2	11.3	5.7	3.1
2.494	55.4	20.3	6.8	3.2	1.7	19.4	5.6	2.3	1.1
2.537	52.4	4.6	0.4	0.1	0.0	4.4	0.3	0.1	0.0
2.900	35.0	2.9	0.2	0.0	0.0	2.8	0.2	0.0	0.0
2.941	33.4	6.0	1.0	0.3	0.1	5.7	0.8	0.2	0.1
2.954	32.8	5.7	0.9	0.3	0.1	5.5	0.8	0.2	0.1
2.973	32.1	8.7	2.2	0.9	0.4	8.4	1.8	0.6	0.3
3.005	30.8	7.8	1.8	0.7	0.3	7.5	1.6	0.5	0.2
3.045	28.8	4.7	0.7	0.2	0.1	4.5	0.6	0.1	0.0
3.056	28.2	4.9	0.8	0.2	0.1	4.7	0.7	0.2	0.1
3.097	26.2	3.2	0.4	0.1	0.0	3.1	0.3	0.1	0.0
3.132	24.9	6.8	1.7	0.7	0.3	6.5	1.5	0.5	0.2
3.156	24.1	18.7	12.6	8.9	6.3	17.9	10.7	6.7	4.2
3.204	22.5	2.1	0.2	0.0	0.0	2.0	0.2	0.0	0.0
3.214	22.1	3.4	0.5	0.1	0.0	3.3	0.4	0.1	0.0
3.245	21.1	3.9	0.7	0.2	0.1	3.8	0.6	0.2	0.1
3.260	20.6	3.7	0.6	0.2	0.1	3.5	0.5	0.1	0.0
3.285	19.7	14.2	8.5	5.1	2.8	13.7	7.3	3.9	1.9
3.317	18.8	12.9	6.9	3.5	1.3	12.4	5.9	2.7	0.9
3.344	18.1	4.2	0.9	0.3	0.1	4.1	0.8	0.2	0.1
3.403	16.5	12.3	7.8	5.1	3.2	11.9	6.7	3.9	2.2
3.450	15.6	12.5	8.9	6.7	5.0	12.0	7.7	5.1	3.5
3.507	14.5	12.5	9.9	8.1	6.7	12.1	8.5	6.2	4.6
3.538	14.2	11.8	8.8	6.9	5.5	11.3	7.6	5.3	3.8
3.573	13.8	10.9	5.4	2.6	1.3	10.5	4.6	2.0	0.9
3.633	13.1	10.8	8.3	6.7	5.5	10.4	7.1	5.2	3.8
3.673	12.6	9.1	6.1	4.6	3.5	8.8	5.3	3.5	2.5
3.696	12.3	10.4	8.2	6.7	5.6	10.1	7.1	5.2	3.9
3.712	12.2	10.9	9.0	7.6	6.5	10.5	7.8	5.9	4.6
3.765	11.5	9.5	7.2	5.9	4.8	9.1	6.3	4.6	3.4
3.812	11.0	8.9	6.7	5.4	4.4	8.6	5.8	4.2	3.1
3.888	10.4	8.1	5.6	4.0	2.9	7.8	4.8	3.1	2.0
3.923	10.1	8.0	5.6	4.2	3.1	7.7	4.9	3.3	2.2
3.948	9.9	7.8	5.5	4.0	3.0	7.6	4.8	3.2	2.1
4.045	9.1	6.7	4.1	2.6	1.5	6.5	3.6	2.0	1.1
Total (W m^{-2})	1353	889.2	448.7	255.2	153.8	800.2	303.1	133.3	63.4

[a] Expressed in W m^{-2} μm^{-1}; H_2O, 20 mm; O_3, 3.4 mm. [From S15.]

TABLE A.2

Monthly Averaged, Daily Horizontal Radiation in Langleys[a,b]

Location	Jan	Yr	Feb	Yr	Mar	Yr	Apr	Yr	May	Yr	Jun	Yr	Jul	Yr	Aug	Yr	Sep	Yr	Oct	Yr	Nov	Yr	Dec	Yr	Ann
AL Annette	63	6	115	6	236	7	364	7	437	6	438	6	438	6	341	6	258	7	122	7	59	7	41	7	243
Barrow	d		38	9	180	8	380	8	513	8	528	8	429	9	255	10	115	10	41	10	d		d		206
Bethel	38	9	108	10	282	9	444	10	457	10	454	10	376	10	252	10	202	10	115	10	44	9	22	9	233
Fairbanks	16	25	71	27	213	25	376	28	461	28	504	29	434	28	317	29	180	29	82	30	26	26	6	26	224
Matanuska	32	6	92	6	242	4	356	7	436	7	462	6	409	6	314	6	198	6	100	6	38	6	15	7	224
AZ Page	300	2	382	3	526	3	618	2	695	2	707	2	680	3	596	3	516	3	402	3	310	3	243	3	498
Phoenix	301	11	409	11	526	11	638	11	724	11	739	11	658	11	613	11	566	11	449	11	334	11	281	11	520
Tucson	315	5	391	5	540	5	655	5	729	5	699	5	626	6	588	6	570	6	442	6	356	6	305	6	518
AR Little Rock	188	9	260	9	353	10	446	9	523	9	559	9	556	8	518	8	439	7	343	8	244	8	187	10	385
CA Davis	174	18	257	17	390	18	528	18	625	18	694	18	682	18	612	18	493	18	347	19	222	19	148	19	431
Fresno	184	31	289	31	427	31	552	31	647	31	702	32	682	32	621	31	510	31	376	32	250	31	161	32	450
Inyokern (China Lake)	306	11	412	11	562	11	683	11	772	11	819	11	772	11	729	10	635	10	467	10	363	11	300	12	568
La Jolla	244	19	302	18	397	19	457	20	506	19	487	21	497	22	464	22	389	22	320	21	277	20	221	20	380
Los Angeles WBAS	248	10	331	10	470	10	515	10	572	9	596	9	641	9	581	10	503	10	373	10	289	10	241	10	463
Los Angeles WBO	243	9	327	9	436	9	483	9	555	9	584	9	651	9	581	9	500	10	362	9	281	10	234	9	436
Riverside[e]	275	8	367	8	478	9	541	9	623	9	680	9	673	9	618	9	535	9	407	9	319	9	270	9	483
Santa Maria	263	11	346	11	482	11	552	11	635	11	694	11	680	11	613	11	524	11	419	11	313	11	252	11	481
Soda Springs	223	4	316	4	374	4	551	4	615	4	691	4	760	3	681	3	510	3	357	4	248	4	182	4	459
CO Boulder	201	5	268	5	401	4	460	4	460	4	525	5	520	5	439	5	412	4	357	3	222	4	182	4	367
Grand Junction[c]	227	8	324	9	434	8	546	8	615	9	708	8	676	8	595	8	514	8	373	10	260	10	212	10	456
Grand Lake (Granby)	212	6	313	7	423	7	512	8	552	8	633	8	600	8	505	7	476	6	361	7	234	6	184	7	417
DC Washington (C.O.)	174	3	266	3	344	2	411	2	551	2	494	2	536	2	446	3	375	3	299	3	211	3	166	3	356
American University	158	39	231	39	322	39	398	39	467	39	510	39	496	39	440	38	364	38	278	38	192	39	141	39	333
Silver Hill	177	7	247	6	342	7	438	7	513	7	555	7	511	7	457	7	391	7	293	8	202	7	156	6	357
FL Apalachicola	298	10	367	10	441	10	535	10	603	9	578	9	529	9	511	9	456	9	413	10	332	10	262	10	444
Belle Isle	297	11	330	10	412	10	463	10	483	10	464	10	488	11	461	10	400	9	366	11	313	11	291	10	397
Gainesville	267	11	343	10	427	12	517	12	579	12	521	10	488	10	483	8	418	9	347	8	300	10	233	10	410
Miami Airport	349	10	415	9	489	9	540	10	553	10	532	10	532	10	505	10	440	10	384	10	353	10	316	10	451
Tallahassee	274	2	311	2	423	3	499	3	547	3	521	3	508	3	542	2	c		299	2	292	2	230	2	g
Tampa	327	8	391	8	474	9	539	8	596	8	574	8	534	9	494	9	452	9	400	9	356	9	300	9	453
GA Atlanta	218	11	290	11	380	11	488	11	533	11	562	11	532	11	508	10	416	10	344	11	268	11	211	11	396
Griffin	234	9	295	9	385	10	522	11	570	11	577	11	556	11	522	11	535	11	368	11	283	11	201	11	413
HI Honolulu	363	4	422	4	516	4	559	5	617	5	615	5	615	5	612	5	573	5	507	5	426	5	371	5	516
Mauna Loa Obs.	522	2	576	2	680	2	689	3	727	3	c		703	3	642	2	602	2	560	2	504	2	481	3	g
Pearl Harbor	359	5	400	4	487	4	529	5	573	5	566	5	598	5	567	5	539	5	466	5	386	5	343	5	484
ID Boise	138	10	236	9	342	9	485	9	585	10	636	9	670	10	576	10	460	10	301	11	182	11	124	11	395
Twin Falls	163	20	240	20	355	20	462	21	552	20	592	18	602	20	540	20	432	19	286	20	176	20	131	19	378
IL Chicago	96	19	147	19	227	19	331	19	424	19	458	18	473	19	403	18	403	19	207	20	120	20	76	20	273
Lemont	170	6	242	6	340	6	402	6	506	6	553	6	540	6	498	6	398	5	275	5	165	5	138	6	352
IN Indianapolis	144	10	213	10	316	10	396	10	488	9	543	9	541	11	490	11	405	11	233	11	177	11	132	11	345

Continued table of monthly mean daily solar-radiation type values by station. Each station row lists twelve successive values followed by an annual value (a script "c" marks a missing entry).

State	Station	(1)	(2)	(3)	(4)	(5)	(6)	(7)	(8)	(9)	(10)	(11)	(12)	(13)
IA	Ames	174	253	326	403	480	541	436	460	367	274	187	143	345
KS	Dodge City	255	316	418	528	568	650	642	592	493	380	285	234	447
	Manhattan	192	264	345	433	527	551	531	526	410	292	227	156	371
KY	Lexington	172	263	357	480	581	628	617	563	494	357	245	174	411
LA	Lake Charles	245	306	397	481	555	591	526	416	449	357	300	250	347
	New Orleans	214	259	335	412	449	443	417	528	383	354	278	198	400
	Shreveport	232	292	384	446	558	557	578	448	414	354	254	205	316
ME	Caribou	133	231	364	400	476	470	508	488	336	212	157	107	350
	Portland	152	235	352	409	514	539	561	449	383	278	152	137	c
MA	Amherst	116	c	300	c	431	514	502	411	354	c	162	124	328
	Blue Hill	153	228	319	389	469	510	486	464	334	266	136	135	301
	Boston	129	194	290	350	445	483	411	436	367	235	164	115	322
	Cambridge	153	235	323	400	420	476	464	432	365	253	163	124	322
	East Wareham	140	218	305	385	452	508	495	466	341	258	135	140	317
	Lynn	118	209	300	394	483	549	528	472	373	241	136	107	311
MI	East Lansing	121	210	309	359	523	547	540	486	322	255	105	108	311
	Sault Ste. Marie	130	225	356	416	496	557	573	522	366	216	146	96	333
MN	St. Cloud	168	260	368	426	530	535	557	509	453	237	225	124	348
MO	Columbia (C.O.)	173	251	340	434	501	574	574	531	417	322	177	158	380
	University of Missouri	166	248	324	429	568	560	583	532	410	324	154	146	365
MT	Glasgow	154	258	385	466	528	605	645	510	407	267	154	116	388
	Great Falls	140	232	366	434	462	583	639	484	354	264	102	112	366
	Summit	122	162	268	414	494	493	560	519	396	216	199	76	312
NE	Lincoln	188	259	350	416	516	544	568	518	410	296	204	159	363
	North Omaha	193	299	365	463	625	546	647	394	551	298	289	170	379
NV	Ely	236	339	468	563	702	712	675	627	551	394	289	218	469
	Las Vegas	277	384	519	621	482	748	509	455	385	429	318	258	509
NJ	Seabrook	157	227	318	403	686	527	683	453	278	278	192	140	339
NH	Mt. Washington	117	218	238	618	440	726	515	626	346	264	334	96	c
NM	Albuquerque	303	386	511	334	432	501	459	453	438	438	120	276	512
NY	Ithaca	116	194	272	369	494	470	543	389	331	231	147	96	302
	Central Park	130	199	290	415	413	565	462	462	385	242	186	115	298
	Sayville	160	249	335	338	502	448	543	397	391	289	128	142	352
	Schenectady	130	200	273	428	531	573	441	475	406	218	182	104	282
	Upton	155	232	339	469	635	564	543	485	471	293	243	146	355
NC	Greensboro	200	276	354	569	494	652	544	562	379	322	282	197	383
	Hatteras	238	317	426	466	550	564	562	476	390	358	235	214	443
	Raleigh	235	302	c	447	502	564	617	516	390	307	161	199	c
ND	Bismarck	157	250	356	286	550	590	562	494	278	272	141	124	369
OH	Cleveland	125	183	303	391	502	562	562	477	422	289	176	115	335
	Columbus	128	200	297	386	471	562	542	487	382	286	144	129	340
	Put-in-Bay	126	204	302	386	468	544	561	487	382	275	144	109	332

(continued)

TABLE A.2 (Continued)

	Location	Jan	Yr	Feb	Yr	Mar	Yr	Apr	Yr	May	Yr	Jun	Yr	Jul	Yr	Aug	Yr	Sep	Yr	Oct	Yr	Nov	Yr	Dec	Yr	Ann
OK	Oklahoma City	251	10	319	10	409	9	494	10	536	10	615	7	610	8	593	8	487	9	377	10	291	9	240	9	436
	Stillwater	205	8	289	8	390	9	454	9	504	9	600	10	596	10	545	10	455	11	354	11	269	11	209	8	405
OR	Astoria	90	7	162	7	270	8	375	8	492	8	469	8	539	8	461	8	354	7	209	8	111	8	79	8	301
	Corvallis	89	2	c		287	3	406	3	517	3	570	3	676	3	558	4	397	4	235	4	144	4	80	4	9
	Medord	116	11	215	11	336	11	482	11	592	11	652	11	698	11	605	11	447	11	279	11	149	11	93	11	389
PA	Pittsburgh	94	6	169	5	216	6	317	6	429	6	491	6	497	7	409	6	339	6	207	5	118	6	77	5	280
	State College	133	19	201	19	295	20	380	20	456	20	518	20	511	20	444	20	358	20	256	20	149	20	118	20	318
RI	Newport	155	23	232	22	334	23	405	23	477	23	527	24	513	24	455	24	377	24	271	24	176	24	139	24	338
SC	Charleston	252	11	314	11	388	11	512	11	551	11	564	11	520	11	501	11	404	11	338	11	286	11	225	11	404
SD	Rapid City	183	11	277	11	400	11	482	11	532	11	585	11	590	11	541	11	435	11	315	10	204	10	158	10	392
TN	Nashville	149	18	228	19	322	19*	432	19	503	19	551	18	530	17	473	17	403	17	308	18	208	18	150	19	355
TX	Oak Ridge	161	11	239	11	331	11	450	11	518	11	551	11	526	11	478	11	416	11	318	11	213	11	163	11	364
	Brownsville	297	10	341	10	402	10	456	10	564	10	610	9	627	8	568	8	475	10	411	10	296	11	263	10	442
	El Paso	333	11	430	11	547	11	654	11	714	11	729	11	666	11	640	10	576	11	460	11	372	11	313	11	536
	Ft. Worth	250	11	320	11	427	11	488	11	562	11	651	11	613	11	593	11	503	11	403	11	306	11	245	11	445
	Midland	283	7	358	8	476	9	550	8	611	8	617	9	608	7	574	8	522	9	396	9	325	9	275	8	466
	San Antonio	279	9	347	9	417	9	445	9	565	9	612	9	639	9	585	9	493	10	398	9	295	10	256	8	442
UT	Flaming Gorge	238	2	298	2	443	2	522	2	565	2	650	2	599	3	538	3	425	3	352	3	262	3	215	3	426
	Salt Lake City	163	8	256	8	354	8	479	8	570	7	621	7	620	6	551	6	446	8	316	8	204	8	146	9	394
VA	Mt. Weather	172	2	274	2	338	2	414	2	508	2	525	3	510	3	430	3	375	3	281	2	202	2	168	2	350
WA	North Head	c		167	7	257	3	432	2	509	3	487	3	486	3	436	3	321	3	205	3	122	3	77	3	9
	Friday Harbor	87	8	157	7	274	8	418	8	514	9	578	9	586	10	507	11	351	8	194	10	102	10	75	8	320
	Prosser	117	4	222	4	351	4	521	5	616	4	680	4	707	4	604	4	458	4	274	4	136	4	100	4	399
	Pullman	121	8	205	9	304	9	462	9	558	10	653	10	699	9	562	5	410	4	245	5	146	5	96	5	372
	University of Washington	67	9	126	9	245	10	364	9	445	9	461	10	496	11	435	10	299	8	170	10	93	9	59	9	272
	Seattle-Tacoma	75	9	139	9	265	9	403	9	503	9	511	9	566	9	452	10	324	10	188	10	104	9	64	10	300
WI	Spokane	119	8	204	8	321	8	474	8	563	9	596	9	665	9	556	9	404	10	225	9	131	9	75	7	361
	Madison	148	46	220	46	313	45	394	47	466	47	514	47	531	47	452	47	348	47	241	47	145	44	115	46	324
WY	Lander	226	8	324	9	452	9	548	9	587	11	678	11	651	11	586	10	472	8	354	8	239	9	196	9	443
	Laramie	216	3	295	3	424	3	508	3	554	3	643	3	606	3	536	3	438	3	324	3	229	3	186	3	408
Island Stations																										
	Canton Island	588	9	626	7	634	7	604	9	561	9	549	8	550	9	597	9	640	9	651	9	600	8	572	8	597
	San Juan, P.R.	404	5	481	4	580	4	622	4	519	5	536	6	639	5	549	6	531	6	460	6	411	6	411	6	512
	Swan Island	442	6	496	7	615	6	646	6	625	6	544	6	588	6	591	6	535	8	457	7	394	8	382	8	526
	Wake Island	438	7	518	7	577	7	627	7	642	8	656	8	629	7	623	7	587	6	525	6	482	7	421	7	560

[a] Langley is the unit used to denote 1 gram calorie per square centimeter.

[b] From "Climatic Atlas of the United States," U.S. Government Printing Office, 1968.

[c] Only one year of data for the month; no means computed.

[d] Barrow is in darkness during the winter months.

[e] Riverside data prior to March 1952 not used because of instrumental difficulties.

[f] Madison data after 1957 not used due to exposure influences

[g] Indicates no data for the month (or incomplete data for the year).

TABLE A.3

Seasonally Averaged Daily Direct-Normal Solar Flux in kWh/m²[a]

	Spring (M,A,M)	Summer (J,J,A)	Fall (S,O,N)	Winter (D,J,F)
Albuquerque	7.6	8.2	6.9	5.8
Apalachicola	5.0	4.5	4.4	3.3
Bismarck	4.7	6.5	3.9	2.7
Blue Hill	3.6	4.2	3.1	2.2
Boston	3.6	4.2	3.1	2.2
Brounsville	4.2	5.8	4.4	3.2
Cape Hatteras	4.8	4.9	4.0	3.1
Caribou	5.9	6.0	3.4	3.9
Charleston	4.3	3.9	3.7	3.1
Columbia	4.3	5.8	4.1	2.9
Dodge City	5.9	7.1	5.5	4.6
El Paso	8.0	8.1	6.9	6.0
Ely	6.9	8.1	6.7	4.8
Fort Worth	4.5	5.9	4.6	3.7
Great Falls	4.5	6.7	3.9	2.3
Lake Charles	3.8	4.3	3.8	2.7
Madison	4.1	5.1	3.3	2.5
Medford	4.5	7.5	3.7	1.5
Miami	4.3	3.5	3.5	3.9
Nashville	3.9	4.6	3.6	2.4
New York	3.4	3.8	2.8	2.0
Omaha	4.6	6.0	4.1	3.2
Phoenix	7.8	8.0	6.6	5.3
Santa Maria	5.7	6.9	5.2	4.1
Seattle	4.8	7.4	3.2	1.5
Washington, DC	3.8	4.3	3.3	2.4

[a] From (K7).

TABLE A.4

Elliptic Integrals of the Second Kind $E(\phi/\alpha)^a$

$$E(\varphi\backslash\alpha) \quad \int_0^\varphi (1-\sin^2\alpha \sin^2\theta)^{\frac{1}{2}} d\theta$$

$\alpha\backslash\varphi$	0°	5°	10°	15°	20°	25°	30°
0°	0	0.08726 646	0.17453 293	0.26179 939	0.34906 585	0.43633 231	0.52359 878
2	0	0.08726 633	0.17453 185	0.26179 579	0.34905 742	0.43631 608	0.52357 119
4	0	0.08726 592	0.17452 864	0.26178 503	0.34903 218	0.43626 745	0.52348 856
6	0	0.08726 525	0.17452 330	0.26176 715	0.34899 025	0.43618 665	0.52335 123
8	0	0.08726 432	0.17451 587	0.26174 224	0.34893 181	0.43607 403	0.52315 981
10	0	0.08726 313	0.17450 636	0.26171 041	0.34885 714	0.43593 011	0.52291 511
12	0	0.08726 168	0.17449 485	0.26167 182	0.34876 657	0.43575 552	0.52261 821
14	0	0.08725 999	0.17448 137	0.26162 664	0.34866 055	0.43555 106	0.52227 039
16	0	0.08725 806	0.17446 599	0.26157 510	0.34853 954	0.43531 765	0.52187 317
18	0	0.08725 590	0.17444 879	0.26151 743	0.34840 412	0.43505 633	0.52142 828
20	0	0.08725 352	0.17442 985	0.26145 391	0.34825 492	0.43476 831	0.52093 770
22	0	0.08725 094	0.17440 926	0.26138 485	0.34809 262	0.43445 488	0.52040 357
24	0	0.08724 816	0.17438 712	0.26131 056	0.34791 800	0.43411 749	0.51982 827
26	0	0.08724 521	0.17436 353	0.26123 141	0.34773 187	0.43375 767	0.51921 436
28	0	0.08724 208	0.17433 862	0.26114 778	0.34753 510	0.43337 709	0.51856 461
30	0	0.08723 881	0.17431 250	0.26106 005	0.34732 863	0.43297 749	0.51788 193
32	0	0.08723 540	0.17428 529	0.26096 867	0.34711 342	0.43256 075	0.51716 944
34	0	0.08723 187	0.17425 714	0.26087 405	0.34689 050	0.43212 880	0.51643 040
36	0	0.08722 824	0.17422 817	0.26077 666	0.34666 093	0.43168 368	0.51566 820
38	0	0.08722 453	0.17419 852	0.26067 697	0.34642 580	0.43122 748	0.51488 638
40	0	0.08722 075	0.17416 835	0.26057 545	0.34618 625	0.43076 236	0.51408 862
42	0	0.08721 692	0.17413 779	0.26047 261	0.34594 343	0.43029 055	0.51327 866
44	0	0.08721 307	0.17410 700	0.26036 893	0.34569 850	0.42981 431	0.51246 037
46	0	0.08720 920	0.17407 613	0.26026 492	0.34545 266	0.42933 594	0.51163 767
48	0	0.08720 535	0.17404 531	0.26016 110	0.34520 710	0.42885 776	0.51081 454
50	0	0.08720 152	0.17401 472	0.26005 795	0.34496 302	0.42838 212	0.50999 501
52	0	0.08719 774	0.17398 449	0.25995 600	0.34472 162	0.42791 134	0.50918 310
54	0	0.08719 402	0.17395 477	0.25985 574	0.34448 409	0.42744 775	0.50838 287
56	0	0.08719 039	0.17392 571	0.25975 765	0.34425 159	0.42699 368	0.50759 843
58	0	0.08718 686	0.17389 745	0.25966 224	0.34402 529	0.42655 138	0.50683 341
60	0	0.08718 345	0.17387 013	0.25956 996	0.34380 631	0.42612 308	0.50609 207
62	0	0.08718 017	0.17384 388	0.25948 126	0.34359 575	0.42571 097	0.50537 811
64	0	0.08717 704	0.17381 883	0.25939 660	0.34339 465	0.42531 712	0.50469 523
66	0	0.08717 408	0.17379 511	0.25931 640	0.34320 404	0.42494 358	0.50404 700
68	0	0.08717 130	0.17377 283	0.25924 104	0.34302 487	0.42459 224	0.50343 686
70	0	0.08716 871	0.17375 210	0.25917 090	0.34285 805	0.42426 495	0.50286 804
72	0	0.08716 633	0.17373 302	0.25910 634	0.34270 440	0.42396 339	0.50234 359
74	0	0.08716 416	0.17371 568	0.25904 767	0.34256 478	0.42368 913	0.50186 633
76	0	0.08716 223	0.17370 018	0.25899 519	0.34243 984	0.42344 363	0.50143 886
78	0	0.08716 053	0.17368 659	0.25894 917	0.34233 022	0.42322 817	0.50106 351
80	0	0.08715 909	0.17367 498	0.25890 983	0.34223 650	0.42304 389	0.50074 232
82	0	0.08715 789	0.17366 539	0.25887 737	0.34215 915	0.42289 175	0.50047 707
84	0	0.08715 695	0.17365 789	0.25885 195	0.34209 857	0.42277 258	0.50026 923
86	0	0.08715 628	0.17365 250	0.25883 370	0.34205 507	0.42268 700	0.50011 993
88	0	0.08715 588	0.17364 926	0.25882 271	0.34202 889	0.42263 547	0.50003 003
90	0	0.08715 574	0.17364 818	0.25881 905	0.34202 014	0.42261 826	0.50000 000
		$\begin{bmatrix}(-8)4\\3\end{bmatrix}$	$\begin{bmatrix}(-7)3\\4\end{bmatrix}$	$\begin{bmatrix}(-7)9\\4\end{bmatrix}$	$\begin{bmatrix}(-6)2\\5\end{bmatrix}$	$\begin{bmatrix}(-6)4\\5\end{bmatrix}$	$\begin{bmatrix}(-6)7\\5\end{bmatrix}$
5	0	0.08726 562	0.17452 624	0.26177 698	0.34901 329	0.43623 105	0.52342 670
15	0	0.08725 905	0.17447 391	0.26160 165	0.34860 188	0.43543 791	0.52207 785
25	0	0.08724 671	0.17437 500	0.26127 157	0.34782 632	0.43394 028	0.51952 597
35	0	0.08723 006	0.17424 275	0.26082 567	0.34677 648	0.43190 776	0.51605 197
45	0	0.08721 113	0.17409 157	0.26031 693	0.34557 562	0.42957 525	0.51204 932
55	0	0.08719 220	0.17394 015	0.25980 603	0.34436 714	0.42721 938	0.50798 838
65	0	0.08717 554	0.17380 680	0.25935 592	0.34329 797	0.42512 769	0.50436 656
75	0	0.08716 317	0.17370 770	0.25902 064	0.34250 043	0.42356 271	0.50164 622
85	0	0.08715 659	0.17365 493	0.25884 192	0.34207 467	0.42272 556	0.50018 720

Compiled from K. Pearson, Tables of the complete and incomplete elliptic integrals, Cambridge Univ. Press, Cambridge, England, 1934 (with permission). Known errors have been corrected.

TABLE A.4 (Continued)

$$E(\varphi \backslash \alpha) = \int_0^\varphi (1 - \sin^2 \alpha \sin^2 \theta)^{\frac{1}{2}} d\theta$$

$\alpha \backslash \varphi$	35°	40°	45°	50°	55°	60°
0°	0.61086 524	0.69813 170	0.78539 816	0.87266 463	0.95993 109	1.04719 755
2	0.61082 230	0.69806 905	0.78531 125	0.87254 883	0.95978 184	1.04701 051
4	0.61069 365	0.69788 136	0.78505 085	0.87220 183	0.95933 459	1.04644 996
6	0.61047 983	0.69756 935	0.78461 792	0.87162 487	0.95859 083	1.04551 764
8	0.61018 171	0.69713 427	0.78401 409	0.87081 998	0.95755 301	1.04421 646
10	0.60980 055	0.69657 784	0.78324 162	0.86979 001	0.95622 460	1.04255 047
12	0.60933 793	0.69590 226	0.78230 343	0.86853 863	0.95461 005	1.04052 491
14	0.60879 577	0.69511 023	0.78120 308	0.86707 031	0.95271 478	1.03814 615
16	0.60817 636	0.69420 492	0.77994 473	0.86539 034	0.95054 522	1.03542 177
18	0.60748 229	0.69318 999	0.77853 323	0.86350 481	0.94810 878	1.03236 049
20	0.60671 652	0.69206 954	0.77697 402	0.86142 062	0.94541 386	1.02897 221
22	0.60588 229	0.69084 814	0.77527 316	0.85914 545	0.94246 984	1.02526 804
24	0.60498 319	0.68953 083	0.77343 735	0.85668 781	0.93928 709	1.02126 023
26	0.60402 308	0.68812 308	0.77147 387	0.85405 695	0.93587 699	1.01696 224
28	0.60300 616	0.68663 077	0.76939 059	0.85126 295	0.93225 186	1.01238 873
30	0.60193 687	0.68506 023	0.76719 599	0.84831 663	0.92842 504	1.00755 556
32	0.60081 994	0.68341 817	0.76489 908	0.84522 958	0.92441 083	1.00247 977
34	0.59966 035	0.68171 170	0.76250 947	0.84201 414	0.92022 452	0.99717 966
36	0.59846 332	0.67994 830	0.76003 726	0.83868 340	0.91588 234	0.99167 469
38	0.59723 431	0.67813 578	0.75749 309	0.83525 115	0.91140 150	0.98598 560
40	0.59597 897	0.67628 229	0.75488 809	0.83173 189	0.90680 017	0.98013 430
42	0.59470 312	0.67439 630	0.75223 383	0.82814 080	0.90209 742	0.97414 397
44	0.59341 278	0.67248 651	0.74954 234	0.82449 369	0.89731 325	0.96803 899
46	0.59211 406	0.67056 191	0.74682 605	0.82080 700	0.89246 858	0.96184 497
48	0.59081 324	0.66863 167	0.74409 773	0.81709 775	0.88758 513	0.95558 873
50	0.58951 664	0.66670 515	0.74137 047	0.81338 346	0.88268 551	0.94929 830
52	0.58823 065	0.66479 183	0.73865 766	0.80968 217	0.87779 305	0.94300 285
54	0.58696 171	0.66290 130	0.73597 286	0.80601 230	0.87293 184	0.93673 272
56	0.58571 622	0.66104 317	0.73332 979	0.80239 262	0.86812 660	0.93051 931
58	0.58450 056	0.65922 707	0.73074 229	0.79884 217	0.86340 261	0.92439 505
60	0.58332 103	0.65746 255	0.72822 416	0.79538 015	0.85878 561	0.91839 329
62	0.58218 382	0.65575 905	0.72578 915	0.79202 582	0.85430 169	0.91254 821
64	0.58109 497	0.65412 585	0.72345 085	0.78879 839	0.84997 709	0.90689 460
66	0.58006 032	0.65257 197	0.72122 260	0.78571 685	0.84583 811	0.90146 778
68	0.57908 549	0.65110 612	0.71911 737	0.78279 987	0.84191 082	0.89630 323
70	0.57817 584	0.64973 667	0.71714 767	0.78006 562	0.83822 090	0.89143 642
72	0.57733 641	0.64847 154	0.71532 545	0.77753 157	0.83479 335	0.88690 237
74	0.57657 189	0.64731 812	0.71366 196	0.77521 434	0.83165 223	0.88273 530
76	0.57588 663	0.64628 328	0.71216 766	0.77312 952	0.82882 031	0.87896 810
78	0.57528 450	0.64537 322	0.71085 210	0.77129 143	0.82631 879	0.87563 185
80	0.57476 897	0.64459 347	0.70972 381	0.76971 298	0.82416 694	0.87275 520
82	0.57434 302	0.64394 879	0.70879 019	0.76840 644	0.82238 177	0.87036 381
84	0.57400 912	0.64344 316	0.70805 745	0.76737 830	0.82097 770	0.86847 970
86	0.57376 921	0.64307 973	0.70753 050	0.76663 912	0.81996 631	0.86712 068
88	0.57362 470	0.64286 075	0.70721 289	0.76619 339	0.81935 604	0.86629 990
90	0.57357 644	0.64278 761	0.70710 678	0.76604 444	0.81915 204	0.86602 540
	$\begin{bmatrix}(-5)1\\5\end{bmatrix}$	$\begin{bmatrix}(-5)2\\5\end{bmatrix}$	$\begin{bmatrix}(-5)3\\5\end{bmatrix}$	$\begin{bmatrix}(-5)4\\6\end{bmatrix}$	$\begin{bmatrix}(-5)5\\6\end{bmatrix}$	$\begin{bmatrix}(-5)7\\6\end{bmatrix}$
5	0.61059 734	0.69774 083	0.78485 586	0.87194 199	0.95899 964	1.04603 012
15	0.60849 557	0.69467 152	0.78059 397	0.86625 642	0.95166 385	1.03682 664
25	0.60451 051	0.68883 790	0.77247 109	0.85539 342	0.93760 971	1.01914 662
35	0.59906 618	0.68083 664	0.76128 304	0.84036 234	0.91807 186	0.99445 152
45	0.59276 408	0.67152 549	0.74818 650	0.82265 424	0.89489 714	0.96495 146
55	0.58633 563	0.66196 758	0.73464 525	0.80419 500	0.87052 066	0.93361 692
65	0.58057 051	0.65333 844	0.72232 215	0.78723 820	0.84788 276	0.90415 063
75	0.57621 910	0.64678 548	0.71289 304	0.77414 195	0.83019 625	0.88079 972
85	0.57387 732	0.64324 351	0.70776 799	0.76697 232	0.82042 232	0.86773 361

(continued)

TABLE A.4 (Continued)

$$E(\varphi \backslash \alpha) \quad \int_0^\varphi (1 - \sin^2 \alpha \sin^2 \theta)^{\frac{1}{2}} d\theta$$

$\alpha \backslash \varphi$	65°	70°	75°	80°	85°	90°
0°	1.13446 401	1.22173 048	1.30899 694	1.39626 340	1.48352 986	1.57079 633
2	1.13423 517	1.22145 628	1.30867 442	1.39589 024	1.48310 448	1.57031 792
4	1.13354 929	1.22063 443	1.30770 767	1.39477 165	1.48182 929	1.56888 372
6	1.13240 837	1.21926 717	1.30609 916	1.39291 030	1.47970 717	1.56649 679
8	1.13081 573	1.21735 820	1.30385 297	1.39031 062	1.47674 288	1.56316 223
10	1.12877 602	1.21491 274	1.30097 484	1.38697 886	1.47294 312	1.55888 720
12	1.12629 522	1.21193 748	1.29747 215	1.38292 302	1.46831 652	1.55368 089
14	1.12338 066	1.20844 065	1.29335 393	1.37815 292	1.46287 363	1.54755 458
16	1.12004 099	1.20443 195	1.28863 089	1.37268 017	1.45662 693	1.54052 157
18	1.11628 624	1.19992 262	1.28331 541	1.36651 823	1.44959 085	1.53259 729
20	1.11212 778	1.19492 542	1.27742 153	1.35968 233	1.44178 179	1.52379 921
22	1.10757 834	1.18945 465	1.27096 502	1.35218 961	1.43321 813	1.51414 692
24	1.10265 204	1.18352 618	1.26396 337	1.34405 903	1.42392 023	1.50366 214
26	1.09736 439	1.17715 743	1.25643 578	1.33531 146	1.41391 049	1.49236 871
28	1.09173 228	1.17036 745	1.24840 326	1.32596 967	1.40321 335	1.48029 266
30	1.08577 404	1.16317 686	1.23988 858	1.31605 841	1.39185 532	1.46746 221
32	1.07950 942	1.15560 796	1.23091 635	1.30560 436	1.37986 503	1.45390 780
34	1.07295 961	1.14768 469	1.22151 305	1.29463 629	1.36727 328	1.43966 215
36	1.06614 728	1.13943 273	1.21170 705	1.28318 499	1.35411 306	1.42476 031
38	1.05909 660	1.13087 946	1.20152 870	1.27128 343	1.34041 965	1.40923 972
40	1.05183 322	1.12205 408	1.19101 036	1.25896 675	1.32623 066	1.39314 025
42	1.04438 435	1.11298 760	1.18018 648	1.24627 240	1.31158 614	1.37650 433
44	1.03677 875	1.10371 291	1.16909 366	1.23324 019	1.29652 865	1.35937 700
46	1.02904 677	1.09426 484	1.15777 077	1.21991 241	1.28110 340	1.34180 606
48	1.02122 034	1.08468 023	1.14625 899	1.20633 398	1.26535 837	1.32384 218
50	1.01333 305	1.07499 796	1.13460 200	1.19255 255	1.24934 449	1.30553 909
52	1.00542 010	1.06525 908	1.12284 604	1.17861 873	1.23311 580	1.28695 374
54	0.99751 835	1.05550 682	1.11104 010	1.16458 621	1.21672 971	1.26814 653
56	0.98966 632	1.04578 671	1.09923 604	1.15051 210	1.20024 724	1.24918 162
58	0.98190 414	1.03614 663	1.08748 883	1.13645 710	1.18373 339	1.23012 722
60	0.97427 354	1.02663 689	1.07585 669	1.12248 590	1.16725 747	1.21105 603
62	0.96681 780	1.01731 023	1.06440 132	1.10866 752	1.15089 364	1.19204 568
64	0.95958 158	1.00822 192	1.05318 814	1.09507 580	1.13472 145	1.17317 938
66	0.95261 084	0.99942 966	1.04228 653	1.08178 986	1.11882 658	1.15454 668
68	0.94595 256	0.99099 354	1.03176 998	1.06889 476	1.10330 172	1.13624 437
70	0.93965 447	0.98297 583	1.02171 634	1.05648 221	1.08824 773	1.11837 774
72	0.93376 462	0.97544 068	1.01220 781	1.04465 133	1.07377 505	1.10106 213
74	0.92833 088	0.96845 360	1.00333 091	1.03350 951	1.06000 556	1.08442 522
76	0.92340 024	0.96208 074	0.99517 606	1.02317 331	1.04707 504	1.06860 953
78	0.91901 802	0.95638 776	0.98783 670	1.01376 904	1.03513 640	1.05377 692
80	0.91522 691	0.95143 847	0.98140 781	1.00543 295	1.02436 393	1.04011 440
82	0.91206 588	0.94729 297	0.97598 331	0.99831 000	1.01495 898	1.02784 362
84	0.90956 905	0.94400 544	0.97165 228	0.99255 019	1.00715 650	1.01723 692
86	0.90776 445	0.94162 171	0.96849 392	0.98830 025	1.00123 026	1.00864 796
88	0.90667 305	0.94017 677	0.96657 142	0.98568 915	0.99748 392	1.00258 400
90	0.90630 779	0.93969 262	0.96592 583	0.98480 775	0.99619 470	1.00000 000
	$\left[\begin{array}{c}(-5)9 \\ 6\end{array}\right]$	$\left[\begin{array}{c}(-4)1 \\ 7\end{array}\right]$	$\left[\begin{array}{c}(-4)2 \\ 7\end{array}\right]$	$\left[\begin{array}{c}(-4)2 \\ 9\end{array}\right]$	$\left[\begin{array}{c}(-4)3 \\ 9\end{array}\right]$	$\left[\begin{array}{c}(-4)4 \\ 10\end{array}\right]$
5	1.13303 553	1.22001 878	1.30698 342	1.39393 358	1.48087 384	1.56780 907
15	1.12176 337	1.20649 962	1.29106 728	1.37550 358	1.45984 990	1.54415 050
25	1.10005 236	1.18039 569	1.26026 405	1.33976 099	1.41900 286	1.49811 493
35	1.06958 479	1.14359 813	1.21665 853	1.28896 903	1.36076 208	1.43229 097
45	1.03292 660	1.09900 829	1.16345 846	1.22661 050	1.28885 906	1.35064 388
55	0.99358 365	1.05063 981	1.10513 448	1.15755 065	1.20849 656	1.25867 963
65	0.95606 011	1.00378 688	1.04769 389	1.08838 943	1.12673 373	1.16382 796
75	0.92579 978	0.96518 626	0.99915 744	1.02823 305	1.05342 632	1.07640 511
85	0.90857 873	0.94269 813	0.96992 212	0.99022 779	1.00394 027	1.01266 351

^{a} From (A6).

TABLE A.5

Normal Probability Integral

$$F(z) = \frac{1}{\sqrt{2\pi}} \int_{-\infty}^{z} e^{-1/2\,t^2}\, dt$$

z	0.00	0.01	0.02	0.03	0.04	0.05	0.06	0.07	0.08	0.09
0.0	0.5000	0.5040	0.5080	0.5120	0.5160	0.5199	0.5239	0.5279	0.5319	0.5359
0.1	0.5398	0.5438	0.5478	0.5517	0.5557	0.5596	0.5636	0.5675	0.5714	0.5753
0.2	0.5793	0.5832	0.5871	0.5910	0.5948	0.5987	0.6026	0.6064	0.6103	0.6141
0.3	0.6179	0.6217	0.6255	0.6293	0.6331	0.6368	0.6406	0.6443	0.6480	0.6517
0.4	0.6554	0.6591	0.6628	0.6664	0.6700	0.6736	0.6772	0.6808	0.6844	0.6879
0.5	0.6915	0.6950	0.6985	0.7019	0.7054	0.7088	0.7123	0.7157	0.7190	0.7224
0.6	0.7257	0.7291	0.7324	0.7357	0.7389	0.7422	0.7454	0.7486	0.7517	0.7549
0.7	0.7580	0.7611	0.7642	0.7673	0.7704	0.7734	0.7764	0.7794	0.7823	0.7852
0.8	0.7881	0.7910	0.7939	0.7967	0.7995	0.8023	0.8051	0.8078	0.8106	0.8133
0.9	0.8159	0.8186	0.8212	0.8238	0.8264	0.8289	0.8315	0.8340	0.8365	0.8389
1.0	0.8413	0.8438	0.8461	0.8485	0.8508	0.8531	0.8554	0.8577	0.8599	0.8621
1.1	0.8643	0.8665	0.8686	0.8708	0.8729	0.8749	0.8770	0.8790	0.8810	0.8830
1.2	0.8849	0.8869	0.8888	0.8907	0.8925	0.8944	0.8962	0.8980	0.8997	0.9015
1.3	0.9032	0.9049	0.9066	0.9082	0.9099	0.9115	0.9131	0.9147	0.9162	0.9177
1.4	0.9192	0.9207	0.9222	0.9236	0.9251	0.9265	0.9279	0.9292	0.9306	0.9319
1.5	0.9332	0.9345	0.9357	0.9370	0.9382	0.9394	0.9406	0.9418	0.9429	0.9441
1.6	0.9452	0.9463	0.9474	0.9484	0.9495	0.9505	0.9515	0.9525	0.9535	0.9545
1.7	0.9554	0.9564	0.9573	0.9582	0.9591	0.9599	0.9608	0.9616	0.9625	0.9633
1.8	0.9641	0.9649	0.9656	0.9664	0.9671	0.9678	0.9686	0.9693	0.9699	0.9706
1.9	0.9713	0.9719	0.9726	0.9732	0.9738	0.9744	0.9750	0.9756	0.9761	0.9767
2.0	0.9772	0.9778	0.9783	0.9788	0.9793	0.9798	0.9803	0.9808	0.9812	0.9817
2.1	0.9821	0.9826	0.9830	0.9834	0.9838	0.9842	0.9846	0.9850	0.9854	0.9857
2.2	0.9861	0.9864	0.9868	0.9871	0.9875	0.9878	0.9881	0.9884	0.9887	0.9890
2.3	0.9893	0.9896	0.9898	0.9901	0.9904	0.9906	0.9909	0.9911	0.9913	0.9916
2.4	0.9918	0.9920	0.9922	0.9925	0.9927	0.9929	0.9931	0.9932	0.9934	0.9936
2.5	0.9938	0.9940	0.9941	0.9943	0.9945	0.9946	0.9948	0.9949	0.9951	0.9952
2.6	0.9953	0.9955	0.9956	0.9957	0.9959	0.9960	0.9961	0.9962	0.9963	0.9964
2.7	0.9965	0.9966	0.9967	0.9968	0.9969	0.9970	0.9971	0.9972	0.9973	0.9974
2.8	0.9974	0.9975	0.9976	0.9977	0.9977	0.9978	0.9979	0.9979	0.9980	0.9981
2.9	0.9981	0.9982	0.9982	0.9983	0.9984	0.9984	0.9985	0.9985	0.9986	0.9986
3.0	0.9987	0.9987	0.9987	0.9988	0.9988	0.9989	0.9989	0.9989	0.9990	0.9990
3.1	0.9990	0.9991	0.9991	0.9991	0.9992	0.9992	0.9992	0.9992	0.9993	0.9993
3.2	0.9993	0.9993	0.9994	0.9994	0.9994	0.9994	0.9994	0.9995	0.9995	0.9995
3.3	0.9995	0.9995	0.9995	0.9996	0.9996	0.9996	0.9996	0.9996	0.9996	0.9997
3.4	0.9997	0.9997	0.9997	0.9997	0.9997	0.9997	0.9997	0.9997	0.9997	0.9998

TABLE A.6

Absorptance and Emittance Values for Various Materials[a]

Substance	Short-wave absorptance	Long-wave emittance	$\dfrac{\alpha}{\epsilon}$
Class I substances: Absorptance to emittance ratios less than 0.5			
Magnesium carbonate, $MgCO_3$	0.025–0.04	0.79	0.03–0.05
White plaster	0.07	0.91	0.08
Snow, fine particles, fresh	0.13	0.82	0.16
White paint, 0.017 in., on aluminum	0.20	0.91	0.22
Whitewash on galvanized iron	0.22	0.90	0.24
White paper	0.25–0.28	0.95	0.26–0.29
White enamel on iron	0.25–0.45	0.90	0.28–0.5
Ice, with sparse snow cover	0.31	0.96–0.97	0.32
Snow, ice granules	0.33	0.89	0.37
Aluminum oil base paint	0.45	0.90	0.50
White powdered sand	0.45	0.84	0.54
Class II substances: Absorptance to emittance ratios between 0.5 and 0.9			
Asbestos felt	0.25	0.50	0.50
Green oil base paint	0.50	0.90	0.56
Bricks, red	0.55	0.92	0.60
Asbestos cement board, white	0.59	0.96	0.61
Marble, polished	0.5–0.6	0.90	0.61
Wood, planed oak	—	0.90	—
Rough concrete	0.60	0.97	0.62
Concrete	0.60	0.88	0.68
Grass, green, after rain	0.67	0.98	0.68
Grass, high and dry	0.67–0.69	0.90	0.76
Vegetable fields and shrubs, wilted	0.70	0.90	0.78
Oak leaves	0.71–0.78	0.91–0.95	0.78–0.82
Frozen soil	—	0.93–0.94	—
Desert surface	0.75	0.90	0.83
Common vegetable fields and shrubs	0.72–0.76	0.90	0.82
Ground, dry plowed	0.75–0.80	0.90	0.83–0.89
Oak woodland	0.82	0.90	0.91
Pine forest	0.86	0.90	0.96
Earth surface as a whole (land and sea, no clouds)	0.83	—	—

TABLE A.6 (*Continued*)

Class III substances: Absorptance to emittance ratios between 0.8 and 1.0			
Grey paint	0.75	0.95	0.79
Red oil base paint	0.74	0.90	0.82
Asbestos, slate	0.81	0.96	0.84
Asbestos, paper	—	0.93–0.96	—
Linoleum, red-brown	0.84	0.92	0.91
Dry sand	0.82	0.90	0.91
Green roll roofing	0.88	0.91–0.97	0.93
Slate, dark grey	0.89	—	—
Old grey rubber	—	0.86	—
Hard black rubber	—	0.90–0.95	—
Asphalt pavement	0.93	—	—
Black cupric oxide on copper	0.91	0.96	0.95
Bare moist ground	0.90	0.95	0.95
Wet sand	0.91	0.95	0.96
Water	0.94	0.95–0.96	0.98
Black tar paper	0.93	0.93	1.00
Black gloss paint	0.90	0.90	1.00
Small hole in large box, furnace, or enclosure	0.99	0.99	1.00
"Hohlraum," theoretically perfect black body	1.00	1.00	1.00

Class IV substances: Absorptance to emittance ratios greater than 1.0			
Black silk velvet	0.99	0.97	1.02
Alfalfa, dark green	0.97	0.95	1.02
Lampblack	0.98	0.95	1.03
Black paint, 0.017 in., on aluminum	0.94–0.98	0.88	1.07–1.11
Granite	0.55	0.44	1.25
Graphite	0.78	0.41	1.90
High ratios, but absorptances less than 0.80			
Dull brass, copper, lead	0.2–0.4	0.4–0.65	1.63–2.0
Galvanized sheet iron, oxidized	0.80	0.28	2.86
Galvanized iron, clean, new	0.65	0.13	5.00
Aluminum foil	0.15	0.05	3.00
Magnesium	0.30	0.07	4.30
Chromium	0.49	0.08	6.13
Polished zinc	0.46	0.02	23.00
Deposited silver (optical reflector) untarnished	0.07	0.01	

TABLE A.6   (*Continued*)

Class V substances: Selective surfaces			
Plated metals:			
Black sulfide on metal	0.92	0.10	9.20
Black cupric oxide on sheet aluminum	0.08–0.93	0.09–0.21	
Copper (5×10^{-5} cm thick) on nickel or silver-plated metal			
Cobalt oxide on platinum			
Cobalt oxide on polished nickel	0.93–0.94	0.24–0.40	3.90
Black nickel oxide on aluminum	0.85–0.93	0.06–0.1	14.5–15.5
Black chrome	0.87	0.09	9.80
Particulate coatings:			
Lampblack on metal			
Black iron oxide, 47 μm grain size, on aluminum			
Geometrically enhanced surfaces:			
Optimally corrugated greys	0.89	0.77	1.20
Optimally corrugated selectives	0.95	0.16	5.90
Stainless-steel wire mesh	0.63–0.86	0.23–0.28	2.7–3.0
Copper, treated with $NaClO_2$ and NaOH	0.87	0.13	6.69

[a] From (K2).

TABLE A.7

Interest Factors for Various Discount Rates

3% Interest Factors for Discrete Compounding Periods

	SINGLE PAYMENT		UNIFORM SERIES					
	Compound Amount Factor	Present Worth Factor	Capital Recovery Factor	Present Worth Factor	Sinking Fund Factor	Compound Amount Factor	Gradient Factor	
N	$(F/P, 3, N)$	$(P/F, 3, N)$	$(A/P, 3, N)$	$(P/A, 3, N)$	$(A/F, 3, N)$	$(F/A, 3, N)$	$(A/G, 3, N)$	N
1	1.0300	.97087	1.0300	.9709	1.0000	1.0000	.0000	1
2	1.0609	.94260	.52262	1.9134	.49262	2.0299	.4920	2
3	1.0927	.91514	.35354	2.8285	.32354	3.0908	.9795	3
4	1.1255	.88849	.26903	3.7170	.23903	4.1835	1.4622	4
5	1.1592	.86261	.21836	4.5796	.18836	5.3090	1.9401	5
6	1.1940	.83749	.18460	5.4170	.15460	6.4682	2.4129	6
7	1.2298	.81310	.16051	6.2301	.13051	7.6622	2.8809	7
8	1.2667	.78941	.14246	7.0195	.11246	8.8920	3.3440	8
9	1.3047	.76642	.12844	7.7859	.09844	10.158	3.8022	9
10	1.3439	.74410	.11723	8.5300	.08723	11.463	4.2555	10
11	1.3842	.72243	.10808	9.2524	.07808	12.807	4.7040	11
12	1.4257	.70139	.10046	9.9537	.07046	14.191	5.1475	12
13	1.4685	.68096	.09403	10.634	.06403	15.617	5.5863	13
14	1.5125	.66113	.08853	11.295	.05853	17.085	6.0201	14
15	1.5579	.64187	.08377	11.937	.05377	18.598	6.4491	15
16	1.6046	.62318	.07961	12.560	.04961	20.156	6.8732	16
17	1.6528	.60502	.07595	13.165	.04595	21.760	7.2926	17
18	1.7024	.58740	.07271	13.753	.04271	23.413	7.7072	18
19	1.7534	.57030	.06982	14.323	.03982	25.115	8.1169	19
20	1.8060	.55369	.06722	14.877	.03722	26.869	8.5219	20
21	1.8602	.53756	.06487	15.414	.03487	28.675	8.9221	21
22	1.9160	.52190	.06275	15.936	.03275	30.535	9.3176	22
23	1.9735	.50670	.06082	16.443	.03082	32.451	9.7084	23
24	2.0327	.49194	.05905	16.935	.02905	34.425	10.094	24
25	2.0937	.47762	.05743	17.412	.02743	36.457	10.475	25
26	2.1565	.46370	.05594	17.876	.02594	38.551	10.852	26
27	2.2212	.45020	.05457	18.326	.02457	40.707	11.224	27
28	2.2878	.43709	.05329	18.763	.02339	42.929	11.592	28
29	2.3565	.42436	.05212	19.188	.02212	45.217	11.954	29
30	2.4272	.41200	.05102	19.600	.02102	47.573	12.311	30
31	2.5000	.40000	.05000	20.000	.02000	50.000	12.666	31
32	2.5750	.38835	.04905	20.388	.01905	52.500	13.016	32
33	2.6522	.37704	.04816	20.765	.01816	55.075	13.360	33
34	2.7318	.36606	.04732	21.131	.01732	57.727	13.700	34
35	2.8137	.35539	.04654	21.486	.01654	60.459	14.036	35
40	3.2619	.30657	.04326	23.114	.01326	75.397	15.649	40
45	3.7814	.26445	.04079	24.518	.01079	92.715	17.154	45
50	4.3837	.22812	.03887	25.729	.00887	112.79	18.556	50
55	5.0819	.19678	.03735	26.774	.00735	136.06	19.859	55
60	5.8913	.16974	.03613	27.675	.00613	163.04	21.066	60
65	6.8296	.14642	.03515	28.452	.00515	194.32	22.183	65
70	7.9173	.12630	.03434	29.123	.00434	230.57	23.213	70
75	9.1783	.10895	.03367	29.701	.00367	272.61	24.162	75
80	10.640	.09398	.03311	30.200	.00311	321.33	25.034	80
85	12.334	.08107	.03265	30.630	.00265	377.82	25.834	85
90	14.299	.06993	.03226	31.002	.00226	443.31	26.566	90
95	16.576	.06033	.03193	31.322	.00193	519.22	27.234	95
100	19.217	.05204	.03165	31.598	.00165	607.23	27.843	100

(continued)

TABLE A.7 *(Continued)*

8% Interest Factors for Discrete Compounding Periods

	SINGLE PAYMENT		UNIFORM SERIES					
	Compound Amount Factor	Present Worth Factor	Capital Recovery Factor	Present Worth Factor	Sinking Fund Factor	Compound Amount Factor	Gradient Factor	
N	$(F/P, 8, N)$	$(P/F, 8, N)$	$(A/P, 8, N)$	$(P/A, 8, N)$	$(A/F, 8, N)$	$(F/A, 8, N)$	$(A/G, 8, N)$	N
1	1.0800	.92593	1.0800	.9259	1.0000	1.0000	.0000	1
2	1.1664	.85734	.56077	1.7832	.48077	2.0799	.4807	2
3	1.2597	.79383	.38803	2.5770	.30804	3.2463	.9487	3
4	1.3604	.73503	.30192	3.3121	.22192	4.5060	1.4038	4
5	1.4693	.68059	.25046	3.9926	.17046	5.8665	1.8463	5
6	1.5868	.63017	.21632	4.6228	.13632	7.3358	2.2762	6
7	1.7138	.58349	.19207	5.2063	.11207	8.9227	2.6935	7
8	1.8509	.54027	.17402	5.7466	.09402	10.636	3.0984	8
9	1.9989	.50025	.16008	6.2468	.08008	12.487	3.4909	9
10	2.1589	.46320	.14903	6.7100	.06903	14.486	3.8712	10
11	2.3316	.42889	.14008	7.1389	.06008	16.645	4.2394	11
12	2.5181	.39712	.13270	7.5360	.05270	18.976	4.5956	12
13	2.7196	.36770	.12642	7.9037	.04652	21.495	4.9401	13
14	2.9371	.34046	.12130	8.2442	.04130	24.214	5.2729	14
15	3.1721	.31524	.11683	8.5594	.03683	27.151	5.5943	15
16	3.4259	.29189	.11298	8.8513	.03298	30.323	5.9045	16
17	3.6999	.27027	.10963	9.1216	.02963	33.749	6.2036	17
18	3.9959	.25025	.10670	9.3718	.02670	37.449	6.4919	18
19	4.3156	.23171	.10413	9.6035	.02413	41.445	6.7696	19
20	4.6609	.21455	.10185	9.8181	.02185	45.761	7.0368	20
21	5.0337	.19866	.09983	10.016	.01983	50.422	7.2939	21
22	5.4364	.18394	.09803	10.200	.01803	55.455	7.5411	22
23	5.8713	.17032	.09642	10.371	.01642	60.892	7.7785	23
24	6.3410	.15770	.09498	10.528	.01498	66.763	8.0065	24
25	6.8483	.14602	.09368	10.674	.01368	73.104	8.2253	25
26	7.3962	.13520	.09251	10.809	.01251	79.953	8.4351	26
27	7.9879	.12519	.09145	10.935	.01145	87.349	8.6362	27
28	8.6269	.11592	.09049	11.051	.01049	95.337	8.8288	28
29	9.3171	.10733	.08962	11.158	.00962	103.96	9.0132	29
30	10.062	.09938	.08883	11.257	.00883	113.28	9.1896	30
31	10.867	.09202	.08811	11.349	.00811	123.34	9.3583	31
32	11.736	.08520	.08745	11.434	.00745	134.21	9.5196	32
33	12.675	.07889	.08685	11.513	.00685	145.94	9.6736	33
34	13.689	.07305	.08630	11.586	.00630	158.62	9.8207	34
35	14.785	.06764	.08580	11.654	.00580	172.31	9.9610	35
40	21.724	.04603	.08386	11.924	.00386	259.05	10.569	40
45	31.919	.03133	.08259	12.108	.00259	386.49	11.044	45
50	46.900	.02132	.08174	12.233	.00174	573.75	11.410	50
55	68.911	.01451	.08118	12.318	.00118	848.89	11.690	55
60	101.25	.00988	.08080	12.376	.00080	1253.1	11.901	60
65	148.77	.00672	.08054	12.416	.00054	1847.1	12.060	65
70	218.59	.00457	.08037	12.442	.00037	2719.9	12.178	70
75	321.19	.00311	.08025	12.461	.00025	4002.3	12.265	75
80	471.93	.00212	.08017	12.473	.00017	5886.6	12.330	80
85	693.42	.00144	.08012	12.481	.00012	8655.2	12.377	85
90	1018.8	.00098	.08008	12.487	.00008	12723.9	12.411	90
95	1497.0	.00067	.08005	12.491	.00005	18701.5	12.436	95
100	2199.6	.00045	.08004	12.494	.00004	27484.5	12.454	100

TABLE A.7 (*Continued*)

10% Interest Factors for Discrete Compounding Periods

	SINGLE PAYMENT		UNIFORM SERIES					
	Compound Amount Factor	Present Worth Factor	Capital Recovery Factor	Present Worth Factor	Sinking Fund Factor	Compound Amount Factor	Gradient Factor	
N	$(F/P, 10, N)$	$(P/F, 10, N)$	$(A/P, 10, N)$	$(P/A, 10, N)$	$(A/F, 10, N)$	$(F/A, 10, N)$	$(A/G, 10, N)$	N
1	1.1000	.90909	1.1000	.9091	1.0000	1.000	.0000	1
2	1.2100	.82645	.57619	1.7355	.47619	2.0999	.4761	2
3	1.3310	.75132	.40212	2.4868	.30212	3.3099	.9365	3
4	1.4641	.68302	.31547	3.1698	.21547	4.6409	1.3810	4
5	1.6105	.62092	.26380	3.7907	.16380	6.1050	1.8100	5
6	1.7715	.56448	.22961	4.3552	.12961	7.7155	2.2234	6
7	1.9487	.51316	.20541	4.8683	.10541	9.4870	2.6215	7
8	2.1435	.46651	.18745	5.3349	.08745	11.435	3.0043	8
9	2.3579	.42410	.17364	5.7589	.07364	13.579	3.3722	9
10	2.5937	.38555	.16275	6.1445	.06275	15.937	3.7253	10
11	2.8530	.35050	.15396	6.4950	.05396	18.530	4.0639	11
12	3.1384	.31863	.14676	6.8136	.04676	21.383	4.3883	12
13	3.4522	.28967	.14078	7.1033	.04078	24.522	4.6987	13
14	3.7974	.26333	.13575	7.3666	.03575	27.974	4.9954	14
15	4.1771	.23940	.13147	7.6060	.03147	31.771	5.2788	15
16	4.5949	.21763	.12782	7.8236	.02782	35.949	5.5492	16
17	5.0544	.19785	.12466	8.0215	.02466	40.543	5.8070	17
18	5.5598	.17986	.12193	8.2013	.02193	45.598	6.0524	18
19	6.1158	.16351	.11955	8.3649	.01955	51.158	6.2860	19
20	6.7273	.14865	.11746	8.5135	.01746	57.273	6.5080	20
21	7.4001	.13513	.11562	8.6486	.01562	64.001	6.7188	21
22	8.1401	.12285	.11401	8.7715	.01401	71.401	6.9188	22
23	8.9541	.11168	.11257	8.8832	.01257	79.541	7.1084	23
24	9.8495	.10153	.11130	8.9847	.01130	88.495	7.2879	24
25	10.834	.09230	.11017	9.0770	.01017	98.344	7.4579	25
26	11.917	.08391	.10916	9.1609	.00916	109.17	7.6185	26
27	13.109	.07628	.10826	9.2372	.00826	121.09	7.7703	27
28	14.420	.06935	.10745	9.3065	.00745	134.20	7.9136	28
29	15.862	.06304	.10673	9.3696	.00673	148.62	8.0488	29
30	17.448	.05731	.10608	9.4269	.00608	164.48	8.1761	30
31	19.193	.05210	.10550	9.4790	.00550	181.93	8.2961	31
32	21.113	.04736	.10497	9.5263	.00497	201.13	8.4090	32
33	23.224	.04306	.10450	9.5694	.00450	222.24	8.5151	33
34	25.546	.03914	.10407	9.6085	.00407	245.46	8.6149	34
35	28.101	.03559	.10369	9.6441	.00369	271.01	8.7085	35
40	45.257	.02210	.10226	9.7790	.00226	442.57	9.0962	40
45	72.887	.01372	.10139	9.8628	.00139	718.87	9.3740	45
50	117.38	.00852	.10086	9.9148	.00086	1163.8	9.5704	50
55	189.04	.00529	.10053	9.9471	.00053	1880.4	9.7075	55
60	304.46	.00328	.10033	9.9671	.00033	3034.6	9.8022	60
65	490.34	.00204	.10020	9.9796	.00020	4893.4	9.8671	65
70	789.69	.00127	.10013	9.9873	.00013	7886.9	9.9112	70
75	1271.8	.00079	.10008	9.9921	.00008	12709.0	9.9409	75
80	2048.2	.00049	.10005	9.9951	.00005	20474.0	9.9609	80
85	3298.7	.00030	.10003	9.9969	.00003	32979.7	9.9742	85
90	5312.5	.00019	.10002	9.9981	.00002	53120.2	9.9830	90
95	8555.9	.00012	.10001	9.9988	.00001	85556.8	9.9889	95
100	13780.6	.00007	.10001	9.9992	.00001	137796.1	9.9927	100

(*continued*)

TABLE A.7 (*Continued*)

12% Interest Factors for Discrete Compounding Periods

	SINGLE PAYMENT		UNIFORM SERIES					
	Compound Amount Factor	Present Worth Factor	Capital Recovery Factor	Present Worth Factor	Sinking Fund Factor	Compound Amount Factor	Gradient Factor	
N	$(F/P, 12, N)$	$(P/F, 12, N)$	$(A/P, 12, N)$	$(P/A, 12, N)$	$(A/F, 12, N)$	$(F/A, 12, N)$	$(A/G, 12, N)$	N
1	1.1200	.89286	1.1200	.8929	1.0000	1.0000	.0000	1
2	1.2544	.79719	.59170	1.6900	.47170	2.1200	.4717	2
3	1.4049	.71178	.41635	2.4018	.29635	3.3743	.9246	3
4	1.5735	.63552	.32924	3.0373	.20924	4.7793	1.3588	4
5	1.7623	.56743	.27741	3.6047	.15741	6.3528	1.7745	5
6	1.9738	.50663	.24323	4.1114	.12323	8.115	2.1720	6
7	2.2106	.45235	.21912	4.5637	.09912	10.088	2.5514	7
8	2.4759	.40388	.20130	4.9676	.08130	12.299	2.9131	8
9	2.7730	.36061	.18768	5.3282	.06768	14.775	3.2573	9
10	3.1058	.32197	.17698	5.6502	.05698	17.548	3.5846	10
11	3.4785	.28748	.16842	5.9376	.04842	20.654	3.8952	11
12	3.8959	.25668	.16144	6.1943	.04144	24.132	4.1896	12
13	4.3634	.22918	.15568	6.4235	.03568	28.028	4.4682	13
14	4.8870	.20462	.15087	6.6281	.03087	32.392	4.7316	14
15	5.4735	.18270	.14682	6.8108	.02682	37.279	4.9802	15
16	6.1303	.16312	.14339	6.9739	.02339	42.752	5.2146	16
17	6.8659	.14565	.14046	7.1196	.02046	48.883	5.4352	17
18	7.6899	.13004	.13794	7.2496	.01794	55.749	5.6427	18
19	8.6126	.11611	.13576	7.3657	.01576	63.439	5.8375	19
20	9.6462	.10367	.13388	7.4694	.01388	72.051	6.0201	20
21	10.803	.09256	.13224	7.5620	.01224	81.698	6.1913	21
22	12.100	.08264	.13081	7.6446	.01081	92.501	6.3513	22
23	13.552	.07379	.12956	7.7184	.00956	104.60	6.5009	23
24	15.178	.06588	.12846	7.7843	.00846	118.15	6.6406	24
25	16.999	.05882	.12750	7.8431	.00750	133.33	6.7708	25
26	19.039	.05252	.12665	7.8956	.00665	150.33	6.8920	26
27	21.324	.04689	.12590	7.9425	.00590	169.37	7.0049	27
28	23.883	.04187	.12524	7.9844	.00524	190.69	7.1097	28
29	26.749	.03738	.12466	8.0218	.00466	214.58	7.2071	29
30	29.959	.03338	.12414	8.0551	.00414	241.32	7.2974	30
31	33.554	.02980	.12369	8.0849	.00369	271.28	7.3810	31
32	37.581	.02661	.12328	8.1116	.00328	304.84	7.4585	32
33	42.090	.02376	.12292	8.1353	.00292	342.42	7.5302	33
34	47.141	.02121	.12260	8.1565	.00260	384.51	7.5964	34
35	52.798	.01894	.12232	8.1755	.00232	431.65	7.6576	35
40	93.049	.01075	.12130	8.2437	.00130	767.07	7.8987	40
45	163.98	.00610	.12074	8.2825	.00074	1358.2	8.0572	45
50	288.99	.00346	.12042	8.3045	.00042	2399.9	8.1597	50

TABLE A.7 (*Continued*)

15% Interest Factors for Discrete Compounding Periods

	SINGLE PAYMENT		UNIFORM SERIES					
	Compound Amount Factor	Present Worth Factor	Capital Recovery Factor	Present Worth Factor	Sinking Fund Factor	Compound Amount Factor	Gradient Factor	
N	$(F/P, 15, N)$	$(P/F, 15, N)$	$(A/P, 15, N)$	$(P/A, 15, N)$	$(A/F, 15, N)$	$(F/A, 15, N)$	$(A/G, 15, N)$	N
1	1.1500	.86957	1.1500	.8696	1.0000	1.000	.0000	1
2	1.3225	.75614	.61512	1.6257	.46512	2.1499	.4651	2
3	1.5208	.65752	.43798	2.2832	.28798	3.4724	.9071	3
4	1.7490	.57175	.35027	2.8549	.20027	4.9933	1.3262	4
5	2.0113	.49718	.29832	3.3521	.14832	6.7423	1.7227	5
6	2.3130	.43233	.26424	3.7844	.11424	8.7536	2.0971	6
7	2.6600	.37594	.24036	4.1604	.09036	11.066	2.4498	7
8	3.0590	.32690	.22285	4.4873	.07285	13.726	2.7813	8
9	3.5178	.28426	.20957	4.7715	.05957	16.785	3.0922	9
10	4.0455	.24719	.19925	5.0187	.04925	20.303	3.3831	10
11	4.6523	.21494	.19107	5.2337	.04107	24.349	3.6549	11
12	5.3502	.18691	.18448	5.4206	.03448	29.001	3.9081	12
13	6.1527	.16253	.17911	5.5831	.02911	34.351	4.1437	13
14	7.0756	.14133	.17469	5.7244	.02469	40.504	4.3623	14
15	8.1369	.12290	.17102	5.8473	.02102	47.579	4.5649	15
16	9.3575	.10687	.16795	5.9542	.01795	55.716	4.7522	16
17	10.761	.09293	.16537	6.0471	.01537	65.074	4.9250	17
18	12.375	.08081	.16319	6.1279	.01319	75.835	5.0842	18
19	14.231	.07027	.16134	6.1982	.01134	88.210	5.2307	19
20	16.366	.06110	.15976	6.2593	.00976	102.44	5.3651	20
21	18.821	.05313	.15842	6.3124	.00842	118.80	5.4883	21
22	21.644	.04620	.15727	6.3586	.00727	137.62	5.6010	22
23	24.891	.04018	.15628	6.3988	.00628	159.27	5.7039	23
24	28.624	.03493	.15543	6.4337	.00543	184.16	5.7978	24
25	32.918	.03038	.15470	6.4641	.00470	212.78	5.8834	25
26	37.856	.02642	.15407	6.4905	.00407	245.70	5.9612	26
27	43.534	.02297	.15353	6.5135	.00353	283.56	6.0318	27
28	50.064	.01997	.15306	6.5335	.00306	327.09	6.0959	28
29	57.574	.01737	.15265	6.5508	.00265	377.16	6.1540	29
30	66.210	.01510	.15230	6.5659	.00230	434.73	6.2066	30
31	76.141	.01313	.15200	6.5791	.00200	500.94	6.2541	31
32	87.563	.01142	.15173	6.5905	.00173	577.08	6.2970	32
33	100.69	.00993	.15150	6.6004	.00150	664.65	6.3356	33
34	115.80	.00864	.15131	6.6091	.00131	765.34	6.3705	34
35	133.17	.00751	.15113	6.6166	.00113	881.14	6.4018	35
40	267.85	.00373	.15056	6.6417	.00056	1779.0	6.5167	40
45	538.75	.00186	.15028	6.6543	.00028	3585.0	6.5829	45
50	1083.6	.00092	.15014	6.6605	.00014	7217.4	6.8204	50

(*continued*)

TABLE A.7 *(Continued)*

20% Interest Factors for Discrete Compounding Periods

	SINGLE PAYMENT		UNIFORM SERIES					
	Compound Amount Factor	Present Worth Factor	Capital Recovery Factor	Present Worth Factor	Sinking Fund Factor	Compound Amount Factor	Gradient Factor	
N	$(F/P, 20, N)$	$(P/F, 20, N)$	$(A/P, 20, N)$	$(P/A, 20, N)$	$(A/F, 20, N)$	$(F/A, 20, N)$	$(A/G, 20, N)$	N
1	1.2000	.83333	1.2000	.8333	1.0000	1.0000	.0000	1
2	1.4400	.69445	.65455	1.5277	.45455	2.1999	.4545	2
3	1.7280	.57870	.47473	2.1064	.27473	3.6399	.8791	3
4	2.0736	.48225	.38629	2.5887	.18629	5.3679	1.2742	4
5	2.4883	.40188	.33438	2.9906	.13438	7.4415	1.6405	5
6	2.9859	.33490	.30071	3.3255	.10071	9.9298	1.9788	6
7	3.5831	.27908	.27742	3.6045	.07742	12.915	2.2901	7
8	4.2998	.23257	.26061	3.8371	.06061	16.498	2.5755	8
9	5.1597	.19381	.24808	4.0309	.04808	20.798	2.8364	9
10	6.1917	.16151	.23852	4.1924	.03852	25.958	3.0738	10
11	7.4300	.13459	.23110	4.3270	.03110	32.150	3.2892	11
12	8.9160	.11216	.22527	4.4392	.02527	39.580	3.4840	12
13	10.699	.09346	.22062	4.5326	.02062	48.496	3.6596	13
14	12.839	.07789	.21689	4.6105	.01689	59.195	3.8174	14
15	15.406	.06491	.21388	4.6754	.01388	72.034	3.9588	15
16	18.488	.05409	.21144	4.7295	.01144	87.441	4.0851	16
17	22.185	.04507	.20944	4.7746	.00944	105.92	4.1975	17
18	26.623	.03756	.20781	4.8121	.00781	128.11	4.2975	18
19	31.947	.03130	.20646	4.8435	.00646	154.73	4.3860	19
20	38.337	.02608	.20536	4.8695	.00536	186.68	4.4643	20
21	46.004	.02174	.20444	4.8913	.00444	225.02	4.5333	21
22	55.205	.01811	.20369	4.9094	.00369	271.02	4.5941	22
23	66.246	.01510	.20307	4.9245	.00307	326.23	4.6474	23
24	79.495	.01258	.20255	4.9371	.00255	392.47	4.6942	24
25	95.394	.01048	.20212	4.9475	.00212	471.97	4.7351	25
26	114.47	.00874	.20176	4.9563	.00176	567.36	4.7708	26
27	137.36	.00728	.20147	4.9636	.00147	681.84	4.8020	27
28	164.84	.00607	.20122	4.9696	.00122	819.21	4.8291	28
29	197.81	.00506	.20102	4.9747	.00102	984.05	4.8526	29
30	237.37	.00421	.20085	4.9789	.00085	1181.8	4.8730	30
31	284.84	.00351	.20070	4.9824	.00070	1419.2	4.8907	31
32	341.81	.00293	.20059	4.9853	.00059	1704.0	4.9061	32
33	410.17	.00244	.20049	4.9878	.00049	2045.8	4.9193	33
34	492.21	.00203	.20041	4.9898	.00041	2456.0	4.9307	34
35	590.65	.00169	.20034	4.9915	.00034	2948.2	4.9406	35
40	1469.7	.00068	.20014	4.9966	.00014	7343.6	4.9727	40
45	3657.1	.00027	.20005	4.9986	.00005	18281.3	4.9876	45
50	9100.1	.00011	.20002	4.9994	.00002	45497.2	4.9945	50

TABLE A.7 (*Continued*)

25% Interest Factors for Discrete Compounding Periods

	SINGLE PAYMENT		UNIFORM SERIES					
	Compound Amount Factor	Present Worth Factor	Capital Recovery Factor	Present Worth Factor	Sinking Fund Factor	Compound Amount Factor	Gradient Factor	
N	(F/P, 25, N)	(P/F, 25, N)	(A/P, 25, N)	(P/A, 25, N)	(A/F, 25, N)	(F/A, 25, N)	(A/G, 25, N)	N
1	1.2500	.80000	1.2500	.8000	1.0000	1.0000	.00000	1
2	1.5625	.64000	.69444	1.4400	.44444	2.2500	.44444	2
3	1.9531	.51200	.51230	1.9520	.26230	3.8125	.85246	3
4	2.4414	.40960	.42344	2.3616	.17344	5.7656	1.2249	4
5	3.0518	.32768	.37185	2.6893	.12185	8.2070	1.5631	5
6	3.8147	.26214	.33882	2.9514	.08882	11.259	1.8683	6
7	4.7684	.20972	.31634	3.1661	.06634	15.073	2.1424	7
8	5.9605	.16777	.30040	3.3289	.05040	19.842	2.3872	8
9	7.4506	.13422	.28876	3.4631	.03876	25.802	2.6048	9
10	9.3132	.10737	.28007	3.5705	.03007	33.253	2.7971	10
11	11.642	.08590	.27349	3.6564	.02349	42.566	2.9663	11
12	14.552	.06872	.26845	3.7251	.01845	54.208	3.1145	12
13	18.190	.05498	.26454	3.7801	.01454	68.760	3.2437	13
14	22.737	.04398	.26150	3.8241	.01150	86.949	3.3559	14
15	28.422	.03518	.25912	3.8593	.00912	109.687	3.4530	15
16	35.527	.02815	.25724	3.8874	.00724	138.109	3.5366	16
17	44.409	.02252	.25576	3.9099	.00576	173.636	3.6084	17
18	55.511	.01801	.25459	3.9279	.00459	218.045	3.6698	18
19	69.389	.01441	.25366	3.9424	.00366	273.556	3.7222	19
20	86.736	.01153	.25292	3.9539	.00292	342.945	3.7667	20
21	108.420	.00922	.25233	3.9631	.00233	429.681	3.8045	21
22	135.525	.00738	.25186	3.9705	.00186	538.101	3.8365	22
23	169.407	.00590	.25148	3.9764	.00148	673.626	3.8634	23
24	211.758	.00472	.25119	3.9811	.00119	843.033	3.8861	24
25	264.698	.00378	.25095	3.9849	.00095	1054.791	3.9052	25
26	330.872	.00302	.25076	3.9879	.00076	1319.489	3.9212	26
27	413.590	.00242	.25061	3.9903	.00061	1650.361	3.9346	27
28	516.988	.00193	.25048	3.9923	.00048	2063.952	3.9457	28
29	646.235	.00155	.25039	3.9938	.00039	2580.939	3.9551	29
30	807.794	.00124	.25031	3.9950	.00031	3227.174	3.9628	30
31	1009.742	.00099	.25025	3.9960	.00025	4034.968	3.9693	31
32	1262.177	.00079	.25020	3.9968	.00020	5044.710	3.9746	32
33	1577.722	.00063	.25016	3.9975	.00016	6306.887	3.9791	33
34	1972.152	.00051	.25013	3.9980	.00012	7884.609	3.9828	34
35	2465.190	.00041	.25010	3.9984	.00010	9856.761	3.9858	35

TABLE A.8

Conversion Factors

Physical quantity	Symbol	Conversion factor
Area	A	$1 \text{ ft}^2 = 0.0929 \text{ m}^2$
		$1 \text{ in}^2 = 6.452 \times 10^{-4} \text{ m}^2$
Density	ρ	$1 \text{ lb}_m/\text{ft}^3 = 16.018 \text{ kg/m}^3$
		$1 \text{ slug/ft}^3 = 515.379 \text{ kg/m}^3$
Heat, energy, or work	Q or W	$1 \text{ Btu} = 1055.1 \text{ J}$
		$1 \text{ cal} = 4.186 \text{ J}$
		$1 \text{ ft lb}_f = 1.3558 \text{ J}$
		$1 \text{ hp/hr} = 2.685 \times 10^6 \text{ J}$
Force	F	$1 \text{ lb}_f = 4.448 \text{ N}$
Heat flow rate	q	$1 \text{ Btu/hr} = 0.2931 \text{ W}$
		$1 \text{ Btu/sec} = 1055.1 \text{ W}$
Heat flux	q/A	$1 \text{ Btu/h/ft}^2 = 3.1525 \text{ W/m}^2$
Heat-transfer coefficient	h	$1 \text{ Btu/h/ft}^2/\text{F} = 5.678 \text{ W/m}^2/\text{K}$
Length	L	$1 \text{ ft} = 0.3048 \text{ m}$
		$1 \text{ in.} = 2.54 \text{ cm}$
		$1 \text{ mile} = 1.6093 \text{ km}$
Mass	m	$1 \text{ lb}_m = 0.4536 \text{ kg}$
		$1 \text{ slug} = 14.594 \text{ kg}$
Mass flow rate	$\dot{m}$	$1 \text{ lb}_m/\text{h} = 0.000126 \text{ kg/sec}$
		$1 \text{ lb}_m/\text{sec} = 0.4536 \text{ kg/sec}$
Power	$\dot{W}$	$1 \text{ hp} = 745.7 \text{ W}$
		$1 \text{ ft/lb}_f/\text{sec} = 1.3558 \text{ W}$
		$1 \text{ Btu/sec} = 1055.1 \text{ W}$
		$1 \text{ Btu/h} = 0.293 \text{ W}$
Pressure	p	$1 \text{ lb}_f/\text{in.}^2 = 6894.8 \text{ Pa (N/m}^2)$
		$1 \text{ lb}_f/\text{ft}^2 = 47.88 \text{ Pa (N/m}^2)$
		$1 \text{ atm} = 101,325 \text{ Pa (N/m}^2)$
Radiation	l	$1 \text{ langley} = 41,860 \text{ J/m}^2$
Specific heat capacity	c	$1 \text{ Btu/lb}_m/\text{°F} = 4187 \text{ J/kg/K}$
Internal energy or enthalpy	e or h	$1 \text{ Btu/lb}_m = 2326.0 \text{ J/kg}$
		$1 \text{ cal/g} = 4184 \text{ J/kg}$
Temperature	T	$T(\text{°R}) = (9/5)T(\text{K})$
		$T(\text{°F}) = [T(\text{°C})](9/5) + 32$
		$T(\text{°F}) = [T(\text{K}) - 273.15](9/5) + 32$
Thermal conductivity	k	$1 \text{ Btu/h/ft/°F} = 1.731 \text{ W/m/K}$
Thermal resistance	R_{th}	$1 \text{ b/°F/Btu} = 1.8958 \text{ K/W}$
Velocity	V	$1 \text{ ft/sec} = 0.3048 \text{ m/sec}$
		$1 \text{ mile/h} = 0.44703 \text{ m/sec}$
Viscosity, dynamic	μ	$1 \text{ lb}_m/\text{ft/sec} = 1.488 \text{ N/sec/m}^2$
		$1 \text{ cP} = 0.00100 \text{ N/sec/m}^2$
Viscosity, kinematic	ν	$1 \text{ ft}^2/\text{sec} = 0.09029 \text{ m}^2/\text{sec}$
		$1 \text{ ft}^2/\text{h} = 2.581 \times 10^{-5} \text{ m}^2/\text{sec}$
Volume	V	$1 \text{ ft}^3 = 0.02832 \text{ m}^3$
		$1 \text{ in.}^3 = 1.6387 \times 10^{-5} \text{ m}^3$
		$1 \text{ gal (U.S. liq.)} = 0.003785 \text{ m}^3$
Volumetric flow rate	$\dot{Q}$	$1 \text{ ft}^3/\text{min} = 0.000472 \text{ m}^3/\text{sec}$

NOMENCLATURE

B

A	Area, m² (ft²); available energy, kJ (Btu)
a	a dimension, m (ft); a constant
a_s	solar-azimuth angle (positive east of sough), deg
a_c	collector azimuth angle (positive east of south), deg
$\bar{B}$	monthly averaged daily beam radiation, kWh/m² day (Btu/day ft²)
BV	book value of an asset $
C	cost, $, D.M., Fr., £, etc.
C/B	cost to benefit ratio
c	speed of light, m/sec (ft/sec); a constant
c_p	Specific heat at constant pressure, kJ/kg °C (Btu/lb °F)
c_v	specific heat at constant volume, kJ/kg °C (Btu/lb °F)
CC	opaque cloud cover expressed as tenth of sky covered: CC = 0 indicates a clear sky; CC = 10 indicates the sky is fully covered with clouds
COP	coefficient of performance
CR	concentration ratio
d, D	a dimension or diameter, m (ft)
D_h	hydraulic diameter, m (ft)
$\bar{D}$	monthly averaged daily diffuse (scattered) radiation, kWh/m² day (Btu/day (ft²)
e	internal energy; eccentricity of earth orbit; electron charge; vapor pressure
E	energy
$E_{b\lambda}$	spectral irradiance of a blackbody in a small bandwidth interval centered at λ, W/m² μm (Btu/hr ft² μm)
ET	equation of time, min
F	fin efficiency; fuel
F'	plate efficiency
F_{ij}	radiation shape factor between surfaces i and j
F_R	heat removal factor
f	focal length, m (ft); friction factor; f-ratio of a lens or mirror

331

f_s	fraction of energy demand delivered by solar system
g	gravitational acceleration, m/sec² (ft/sec²); gap dimension, m (ft)
Gr	Grashof number
G	mass flow per unit area, superficial velocity kg/m² hr (lb/ft² hr)
h_s	local solar-hour angle (measured from local solar noon, 1 h = 15°), positive before solar noon, negative after
$\bar{H}_{o,h}$	monthly averaged daily solar flux on a horizontal extraterrestrial surface, kW hr/m² day, (Btu/ft² day)
$\bar{H}_h$	monthly averaged daily total radiation on a horizontal surface, kW hr/m² day (Btu/ft² day)
h_{ss}, h_{sr}	hour angle between sunset (or sunrise) and local solar noon.
h	hour angle, deg; enthalpy, kJ/kg (Btu/lb); Planck's constant, Js (Btu sec)
h_c, h_f	coefficient of convection heat transfer from a surface to a fluid, W/m² °C (Btu/hr ft² °F)
I	instantaneous or hourly solar radiation on a surface, W/m² (Btu/hr ft²); use subscripts to denote bean, diffuse, reflected, horizontal, tilted, direct normal, etc.
$\bar{I}$	daily total solar radiation on a surface, W/m² day (Btu/day ft²); use subscripts to denote beam, diffuse, reflected, horizontal, tilted, direct normal, etc.
I_{sc}	solar constant, i.e., the amount of solar energy received by a unit area of surface placed perpendicular to the sun's rays outside the earth's atmsophere at the earth's mean distance from the sun, W/m² (Btu/hr ft²)
i	incidence angle, deg; interest or discount rate, decimal percent
I	current, A
ITC	investment tax credit
K	extinction coefficient, m⁻¹ (ft⁻¹)
k	thermal conductivity W/m °C (Btu/hr ft °F); specific heat ratio; Boltzmann's constant, J/K (Btu/°R); a constant
k_t	˙instantaneous or hourly ratio of horizontal total radiation on a terrestrial surface to that on the corresponding extraterrestrial surface, i.e., the instantaneous or hourly clearness index; sometimes called percent of possible radiation
K_T	daily ratio of horizontal total radiation on a terrestrial surface to that on the corresponding extraterrestrial surface, i.e., the daily clearness index
$\bar{K}_t$	ratio of monthly averaged horizontal total radiation on a terrestrial surface to that on the corresponding extraterrestrial surface, i.e., the monthly averaged clearness index
L, L	latitude, rad (deg); length; thermal load or demand
l	a length m (ft); a constant
m	air mass ratio, i.e., the distance through which radiation travels from the outer edge of the earth's atmosphere to a recovery point on the earth divided by the zenith distance radiation would travel to the same point
m	mass, kg (lbₘ)
$\dot{m}$	mass flow rate, kg/sec (lbₘ/hr)
n	a number
n_r, n	index of refraction
N	day number counted from January 1
NTU	number of transfer units, dimensionless
Nu	Nusselt number
r_T, r_d	functions to convert solar flux on a horizontal surface to that on a nonhorizontal surface (Table 2.11) for daily or monthly time scales

P	pressure, N/m² (lb/in.²)
Pr	Prandtl number
q	heat flux, W/m² (Btu/hr ft²)
Q	quantity of energy or heat, kJ or kW hr (Btu)
r, R	radius or radial coordinate, m (ft)
r_p	pressure ratio
R	thermal resistance, °C m²/W (°F ft² hr/Btu)
Ra	Rayleigh number
Re	Reynolds number
Ri	Richardson number
S	entropy, J/K (Btu/°R)
T	temperature, K or °C (°F or °R); tax or tax rate
t	time, sec (hr); thickness, m (ft)
U	overall heat transfer coefficient, W/m² °C (Btu/hr ft² °F)
v, V	velocity, m/sec (ft/sec)
V	volume, m³ (ft³)
V	voltage, V
W	work, kJ or kW hr (ft lb); with m (ft)
x	a Cartesian coordinate
X	critical intensity ratio
y	a Cartesian coordinate
z	zenith angle, deg; altitude above mean sea level, m (ft)

SUBSCRIPTS

a	air; ambient conditions; aperture; absorber
amb	ambient
ann	annual
atm	atmospheric
aux	auxiliary
b	beam; particle bed
C	Carnot cycle
c	collector; convection
coll	collection, collector
d	diffuse (scattered)
day	day
e	environmental; electrical
eff	effective
exp	expansion
f	fluid, fin; fuel
g	glass
h	horizontal surface
hw	hot water
hx	heat exchanger
i	infiltration; inlet

inc	income; incidence
ir	infrared
k	conduction
l	loss; liquid
m	mirror; monthly totals; maintenance; motor
max	maximum
min	minimum
n	normal
o	optical (0 for: reference or standard; extraterrestrial)
opt	optimum
out	outlet or outside
⊥	perpendicular
‖	parallel
p	pump
R	Rankine cycle
r	reflected; roof; refracted; receiver
rad	radiation or radiative
s	solar; surface; saturation
sr	sunrise
ss	sunset
sky	sky
t	tilted surface; top; tubular; thermal
terr	terrestrial
th	thermal
tot	total
T	total
u	useful
v	vapor
v	constant volume
w	wind; wall; water
y	yearly
z	zenith

GREEK SYMBOLS (MAY BE SUBSCRIPTS)

α	absorptance; solar altitude angle
β	collector tilt angle from horizontal plane
γ	solar profile angle; an angle
δ	declination; optical intercept factor
Δ	difference, change in
ϵ	emittance; heat exchanger effectiveness; surface roughness
ϵ_v	void fraction
η	efficiency; effectiveness
η_1	first law efficiency
η_2	second law efficiency

θ, ψ	angles
λ	wavelength; Lagrange multiplier
λ_m	molecular mean free path, cm
μ	dynamic viscosity; mean value
ν	frequency; kinematic viscosity
ρ	density; reflectance
σ	Stefan–Boltzmann constant; standard deviation; stress; surface tension
τ	transmittance; shear stress
ϕ	relative humidity; an angle; utilizability
ω	solid angle

Note: Symbols used infrequently are defined in the text where used.

INDEX

A

Absorber area, definition, 101
Absorber assembly, heat loss from, 74–77
Absorber–reflector surface, 93
Absorptance, total, 54
Absorptance values for various materials, 316–318
Accelerated depreciation, *see* Depreciation
Acceptance angle, 103
Air, thermal conductivity of, 72
Air conditioning, second law efficiency of, 66
Air mass, 14–16
  value calculation, 16
  Air mass ratio, 14, 16
  standard irradiance curves for, 17–19
Air pollution, 8–9
Aluminum
  reflectance and dispersion data for, 61
  thermal conductivity of, 73
Analemma, 25
Analytic production function, 293–296, *see also* Production function; Tabular production function
Annual revenue, 291
Annual worth, 281
Aperture area, definition, 101
Atmospheric transmittance, *see* Clearness index
Attenuation coefficients, 16–17
Available energy, 64, *see also* Thermodynamics; Energy
Average altitude angle, evaluation of, 46

B

Beam radiation, *see* Extraterrestrial radiation; Terrestrial beam radiation
Beer's law, 14, 52
Before-tax analysis, 291–292
Benzene, thermal conductivity of, 72
Bifurcated fin receiver, 126
Blackbody radiation functions, 87
Blast furnace, 4
Boeing receiver, 246
Boiling heat transfer, 82–83, *see also* Heat transfer
Boltzmann's constant, 86
Bond conductance, 76
Book value, 284, 285, 286, 287
Bouger's law, 14
Brayton cycle, 67–69
  compared with Rankine cycle, 255–256
  first law efficiency, 69
  second law efficiency, 69

337